OPTIMIZATION OF INDUSTRIAL UNIT PROCESSES

SECOND EDITION

OPTIMIZATION OF INDUSTRIAL UNIT PROCESSES

SECOND EDITION

Béla G. Lipták

CRC Press
Taylor & Francis Group
Boca Raton London New York

CRC Press is an imprint of the
Taylor & Francis Group, an **informa** business

CRC Press
Taylor & Francis Group
6000 Broken Sound Parkway NW, Suite 300
Boca Raton, FL 33487-2742

First issued in paperback 2019

ISBN-13: 978-0-8493-9873-5 (hbk)
ISBN-13: 978-0-367-40026-2 (pbk)
Library of Congress Card Number 98-20091

Library of Congress Cataloging-in-Publication Data

Lipták, Béla G.
    Optimization of industrial unit processes / Béla Lipták. — 2nd
  ed.
      p. cm.
    Includes bibliographical references and index.
    ISBN 0-8493-9873-8 (alk. paper)
    1. Manufacturing processes. 2. Industrial equipment.
  3. Industrial efficiency. I. Title.
  TS183.L57 1998
  670.42—dc21
                                                    98-20091
                                                       CIP

Visit the Taylor & Francis Web site at
http://www.taylorandfrancis.com

and the CRC Press Web site at
http://www.crcpress.com

Preface

In this volume, the term "optimization" is used as that combination of process control strategies that will safely maximize productivity while minimizing operating costs. In other words, this is not a theoretical or esoteric book. It deals with real-world processes, where pipes leak, sensors plug, and pumps cavitate. It suggests practical solutions to real problems.

In a fully optimized plant, levels, temperatures, pressures, or flows should only be constraints, as these variables float within their allowable limits, while the efficiency or productivity of the plant is continuously maximized. This control philosophy differs from the earlier mentality, where levels and temperatures were controlled at constant values and plant productivity was only an accidental and uncontrolled consequence of those controlled variables.

Therefore, in this volume, the single loops serve only as constraints in forming the multivariable control envelopes. The control envelope is a polygon, with its sides representing the various constraints, while inside the envelope the process is continuously moved to maximize efficiency and productivity. Common sense is the basis for multivariable optimization. It is also nature's way of control. For one who understands the process, this control method is both natural and simple. However, for multivariable optimization, it is essential to fully understand the process; this book will discuss the nature, the time contants, and the unique characteristics of each process.

The most important characteristic of the control systems discussed is that they work. Each chapter is a snapshot of the state of the art of controlling a particular unit operation. I use the term "snapshot" because our understanding of process control is improving every day. The control systems described in this book have evolved through many decades and through the contributions of many control experts. Further advances are forthcoming, and the next edition of this book will report on them.

To serve progress, technology must not be value free, it must not respond only to market forces, it must also serve higher human goals and values. Optimization means more than production at minimum cost. One can also optimize to maximize safety and useful life or to minimize pollution and energy consumption. This book is dedicated to those process control engineers who will use the technology of optimization to improve the quality of our lives.

Béla Lipták July 4, 1997

Contents

Chapter 6   CSTR and BSTR Chemical Reactors

1 Boilers

Each chapter in this book describes an industrial unit operation from three different perspectives. The first part of the chapter introduces the reader to the particular process, its characteristics, and those of the associated equipment. The second part briefly describes the traditional control strategies. The third part concentrates on maximizing the efficiency or productivity of the unit operation; in other words, it is devoted to the subject of optimization. This chapter is structured the same way, and the discussion therefore begins with a description of the boiler equipment and the boiling process.

I. THE PROCESS

In the various industries, steam boilers are used to provide heat and/or energy to the various processing operations. As illustrated in Figure 1.1, boilers consist of burners where air and fuel are mixed and ignited, a furnace where the flames release their heat of combustion as the fuel is burned into combustion products, and a water system that receives the heat generated by combustion.

Gas fuels need no special preparation, while the particle size of oil is reduced by atomization and the particle size of coal by pulverization. Good combustion efficiency depends on the three Ts (time, temperature, and turbulence); the fuel needs time to make contact with the air, needs to be at ignition temperature, and needs turbulence for thorough mixing for each carbon molecule to find two oxygen molecules to form CO_2. Under perfect combustion conditions, each pound of carbon will produce 3.67 lb of CO_2.

In a drum-type boiler, the water, through natural circulation in the boiler tubes, returns as a hot-water and steam mixture to the steam drum, while the solids accumulate in the mud drum. The steam leaves the steam drum, and fresh feedwater is added to maintain the water reserve (level). Large, supercritical, once-through boilers do not use recirculation at full load.

The feedwater must be deaerated (heated in a vented tank to remove oxygen) and demineralized, because otherwise the hard water would form scale deposits, while low-alkaline water in the presence of oxygen would cause corrosion. Chemicals are used to completely remove oxygen (sodium sulfite), to precipitate hardness (phosphate salts), or to prevent the formation of scale and sludge (chelating chemicals such as EDTA).

If superheated steam is needed, the saturated steam from the drum is returned into the furnace to pick up additional heat. If the superheat tubes are exposed to the flames, the superheater is called *radiant*; if the superheater receives the heat only from the hot combustion gases, it is called a *convection superheater*. If both sections are provided, such as in Figure 1.1, the design is called a *combination* superheater design. In this case, as the saturated steam leaves the steam drum, it first passes through the convection and then the radiant sections. The received amount of superheat can be adjusted by the dampers (#1 and #2 in Figure 1.1), which can bypass some or all the hot gases around the convection section.

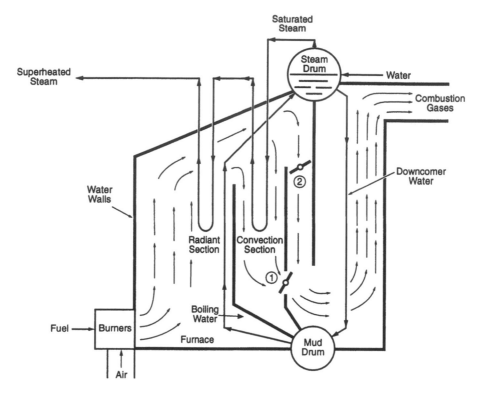

FIGURE 1.1   The boiler. The design shown is a single-drum, water tube boiler with combination superheat sections.

A. BOILER TYPES AND COMPONENTS

The furnace can be integral or separate from the boiler, and these designs are therefore called *internally* or *externally* fired. The furnace can be *pressurized* or *balanced draft*; the burners can be *horizontal* or *tangential firing* types.

The water can be in- or outside the boiler tubes, and such boilers are called *water-tube* (Figure 1.1) or *fire-tube* boilers. The tubes can be *horizontal* or *vertical* and can be arranged in *single* or *multiple passes*. The water circulation can be *natural* or *pump controlled*. The *once-through* design uses no recirculation at full load.

The fuel can be gas, oil, coal, or waste. Oil can be atomized by *air, steam* or *pressure. Multifuel* burners can burn both gas and pulverized coal or gas and oil. When burning coal, the boiler can be *stoker* fired or burn *pulverized coal*. The pulverizers can be *ball and race, roll and race, ring and roller,* or *bowl* types. The stoker-fired boilers can utilize *chain, traveling-grate, underfeed,* or *spreader* stokers. In *fluidized-bed* boilers, limestone can be added directly into the combustion zone to remove sulfur. If the ash is removed in the dry state, the boiler is called a *dry-bottom* design, and, if it is taken out in the liquid form, it is called a *wet-bottom* boiler.

Fire-tube boilers are usually smaller, under 25,000 lb/hr (11,000 kg/hr), and generate steam at pressures below 250 PSIG (17 barg). They are also less expensive and more compact. On the other hand, their disadvantages are that their joints are exposed to furnace heat, the thermal gradients cause stresses through unequal expansion, their response is slow due to the large water volume, and they are limited in capacity, pressure, and temperature.

Water-tube boilers, particularly with small-diameter tubes, are more responsive and more efficient. On the other hand, their initial costs are also higher, but because, during the lifetime of a boiler the fuel cost is many times more than the capital cost, their total cost can still be lower.

Water-tube boilers can be *electric* or *fired*, the water tubes can be *vertical* or *inclined*, and they can be the *one-*, *two-*, *three-*, or *four*-drum designs. In more modern boilers, usually only one or two drums are used (in addition to the mud drum). Most modern boilers are also provided with membrane-type *water walls*, which serve to increase heat transfer and, through their welded outer casings, also reduce air infiltration.

Prepackaged water-tube boilers are available in capacities up to 180,000 lb/hr (82,000 kg/hr) and generate steam at pressures and temperatures up to 1050 PSIG (72 barg) and 825°F (440°C).

Large utility boilers are the size of a 30-story building. They are often the once-through design. They burn 500 tons of coal every hour and generate 6 to 7 million pounds (3 million kilograms) of steam at 3850 PSIG and 1050°F (265 barg and 565°C). Each pound (0.45 kg) of coal produces about 10 lb of steam, 10 lb of flue gases, and 1 kW of electricity. According to Lammers,[5] the daily consumption of a 500-MW plant is 11 million gallons (42 million liters) of water and 80 railroad carloads of coal. There are 250 such generating units in the United States alone.

1. Heat Recovery Equipment

The greatest heat loss in boilers results from the exit of hot flue gases. This heat can be partially recovered by *economizers* and *air preheaters*. The flue gases should stay above the dew point by about 50°F (28°C) to prevent water condensation, which in the presence of sulfur oxides or chlorine will cause acid formation and corrosion. The dew point of water is around 120°F (49°C), while the dew point of the flue gas varies with the concentration of sulfur trioxide in it.

The *economizer* is usually located in the passage between the boiler and the stack (Figure 1.2). In it, the feedwater is preheated as it passes through *parallel* or *counterflow* economizer tubes. The counterflow units are considered to be more efficient.

The *air preheater* is located downstream of the economizer and can be *tubular* or *regenerative* in its design. In the tubular design, the combustion gases pass through a series of tubes and the cold ambient air travels countercurrently on the outside of these tubes. In the regenerative design, a motor continuously rotates a heat transfer surface through the flue gas and the ambient air at slow speed. When the fuel contains sulfur or chlorine, the air preheater must be protected from corrosion, caused by acid formation if the flue gases are cooled by outside air to the point where acids are condensed. At any point in the preheater, the metal temperature is assumed to be the average of the air and gas temperatures. Therefore, corrosion can be prevented by providing a bypass around the preheater and FD fan, raising the entering air temperature above the level where acids condense.

The heat content of blowdown is recovered by recovering low-pressure flash-steam and by preheating the feedwater.

A coal-fired boiler without heat recovery is expected to operate at 72 to 75% efficiency. The addition of an air preheater alone will increase that to 76 to 83%, and the addition of both an economizer and an air preheater will increase the efficiency to 86 to 89%.

Boiler efficiency is also affected by buildup on the heat transfer surfaces. The cleanliness of the water side of the tubes is protected by automatic blowdown controls, while soot and fly-ash accumulation on the fire side is removed by automatically sequenced, retractable soot blowers.

2. Fans and Draft

The theoretical air required by one unit mass of fuel (A_{th}) can be calculated from the composition of the fuel, where C is the mass percentage of carbon, and H, O, and S are those of hydrogen,

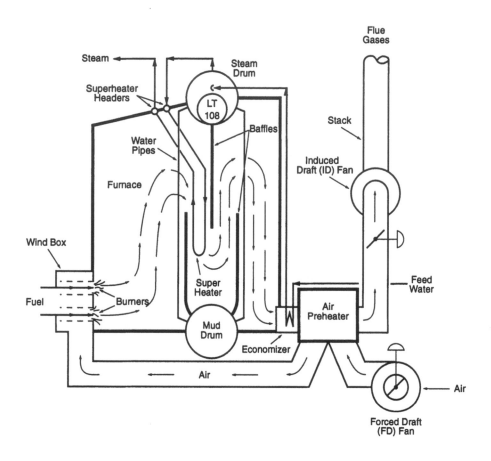

FIGURE 1.2   Boiler accessories: Economizer, air preheater, and forced- and induced-draft fans.

oxygen, and sulfur (Eq. 1). The factor O/8 corrects for the amount of hydrogen that is present in the form of water. On the average, the moisture content of air is 1% by weight.

$$A_{th} = 11.53C + 34.56(H - O/8) + 4.32S \tag{1}$$

The smaller industrial boilers are provided with only a stack and a *forced draft fan* and are therefore called *pressurized designs*, while larger industrial and utility boilers are provided with both *forced-* and *induced-draft fans* and are called *balanced-draft* boilers (Figure 1.2). Induced-draft fans are required when the draft generated by the stack is insufficient to provide the needed pressure drop across the economizers, air preheaters, and the flue-gas filters or scrubbers.

The draft (D in H_2O vacuum) generated by the stack is calculated by using Eq. 2,

$$H = 0.52H \times P(1/Ta - 1/Tg) \tag{2}$$

where H is the stack height in feet, P is atmospheric pressure in PSIA (usually 14.7), Ta and Tg are the absolute temperatures (°F + 460) of the outside air and of the flue gas. The theoretical draft

(*D*) generated by the stack is reduced by friction, so that during operation only about 80% of the theoretical draft (vacuum) is obtained at the bottom of the stack.

The advantage of the pressurized boiler design is that the high maintenance induced-draft fan, located in the hot combustion gases, is eliminated. Its disadvantage is higher flue-gas and fly-ash leakage from the boiler. In the balanced-draft design, the furnace operates under a slight vacuum, referred to as "draft" (0.05" H_2O or 1.2 mm H_2O), and, therefore, if there is leakage, it is into the furnace.

Fans and their controls are discussed in Chapter 4. Fan blades can be radial, forward, or backward curved. The backward-curved blade gives the best efficiency (80%) and is the least sensitive to heat and dirt. Fans are throttled by discharge dampers, inlet vanes, blade pitch positioners, or by varying their speed. Energy is conserved, if, instead of burning up the fan energy in dampers, that energy is not introduced in the first place. Therefore, fan speed or blade pitch throttling is preferred to damper throttling. The resulting energy savings are shown in Figures 9.3, 9.4, and 11.6. On induced-draft fans, the variable-inlet vanes should be considered only if the fan is located downstream of the particulate control equipment.

Figure 1.3 shows the pressure profiles under different load conditions in a pressurized boiler that is not provided with heat-recovery equipment.

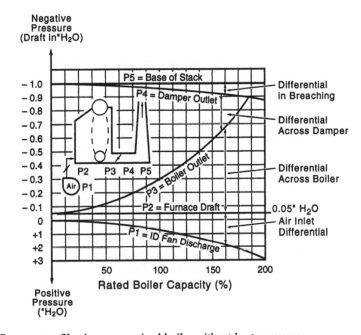

FIGURE 1.3   Pressure profiles in a pressurized boiler without heat recovery.

The fan shaft horsepower (HP) required can be calculated by Eq. 3, where air flow is in actual cubic feet per minute (ACFM), the pressure differential across the fan is in "H_2O, and the fan efficiency is in percentage (%):

$$HP = 5.2(ACFM) ("H_2O)/33,000(\%) \tag{3}$$

3. Dampers and Inlet Vanes

While speed control or blade-pitch positioning give better linearity, less hysteresis, or less dead band or leakage, many boilers still use inlet vanes (Figure 1.4) or discharge dampers (Figures 11.7

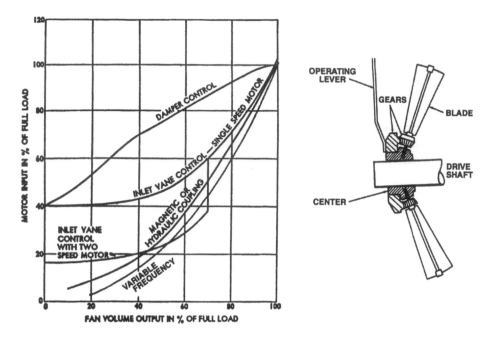

FIGURE 1.4 The design of a prerotation inlet valve and the power consumption of the different volume modulation techniques.

and 11.8) for fan flow control. The prerotation-type inlet vanes are more energy efficient, but the nonlinear characteristics of both vanes and dampers require characterization for good control.

Figure 1.5 describes the nonlinear nature of multiblade dampers. At the bottom, curves A and B relate damper opening and damper flow to their rotational angle. The upper portion of the figure allows one to evaluate the *installed performance* of the same damper. For example, if the damper pressure drop is 25% of the total system drop and the damper blade is throttled to a 45° position (curve 7), the actual air flow in the system will be 88% of maximum (the damper pressure drop will reduce the flow by 12%).

In addition to the dead band and the leakage of dampers, their gain (increase in flow per degree of rotation) is also nonlinear, much larger in the beginning of the damper opening than toward the end (quick opening characteristics). This is undesirable, because for stable control, the gain of the final control element should be constant. New damper designs could provide linear characteristics, while minimizing leakage and dead band (Figure 1.6), but the general practice is only to provide characterizing positioners or gain compensating algorithms on conventional, low-leakage (Figure 11.7) dampers.

4. Pumps

Information on pump features, characteristics, and controls, and the data for calculating the required pump HP are provided in Chapter 12. A number of pumps (oil, condensate, sump, and cooling water) are used on boilers, but the critical one is the feedwater pump because, at steam pressures of 3650 PSIG (251 barg), some of the steam discharge-header pressures have risen to the supercritical region of 6500 PSIG (448 barg).

In a *duplex pump*, the pressure of steam (or other media) moves one end of a shaft, which at its other end operates a piston that displaces and thereby pumps the water. The energy source is usually steam in the *simplex pumps* also. In *power pumps*, the pump is connected by gears or belts or directly to the energy source, which can be a gas engine or an electric motor. *Metering pumps*

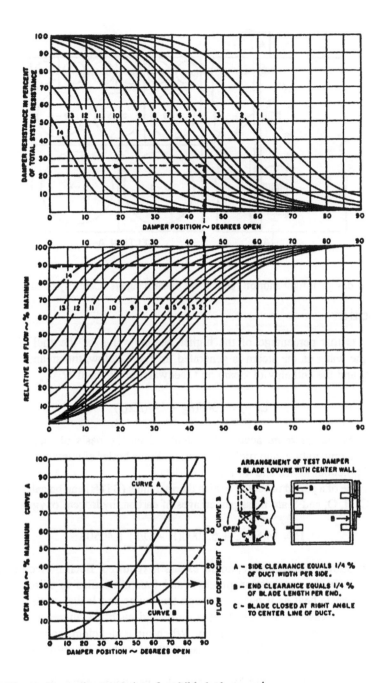

FIGURE 1.5 The nonlinear characteristics of multiblade dampers.[4]

can generate pressures up to 7500 PSIG (517 barg) and can operate with turndown ratios exceeding 10:1. *Rotary, centrifugal,* and *multistage centrifugal* pumps are all used in boiler applications.

Boiler feedwater pumps are sized to handle 2 or 2.5 times the actual boiler capacity. As boiler capacities are often expressed in boiler HP (the energy required to evaporate 34.5 lb of water at 212°F or 15.7 kg at 100°C), the required feedwater pump capacity is 2.5 × 34.5 = 86 lb of water for each boiler HP.

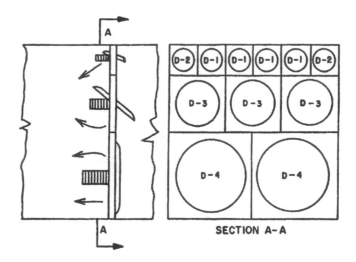

FIGURE 1.6   The design of tight shut-off, potentially linear dampers (Courtesy of Mitco Corporation).

B. BOILER EFFICIENCY

Boiler efficiency is that percentage of the higher heating value of the fuel (H) that has been transferred into the water as the steam is generated. In Eq. 4 below, the steam and fuel flows are noted as S and Q and the enthalpy of steam and feedwater are hg and hf.

$$E(\%) = 100S(hg - hf)/(HxQ) \qquad (4)$$

Boiler efficiency can be more accurately calculated on the basis of measuring the flue gas composition, steam temperature, and fuel composition. Some boiler efficiency calculations consider seven sources of potential losses and 44 pieces of data.[3]

The heating value of a pound of fuel in British thermal units (BTUs) can be calculated by Dulong's formula, where C, H, O, and S are mass percentages of carbon, hydrogen, oxygen, and sulfur:

$$H = 14,540C + 62,000(H - O/8) + 4050S \qquad (5)$$

For the combustion to be complete, an air excess is required. Each molecule of oxygen that does not participate in the combustion process (and four molecules of nitrogen, which enters with it), will only waste heat as it passes through the boiler and leaves with the hot flue gases. Each 50°F (28°C) increase in flue gas temperature[1] and each 2% increase in excess oxygen[2] results in about a 1% drop in boiler efficiency. When air is deficient, some of the carbon will not burn into CO_2 but only into CO. In that case, 70% of its heating value is lost as CO passes out of the stack unburned.

Boiler efficiency can be increased by reducing flue gas temperature, by keeping boiler tubes clean on both the water and the fire sides, by sealing against air infiltration, by using flue gases to preheat both the water and the air, and by operating the boiler at a capacity that corresponds to the boiler's maximum efficiency point. Under these conditions, 88 to 89% efficiency can be obtained from a coal-fired boiler, when operated at 4% excess oxygen, 87 to 87.5% can be obtained from oil firing at 3% excess oxygen, and 82 to 82.5% from gas firing at 1.5% excess oxygen. In the case of a 100,000-lb/hr (45,454-kg/hr) boiler, a 1% drop in efficiency increases the yearly operating cost by about $20,000.

As can be seen in Figure 1.7, the efficiency of each boiler varies with load. When multiple boilers are used to meet the total load, it is frequently advisable to let one boiler respond to the

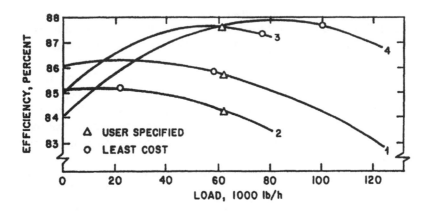

FIGURE 1.7 Boiler efficiencies vary with both the design of the boiler and its loading.

load variations, while the capacities of the others are fixed at their maximum efficiency points. Another common technique is to assign all increases in load to the most efficient and all decreases to the least efficient boiler in operation.

C. BOILER DYNAMICS

The boiler must respond to both changes in the load (process demand for steam) and changes in the heating value of the fuel, which is likely to vary if the fuel used is waste, such as in incinerators. The key dynamic characteristics of boilers (or of any other process) are their *dead time* (T_d) (the time it takes to start responding to a change) and their *time constant* (T_c) (the speed at which they respond). It is the ratio of T_d/T_c that determines the quality of control. The usual goal is to keep the dead time under 10% of the time constant. In fast processes, such as boilers, where the time constant is under a minute, this limits the allowable dead time to a few seconds.

The dead time of a boiler control loop is the sum of its transportation lags, mixing lags, and instrument lags. The fuel transportation lags are the largest with solid fuels, less with liquid or gas. There is also a dead time due to transportation as the combustion gases displace the furnace. Because the volume to be displaced is constant, the furnace displacement dead time drops as the load (flow rate of flue gases) rises. Instrument dead times can be caused by sensors, final control elements, or the cycle time of the Distributed Control System (DCS) itself. Sampling analyzers in the flue gas stream (O_2 or CO) can introduce more than a minute of dead time because of their location, fly-ash filter, and sampling tube volume. This can make them unuseable for closed loop control. Air vanes or dampers can also introduce several seconds of dead time due to their dead band and hysteresis. When cross-limited firing controls are used to guarantee that air always "leads" fuel, the dead time of the air controls further slows the fuel response. The cycle time of the DCS system (usually more than a second) is pure dead time, since the system is not even looking at the controlled variable until the cycle time passes. For all liquid pressure and flow loops and for most level and gas pressure loops, the dead times introduced by valve dead bands, smart transmitters, field bus update times, and DCS cycle times are long enough to deteriorate performance.

Loops oscillate with a period of three-to-five dead times, and the integral and derivative settings are adjusted to correspond to the loop dead times. Because loops are tuned for quarter amplitude damping, a boiler will go through three or four cycles after each upset before reaching its new steady state. Therefore, by minimizing dead time, one is minimizing the time during which the boiler cycles after an upset.

The loop dead time between steam pressure and firing rate controls (in well-designed boilers) is approximately 5 to 10 seconds: after a change in fuel valve opening, it takes 5 to 10 seconds for the steam pressure to *start* responding. The period of oscillation of such a loop is under 1 minute. In not-so-well-designed boilers, dead times and oscillating periods can be an order of magnitude greater. The dead time of the excess oxygen control loop is usually the highest (often over a minute) and it also varies (drops) with load. The water side (feed water and drum level) usually has very little dead time (less than a second).

Feedforward is a very effective means to reduce the dead time of loops. One can anticipate and respond to the changes before they occur by realizing that, for example, steam flow responds faster to load variations than steam pressure, that changing air flow into the furnace (FD fan) will require a change if flue gas flow out of it (ID fan) a few seconds later, or that the "shrink/swell effect" in the drum causes the drum level signal to misrepresent water inventory for about 10 seconds after each upset.

The time constant (the time it takes to reach 63% response to a step change) sets the speed of response of the boiler. This speed is a function of boiler design, load, and instrumentation. Some of the design features that affect the responsiveness include (1) the speed at which the flame volume can grow or shrink, and (2) the total mass of water that is to be heated (more in fire-tube, less in water-tube designs). The mass of water also varies with load: it is the largest at zero load (no bubbles) and the minimum at 100% load. (3) As the heat-transfer area is relatively constant, the larger the load, the less efficient the heat transfer. This is why the flue gas temperature has to rise (to increase ΔT) with load.

As the goal of tuning for quarter amplitude damping is to set the loop gain at a constant value of about 0.5, the higher the process gain (the more responsive the process), the lower the controller gain (wider the proportional band) has to be to result in a constant loop gain product. Therefore, the pressure controller on a responsive boiler would be tuned for a low gain (e.g., Proportional Band (PB) = 100%), but provided with some derivative, which is set to match the dead time of the loop (5 to 10 seconds) to increase the loop response to upsets. Most boilers are fast enough to respond to load changes that occur at a rate of 20 to 100% of full load per minute.

II. BASIC CONTROLS

In this section, a brief summary is given of the sensors and safety devices used, followed by a discussion of pressure, fuel, air, draft, water, flue gas, and steam superheat temperature controls.

A. Sensors

The normal "on-line" requirements for steam boilers are to control steam pressure within ±1% of desired pressure; fuel-air ratio within ±2% of excess air (±0.4% of excess oxygen), based on a desired "load" versus "excess air" curve; steam drum water level within ±1 inch of desired level; and steam temperature (where provision is made for its control) within ±10°F (5.6°C) of desired temperature. In addition, the efficiency of the boiler should be monitored within ±1%. To reach these performance goals, it is necessary to install accurate sensors.

Figure 1.8 shows the in-line instruments used on a boiler. The most important sensors are the flow detectors, which provide the basis for both material and heat balance controls. For successful control of the air-fuel ratio, combustion airflow measurement is important. In the past it was impossible to accurately measure air flow. Therefore, the practice was to provide some device in the flow path of combustion air or combustion gases and to field calibrate it by running combustion tests on the boiler.

These field tests, carried out at various boiler loads, used fuel flow measurements (direct or inferred from steam flow) and measurements of percent of excess air by gas analysis; they also used the combustion equations to determine air flow. Because what is desired is a relative measurement with respect to fuel flow, the airflow measurement is normally calibrated and presented on a relative basis. Flow versus differential pressure characteristics, compensations for normal variations in temperature, and variations in desired excess air as a function of load—are all included

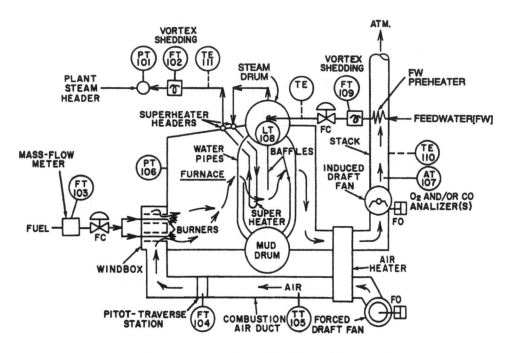

FIGURE 1.8   The main in-line instruments are shown here for a drum-type boiler.

in the calibration. The desired result is to have the airflow signal match the steam or fuel flow signals when combustion conditions are as desired.

The following sources of pressure differential are normally considered:

> Burner differential (windbox pressure minus furnace pressure)
> Boiler differential (differential across baffle in combustion gas stream)
> Air heater differential (gas side differential)
> Air heater differential (air side differential)
> Venturi section or flow tube (installed in stack)
> Piezometer ring (at forced draft fan inlet)
> Venturi section (section of forced draft duct)
> Orifice segments (section of forced draft duct)
> Air foil segments (section of forced draft duct)

Of these, the most desirable are the last four, because they use a primary element designed for the purpose of flow detection and measure flow on the clean-air side. None of these sensors meets the dual requirement of high accuracy and rangeability. Actually, they are of little value at 30% flow or less. Table 1.1 lists some better flow sensors, such as the multipoint thermal flow probe or the area-averaging pitot stations provided with "hexcel"-type straightening vanes and with membrane-type pressure balancing d/p cells. These are major advances in combustion airflow detection. This table shows the measurement errors that can be anticipated. Unfortunately, as the flow is reduced, the error—in percentage of actual measurement—increases in all cases except the first two. With linear flow meters, the error increases linearly with turndown. Therefore, the thermal flowmeter error increases tenfold at a turndown of 10:1. In the case of nonlinear flowmeters, the error increases exponentially. Therefore, at a turndown of 10:1, the orifice or pitot error increases 100-fold and causes these devices to become useless. This situation can be alleviated somewhat by the use of two d/p cells on the same element or by the use of "smart" units.

TABLE 1.1
Flow Sensor Errors on Boilers

Flow Streams Measured	Type of Flowmeter	Inaccuracy (% of Flow)			Rangeability	Limitation
		At 10%	33%	100%		
Fuel (Oil)	Coriolis mass flow	0.5	0.5	0.5	20:1	Not for gases
Steam and Water	Vortex shedding					
	Steam	1–1.5	1–1.5	1–1.5	10:1	Min. R_e = 20,000
	Water	0.5–1	0.5–1	0.5–1		Max. temp = 750°F (400°C)
	Orifice, venturi or flow nozzle	NG*	2–5	0.5	3:1	Rangeability can be increased if using two conventional d/p cells or a "smart" d/p cell
Air	Area averaging pitot traverse station	NG	2–10	0.5–2	3:1	
	Multipoint thermal	5–20	2–5	1–2	10:1	Dual range unit required
	Piezometer ring, orifice segment Venturi section airfoil section	NG	3–20	2–3	3:1	Cannot be used below 25% of max. flow

Note: *NG = Not Good

Based on the data in Table 1.1, if the boiler efficiency is to be monitored on the basis of time-averaged fuel and steam flows, the lowest error that can be hoped for is approximately 1%.

Similarly, the air-fuel ratio cannot be measured to a greater accuracy than the airflow. At high turndown ratios, this error can be very high. Considering that a 2% reduction in excess oxygen will increase the boiler efficiency by 1%, both the accurate measurement and the precise control of airflows is essential in boiler optimization.

The demand for sensor accuracy elsewhere is not so stringent. Standard instrumentation allows for the control of steam pressure within ±1%, furnace pressure within ±0.1 in. H_2O, (2.5 millibar), water level within ±1 in. of desired level, and steam temperature to within ±10°F (5.6°C).

B. SAFETY INTERLOCKS

Operations relating to safety, startup, shutdown, and burner sequencing are basically digital in nature and operate from inputs such as contact closures.

The basic safety interlocks are as follows:

Purge interlock — Prevents fuel from being admitted to an unfired furnace until the furnace has been thoroughly air purged.
Low airflow interlock and/or fan interlock — Fuel is shut off upon loss of airflow and/or combustion air fan or blower.
Low fuel supply interlock — Fuel is shut off upon loss of fuel supply that would otherwise result in unstable flame conditions.
Loss of flame interlock — All fuel is shut off upon loss of flame in the furnace and/or fuel to an individual burner is shut off upon loss of flame to that burner.
Fan interlock — Stop forced draft upon loss of induced draft fan.
Low water interlock (optional) — Shut off fuel on low water level in boiler drum.
High combustibles interlock (optional) — Shut off fuel on highly combustible content in the flue gases.

Where fans are operated in parallel, an additional interlock is required to close the shutoff dampers of either fan when it is not in operation. This is necessary to prevent air recirculation around the operating fan.

Automatic startup sequencing for lighting the burners and for sequencing them in and out of operation is common. This is accomplished at present by either relays or solid-state hard-wired logic systems. These functions are frequently accomplished by Programmable Logic Controllers (PLCs).

Design and specification of the various safety interlocks are specialized as a result of insurance company regulations, NFPA (National Fire Protection Association), and state regulations.

C. STEAM PRESSURE CONTROL

Figure 1.9 shows one set of possible basic controls for an oil-fired, balanced, water boiler. Some of the basic control loops, starting with steam pressure controls, are described below.

The thermal power required from a boiler is related to the square root of the h/p ratio, where h is the pressure differential through a venturi flow element and p is steam pressure.[6] While steam drum pressure alone is a good indicator of the balance between firing rate and steam demand, the response of this pressure to a load (or firing rate) change is slower than the response of steam flow (h) (Figure 1.9). Therefore, the addition of a feedforward signal based on steam flow (FT-102) will make the loop more responsive. The feedback trim by pressure control (PIC-101) is still required because control of steam flow alone would contain a positive feedback, where an increase in flow increases firing rate, which in turn would further increase steam flow.

The multiplier in Figure 1.10 (PY-101) also adjusts the control loop gain in proportion to the load (h). This is a desirable feature because at low loads the process gain is high (a large heat transfer area is used to transfer little heat), and, therefore, to make the loop gain (the product of the dynamic gains of all loop components including the process) constant, the controller gain has to drop.

As shown in Figure 1.10, a lead/lag compensator (FY-102) is provided to simulate the time delay between a change in firing rate and the resulting change in the rate of steam generation.

In the scheme shown in Figure 1.10, an increase in load (steam demand) causes flow (h) to rise and pressure (p) to fall. Both of these effects cause the h/p ratio to rise, which increases the firing rate. In this system, therefore, load changes cause both h and p to move in *opposite* directions. In the case of an upset on the heat supply side, e.g., an increase in the heating value of the fuel, both pressure (p) and flow (h) will rise (they move in the *same* direction), and therefore the value of h/p and the feedforward signal itself stays constant (as the load is constant). PIC-102 corrects for this type of an upset in the feedback by reducing the fuel-to-steam ratio setting of the multiplier (PY-101). This is a valuable feature when burning solid fuels or when operating heat-recovery incinerators.

PIC-101 does not need to be tuned very tightly because, if the feedforward part of the loop is properly adjusted, the feedback controller will have very little to do other than correcting for sensor errors and fuel variations. The settings for PIC-101 are a function of the volume of the steam drum and distribution headers, but typically can be set for PB = 100%, I = 1 to 5 repeats per minute, and D = 0 to 10 seconds. If implemented in DCS systems, the scan or cycle period should not exceed 0.1 seconds. If the PIC is used directly to set the firing rate (without feedforward correction), the loop has to do much more work (tuned more tightly), e.g., to PB = 15 to 20%, and the integral correction must be reduced, e.g., to I = 0.2 repeats per minute. Some authorities recommend the elimination of the integral under such conditions and the use of derivative instead[7] (D = 0.05 to 0.1 minutes).

D. FUEL CONTROLS

The firing rate signal from Figure 1.10 is the set point for the fuel controls. To maintain a linear relationship (constant gain) between set point and fuel flow on a mass basis, the use of a complete flow control loop is recommended.

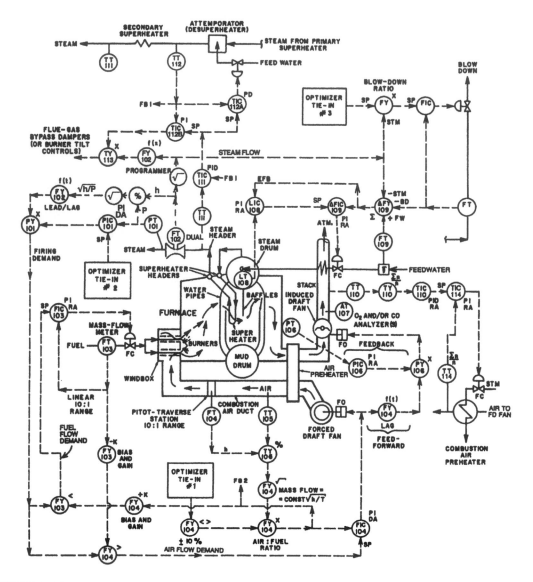

FIGURE 1.9 Basic controls for an oil-fired steam pressure control without optimization.

1. Gas Fuel

When the mass flow of a gas fuel is to be controlled (Figure 1.11), sometimes it is necessary to control the inlet gas pressure because of burner considerations or because the flowmeter requires it.[8] When pressure regulation is also required, one might use a self-contained regulator or a whole pressure control loop. If the pressure drop across the pressure regulator (V1 in Figure 1.11) is smaller than the pressure drop assigned to the flow control valve (V2) that is an acceptable configuration. On the other hand, if the ΔP across V1 is larger (V1 is smaller than V2), then the use of a regulator will result in substantial interaction between flow and pressure controls. This is because the relative gain of the flow loop[6] is (P2-P3)/(P1-P3) and under such conditions the relative gain of the flow loop would be under 0.5. If that is the case, V1 must be assigned to the FIC and V2 to the PIC in Figure 1.11, and this is only possible if a full pressure control loop is substituted for the regulator.

As gas flow rises, the burner back pressure (P3) increases, which leaves less ΔP for the flow control valve. Therefore, if the valve ΔP varies by more than 2:1 as the load changes from minimum

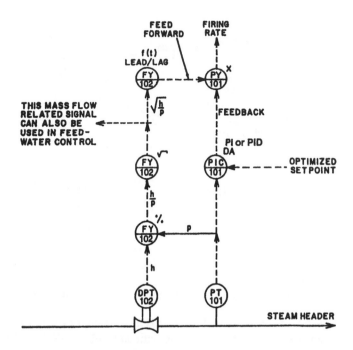

FIGURE 1.10 Firing rate determination using feedforward loop with feedback trim.

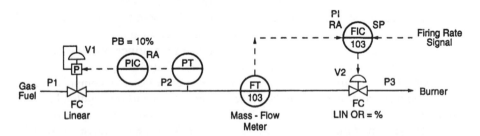

FIGURE 1.11 Gas fuel controls. FIC-103 should control the smaller of the two values (V1 or V2).

to maximum (to keep the loop gain relatively constant), an equal percentage valve should be used. For gas pressure control, a linear valve is generally acceptable if the ΔP variation is under 2:1, while for regulators with a gain over 20 (PB < 5%), the standard characteristics offered by the manufacturer are acceptable. A valve positioner will improve the performance of the PIC loop, but might contribute to cycling in the FIC loop. Therefore, if the positioner cannot be removed, or bypassed, the FIC output should be lagged, thereby making the valve response artificially faster than the process variable (flow).

The PIC requires proportional mode only with a gain of 10 or even 20 (PB = 5 to 10%). The FIC should be set for a low gain, such as 0.3 (PB = 300%), with much integral action (10 to 20 repeats per minute). The measurement to both controllers should be filtered to remove noise.

2. Liquid Fuel

To guarantee good mixing with the combustion air, liquid fuels must be atomized. The atomizing air or steam flow is ratio controlled off the firing rate signal or off the fuel flow signal (Figure 1.12). The amount of steam used for atomization is about 1% of the total steam generated by the

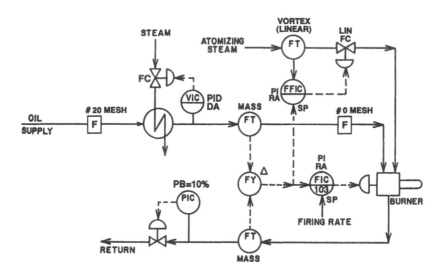

FIGURE 1.12 Oil flow control for recirculating burner provided with steam atomization.

boiler. The steam valve can be linear if the flowmeter is also linear, or quick opening, if an orifice-type flow sensor is used. The FFIC settings are approximately PB = 300% and I = 10 to 20 r/m.

For the atomizer to operate properly, particularly when heavy residual oils are burned, it is necessary to heat the oil in order to keep its viscosity low and constant. For proper burner operation and to prevent flashing, the back pressure on the burner must also be controlled, as the heavy residual oil fuel is continuously circulated past the burner. A #20-mesh strainer is used upstream of the heater and a #60-mesh before the burner. The net fuel flow to the burner is signaled by the difference between the two mass flowmeters. They usually are of the Coriolis type. Firing rate provides the set point to FIC-103, while its output modulates the orifice opening in the burner.

The tuning requirements for FIC-103 are similar to those listed for gas fuel control in Figure 1.11 (PB = 300%, I = 10 to 20 r/m), except that the loop is even faster (incompressible fluid) and therefore the dead band contributions (of the burner actuator, the smart transmitter, long pneumatic transmission lines, field bus update time, and DCS cycle time) all become more critical and must be minimized for good performance. In addition, the noise from the measurement signal must be filtered by blocking any flow changes that occur faster than the response speed of the burner.

3. Multiple or Waste Fuels

When more than one fuel is used, it is necessary to total their flows on the basis of their individual heating values before the total is compared to the firing rate signal. The total heating value (calories or BTUs per unit of fuel mix) is not necessarily the same as the total combustion air required to burn that mixture. The theoretical air requirement (on a mass basis) to burn a unit of carbon is 11.53, for hydrogen it is 34.56, and for sulfur it is 4.32. Therefore, separate summers are needed to determine the total heat input or air requirement.[5]

In Figure 1.13, a waste gas is burned. If high waste concentrations could cause flameout, then FY-1 is set for less than 100% to prevent that. If the boiler can be operated on 100% waste, FY-1 can be left out. The subtractor (PY-2) must be scaled on the basis of the heating values of the two fuels, and the scaling must also consider the transmitter ranges. If waste fuel availability becomes limited, the gas supply pressure drops and the FY-3 selector will transfer control over to the PIC. As a consequence, FY-2 will increase the set point of the primary fuel FIC-2. The summer FY-4 should be scaled on an "air-required" basis because its output is used in the air/fuel ratio control loop.

Particularly in multiple fuel burner controls, it is advisable to provide all valves and transmitters with linear characteristics so that signals can be freely combined, subtracted, multiplied, or divided.

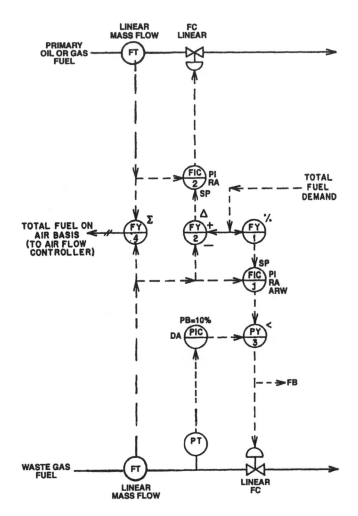

FIGURE 1.13 Automatic control system for burning limited-availability waste up to preset maximum percentage in total mixture.

The tuning requirements for the flow and pressure controllers are similar to the ones in Figures 1.11 and 1.12. Because of the selective control over the waste fuel valve, FIC-1 must be provided with antireset wind-up (ARW) from the external feedback (FB) of the output signal of PY-3.

4. Coal

When burning solid fuels such as coal, bark, bagasse, or municipal solid waste, the control system is a function of the capacity of the mill, which is used to reduce the particle size of the fuel. When hammer mills are used, this capacity is small, there is only an equivalent of a few seconds' coal supply in the mill, and therefore a change in the rate of feed to the mill results in an almost immediate change in pulverized coal flow to the burner. In such installations in earlier designs, the power requirement of the pulverizer was used as an indication of coal flow.

When the coal feed to the mill is measured directly (Figure 1.14), the lag time of transportation is compensated by FY-1. If the fuel demand set point rises or drops, FIC-103 will detect this only after the lag time has passed and therefore the mill will be temporarily over or underfed (to improve its response) before a new steady-state is reached.

When ball or roller mills are used, the mill capacity is relatively large, the mill delay exceeds 1 minute, and therefore neither pulverizer power nor coal inlet flow are acceptable means of

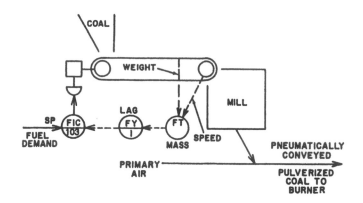

FIGURE 1.14 Compensation for mill capacity when coal flow is measured.

detecting burner feed rate. In that case, the primary air flow that conveys the pulverized coal is manipulated as a function of coal demand. As the demand for coal rises, the resulting increase in air flow also increases the coal feed to the pulverizer.

Because the coal loading of the air is not uniform, and because the heating value of the coal varies, the total air flow to all the pulverizers (FY-2) does not accurately reflect the total coal feed to the burners (Figure 1.15). Therefore, it is advisable to apply some feedback correction to this signal by calculating the *actual* heat release from the boiler (FY-1) based on the flow and pressure of the steam generated. Shinskey recommends bringing the estimated heat flow (air based) and actual heat release together by inserting the optimizing controller (OC) provided with a slow integral action. It is important that the integral time of this controller be *longer* than the boiler time constant so that an increase in steam flow will not cause a drop in fuel flow through the integral action of the OC.

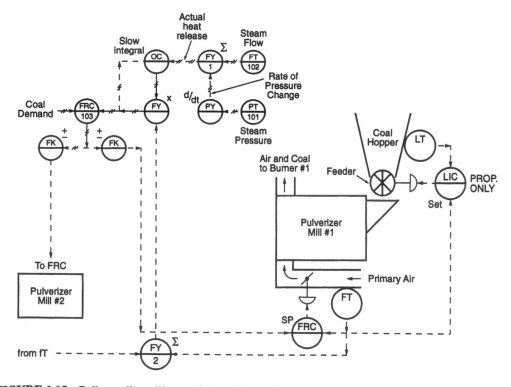

FIGURE 1.15 Ball or roller mill control through steaming rate-corrected airflow modulation.

E. Air Controls

In the early designs of boiler air controls, the inaccurate sensors, leaking and nonlinear dampers, dead band, and hysteresis in linkages or fan inlet vanes made air control unreliable. This often resulted in disabling the air-fuel ratio trim, when the load dropped below 25%. In these early systems, primitive air sensors were field calibrated by tests that were based on fuel (or steam) flow, excess oxygen in the flue gases, and the combustion equations. These field-calibrated sensor signals were quite noisy, and therefore their signals had to be dampened.

In today's installations, the quality of the sensors has improved (Table 1.1) and the air flow is controlled in self-linearizing closed loops. To increase the turndown (rangeability) of the system on steam-turbine driven fans, both the damper opening and the fan speed are throttled in a split-range configuration: as the demand for air flow drops from 100 to 60%, the damper is throttled from full open to its minimum opening, and, as the demand for air drops from 70 to 0% (or the minimum allowable), the fan speed is reduced from its maximum to minimum value.

Figure 1.16 describes an air control loop, which is configured to measure the mass flow of air ($\sqrt{h/T}$) over a 10:1 range. It is best to measure and control air on the same side of the boiler (to minimize interaction with draft controls); therefore, FIC-104 is throttling the inlet vanes (and/or speed) of the forced draft (FD) fan(s). The tuning requirements of FIC-104 are likely to be similar to the ones described for FIC-103 in Figure 1.11.

1. Draft Controls

In balanced-draft boilers, a slight negative pressure ($0.05"H_2O = 1.25$ mm H_2O) is maintained at the top of the furnace so that the leakage that might exist will result in the atmospheric air entering (and not unburned fuel escaping from) the furnace. If the draft is detected below the top of the furnace, it is necessary to compensate for the chimney effect of the hot gases and increase the draft (vacuum) setting by about $0.01"H_2O$ for each foot (about 1 mm H_2O per meter) as the sensing tap is lowered. The pressure tap should be large, at least 1 in (25 mm) in diameter. Because draft measurement is critical, in larger installations three transmitters are used. They are configured in voting or median-selector configurations to increase system reliability.

As the flow and pressure loops (FIC-104 and PIC-106) do interact and because both the FD and the ID fan affect the draft, the measurement tends to be noisy. This can be reduced by using a wide transmitter span or by filtering. The controller (PIC-106) can be provided with inverse-derivative action to filter out the noise or can be provided with wide proportional and high integral settings (several repeats per minute), which give essentially floating control.

Because every time the air inflow to the furnace (FIC-104) changes, the furnace pressure will also be upset, unless the material balance is reestablished by also increasing the outflow; therefore, a feedforward loop is provided in Figure 1.16 to decouple this potential source of interaction. If the draft is measured near the top of the furnace, it will respond faster to a change in outlet damper opening, and therefore a dynamic lag (FY-104C) is provided in the feedforward path. This feature leaves much less work to be done by the draft controller, which provides the feedback trim.

A multiplier (PY-106) in the decoupling loop is preferred to a summer, because the multiplier automatically adjusts the gain of the feedforward decoupler in proportion with flow. This is desirable. On the other hand, PY-106 also causes the PIC-106 feedback loop gain to also vary directly with flow. This should be compensated: to keep the loop gain constant, it is recommended[6] to also automatically reduce the controller (PIC-106) gain as the flow rises.

2. Air-Fuel Ratio

Efficient boiler operation requires the continuous matching of the fuel and air flows while maintaining a slight air excess. The air and fuel flowmeters are scaled so they both need to be at the

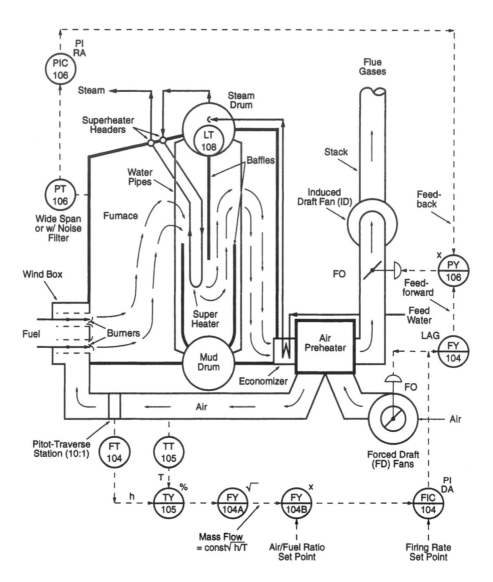

FIGURE 1.16 Airflow and draft controls.

same percentage to give the right air-fuel ratio. A well-designed air-fuel ratio system will guarantee that air is "leading" all the time. This means that on a load increase, the air flow increases faster than the fuel and on a load decrease, the air flow is "lagging" (its flow drops slower) than that of the fuel. The high and low selectors (FY-104A and FY-103 in Figure 1.17) provide this feature and guarantee that no "fuel-rich" mixture can exist, even for a few seconds. The system also protects from the consequences of a sticking fuel valve or a fan failure. The cross-limited parallel air-fuel ratio controls guarantee that (1) fuel flow is limited to actual airflow, (2) minimum fuel required for stable flame is guaranteed, and (3) the minimum air flow required to match minimum fuel is always provided (FY-104B in Figure 1.17).

An undesirable consequence of cross-limiting is that the boiler response is reduced to match the response of the slower of the two final control elements (fuel valve or air inlet vane). When the change in firing rate is small (less than 1%), this effect is eliminated by the "bias and gain" elements (FY-103 and 104), because they bias the high and low selectors as if the air flow were

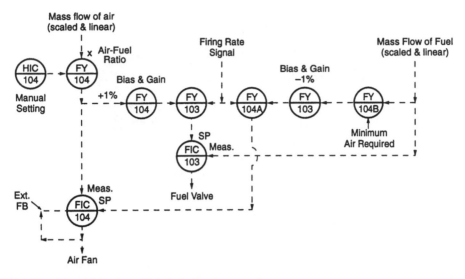

FIGURE 1.17 Cross-limited parallel air-fuel ratio controls.

1% more and the fuel flow 1% less than its actual value. Therefore, the small changes pass through immediately (which is safe, because of the excess air), while larger changes are cross-limited. By limiting the set point of the air flow (FY-104A and B), one could cause integral wind-up. Therefore, ARW protection is added to block the integral action in the airflow controller (through external feedback), during periods of "air limiting."

F. Feedwater Controls

It is necessary to control the water level in the steam drum because a drop in level could result in the uncovering and thereby overheating of boiler tubes, and a rise in level will interfere with the operation of moisture separators. The apparent water level and the water inventory (mass of water in the drum) do not coincide because, if the load rises (steam flow increases and steam pressure drops), the level tends to "swell" due to the increased volume taken up by bubbles. Inversely, a drop in load, resulting in an increase in steam pressure (or an increase of feedwater flow, resulting in a drop in drum temperature), causes bubbles to collapse and an apparent drop ("shrink") in the level (Figure 1.18). The inverse response of the level controller therefore upsets the control of feedwater. At zero load, there are no bubbles; so, the water inventory (mass) is maximum. As the load rises and part of the liquid volume is displaced by bubbles, the liquid inventory drops, and some of the water must be removed to keep the level constant.

In small boilers with large water-reserve capacities, level control can still be used in spite of the effect of the inverse response. In such cases, if the level controller has only a proportional control mode, the setting can be PB = 25% (gain = 4). In intermediate-size boilers, a two-element (level + steam flow) loop might suffice, where the steam flow and the level controller output are summed and the summer is so scaled that on a load change the two changes are equal: the change in level controller signal in one direction and the change in steam flow transmitter signal in the opposite direction are equal.

In larger boilers, more sophisticated dynamic compensation is needed. One approach is to measure *four related variables*: the level and pressure in the drum and the flows of both steam and water. The dynamic "swell or shrink" period following a load change usually lasts for less than a minute. It is during this period that the level signal is inversed and a compensation is needed. After that, a new steady state is reached, with a new water inventory corresponding to the controlled level because the volume taken up by bubbles has changed. A change in load always causes the

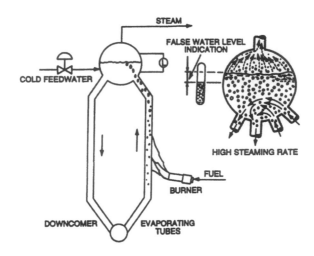

FIGURE 1.18 Partial vaporization in the evaporating tubes causes drum level to shrink when feedwater flow increases and when pressure rises. On the other hand, an increase in the demand for steam causes the level to "swell."

steam pressure and steam flow to move in opposite directions. When this occurs, the level measurement should temporarily be overruled. If steam pressure and flow move in the same direction, an increase in the heating value of the fuel is indicated. If pressure and level measurements are moving in opposite directions, a load change (shrink or swell period) has occurred. If they move in the same direction, an imbalance in the water and steam flows has occurred. Based on the knowledge of the above relationships, the measurement of the four variables can be used to control the dynamic periods of rapid load change.

Another control choice is the material-balance control scheme shown in Figure 1.9. There the two flowmeters (FT-102 and FT-109) have identical ranges on a mass-flow basis. When the two flow signals are equal, ΔFY-109 sends a 50% signal to the flow-difference controller (ΔFIC-109). LIC-108 trims the ΔFIC set point to correct for flowmeter errors, blowdown losses, and variations in water inventory at different loads. The integral setting of the LIC is an order of magnitude slower than that of the ΔFIC; therefore, the LIC balances the system during periods of steady load conditions. LIC-108 is provided with an external feedback signal (EFB) to protect against reset wind-up, when the feedwater flow is under manual control or is otherwise limited.

Figure 1.19 illustrates a feedwater control system that takes into consideration both the swell—shrink effect and the water inventory variation with load. The inverse response of the level is corrected by the addition of an impulse signal to the level transmitter output in a summer (LY-108), which is equal in size (set by K) but of opposite sign than the shrink—swell error. The value of K is lower for high-pressure boilers because the shrink—swell effect is reduced with pressure. The lag time of the impulse is set for about 20 seconds. After this lag time, the effect of the dynamic compensator (LY-108) is automatically eliminated, as the impulsed and the bypass signals are equal and of opposite signs. Steam flow is sent as a feedforward signal to the integral port of LIC-108. This corrects for the drop in water inventory with load. The P and I settings of LIC-108 are approximately PB = 25%, I = > 1 min per repeat.

The feedwater valve usually handles very pure water, which is "metal hungry" at high pressure drops. Chrome-molybdenum steel valves can withstand the corrosion erosion that results, while cavitation must be eliminated. The valve should have a constant gain so that the loop gain does not vary with load. This can be provided by linear characteristics or by a linearizing positioner. Oversizing is a common problem because these valves are often sized for the sum of boiler capacity *plus* relief valve capacity, which is unnecessary.

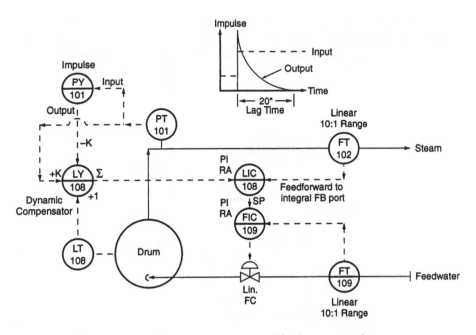

FIGURE 1.19   Inverse response and inventory-compensated feedwater controls.

G. STEAM SUPERHEAT CONTROLS

In central stations, the goal is to generate steam at 1050°F (566°C). This is the maximum temperature at which the life of the generator turbine blade is still likely to be acceptable. As shown in Figure 1.20, the uncontrolled steam temperature tends to rise with load because the hottest gases tend to propagate further at higher loads. At low loads (Zone 1 in Figure 1.20), the only way to increase steam temperature is to either use recirculation blowers or to tilt the burners (the blowers and the tilt mechanisms are not shown in Figure 1.21) so that the flames will be more aimed at the superheat tubes.

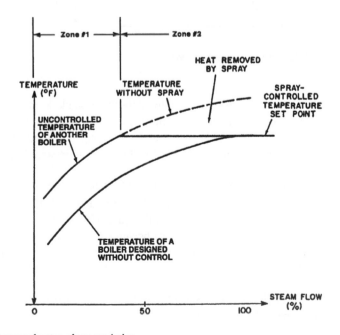

FIGURE 1.20   Desuperheater characteristics.

At high loads (Zone 2 in Figure 1.20), the rise in steam temperature is prevented by either bypassing the convection section of the superheater (closing damper 1 and opening damper 2) or by attemperation with water. The required burner and damper positions corresponding to each load are well established for most boilers, and so are often feedforward "programmed" directly from the load, based on either steam (FY-102 in Figure 1.21) or air flow.

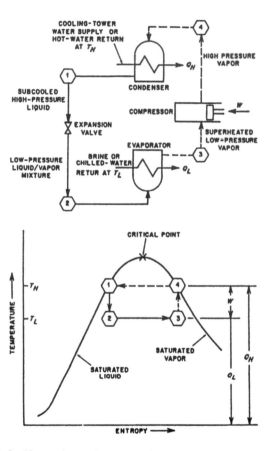

FIGURE 1.21   Cascaded, feedforward superheat controls.

Figure 1.21 describes a feedforward-corrected temperature cascade loop with two slaves (TIC-112A and B). The master, TIC-111, detects the final superheat temperature and is provided with an external feedback (from TT-112) so that its integral mode will not wind up if the temperature is below set point. This can occur because at low loads the boiler is incapable of maintaining high steam temperature (Zone 1 in Figure 1.20).

The two slaves detect the steam temperature (TT-112) downstream of the attemperator. The faster slave, TIC-112A, controls the attemperator and is provided with PD control modes and a bias. The use of the PD modes avoids conflict with the other (PI) slave and minimizes the usage of water (an irreversible waste of boiler efficiency) because the derivative mode shuts the valve before the set point is reached. The bias is provided to keep the attemperator responsive by delivering a minute amount of water even when TIC-112A is on set point and would otherwise call for no water at all.

The slower slave (TIC-112B) is a PI controller and modulates the tilt of burners or the flow through recirculation or bypass dampers. Because at low loads the integral action could wind up

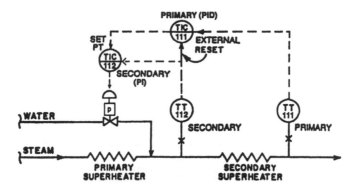

FIGURE 1.22 Cascade control with external reset.

if the steam temperature remained below the set point, a batch switch is provided with this controller. The batch switch prevents wind-up by sending a preload signal to the integral mode whenever the controller output exceeds its preset limit. The amount of work that TIC-112B has to do is minimized by the feedforward signal generated by FY-102, which anticipates the required tilt or bypass settings based on the prevailing load. The use of a multiplier (FY-113) increases the loop gain with load, which is desirable, because the process gain drops with load and therefore these effects tend to cancel.

Figure 1.22 illustrates the controls used when the only means of steam superheat control is water spraying. When the boiler load is low (Zone 1 in Figure 1.20), no spraying is needed because the steam temperature is below the set point. Yet the loop must quickly open the spray valve if a sudden load increase causes the steam temperature to quickly rise. This capability is provided by eliminating reset wind-up through the use of the external feedback provided in Figure 1.22.

III. OPTIMIZING CONTROLS

When the yearly boiler fuel cost is in the millions, even a few percentage points of improved efficiency can justify the costs of added instruments and controls. In the following paragraphs, a number of optimization techniques will be described. The various goals of optimization include

1. Minimize excess air and flue-gas temperature.
2. Minimize steam pressure.
    a. Turbines thereby will open up turbine governors.
    b. Reduce feed pump discharge pressures.
    c. Reduce heat loss through pipe walls.
3. Minimize blowdown.
4. Measure efficiency.
    a. Use the most efficient boilers.
    b. Know when to perform maintenance.
5. Provide accountability.
    a. Monitor losses.
    b. Recover condensate heat.

The first three methods of optimization are achieved by closed-loop process control and can be superimposed upon the overall boiler control system shown in Figure 1.9. The tie-in points for these optimization strategies are also shown in that figure. The benefits of the last two methods

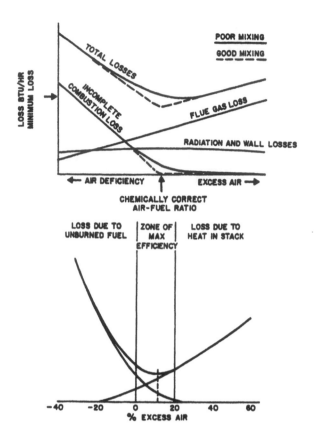

FIGURE 1.23   Boiler losses can be plotted as a function of excess air. (Top) The minimum point of the total loss curve of a boiler is where optimized operation is maintained. (Bottom) Most efficient operation of a boiler occurs when the amount of excess air in the stack balances the losses in unburned fuel. (Adapted from Walsh, T.J., *I&CS*, January 1981.)

(efficiency and accountability) are not obtained in the form of closed-loop control signals, but they do contribute to better maintenance.

A. EXCESS AIR OPTIMIZATION

If a boiler is operating on a particular fuel at a specific load, it is possible to plot the various boiler losses as a function of air excess or efficiency, as shown in Figure 1.23. The sum total of all the losses is a curve with a minimum point. Any process that has an operating curve of this type is an ideal candidate for instrumental optimization. Such process control systems are operated by continuously determining the minimum loss point at that particular load and then shifting the operating conditions until that point is reached.

As shown in Figure 1.23, the radiation and wall losses are relatively constant. Most heat losses in a boiler occur through the stack. Under air-deficient operations unburned fuel leaves, and when there is an air excess, heat is lost as the unused oxygen and its accompanying nitrogen are heated up and then discharged into the atmosphere. The goal of optimization is to keep the total losses at a minimum. This is accomplished by minimizing excess air and by minimizing the stack temperatures (Figure 1.24).

The minimum loss point in Figure 1.23 is not at zero excess oxygen. This is because no burner is capable of providing perfect mixing. Therefore, if only as much oxygen would be admitted into the furnace as is required to convert each carbon molecule into CO_2, some of the fuel would leave

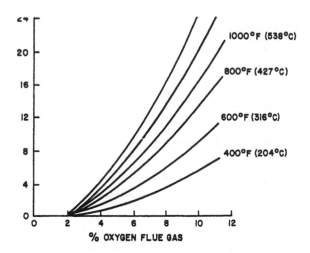

FIGURE 1.24 Fuel-saving potentials in a boiler with optimum conditions corresponding to 400°F stack gas temperature and 2% excess oxygen.

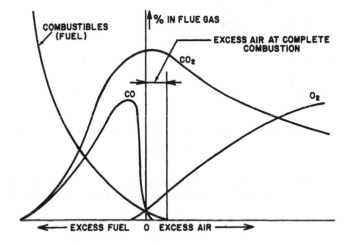

FIGURE 1.25 The major components of flue gas are oxygen, carbon dioxide, carbon monoxide, and unburned hydrocarbons. (Adapted from Walsh, T.J., *I&CS*, January 1981.)

unburned, as not all O_2 molecules would find their corresponding carbon molecules. This is why the theoretical minimum loss point shown by the dotted line in Figure 1.23 is to the left of the actual one. This actual minimum loss or maximum efficiency point is found by lowering the excess oxygen as far as possible, until opacity or CO readings indicate that the minimum has been reached. At this minimum loss point, the flue-gas losses balance the unburned fuel losses.

1. Flue-Gas Composition

Figure 1.25 shows the composition of the flue gas as a function of the amount of air present. The combustion process is usually operated so that enough air is provided to convert all the fuel into CO_2, but not much more. This percentage of excess oxygen is *not* a constant. It varies with boiler design, burner characteristics, fuel type, air infiltration rates, ambient conditions, and load.

The top portion of Figure 1.26 shows that the percentage of excess air must be increased as the load drops off. This is because at low load the burner velocities drop off and the air flow is reduced, while the furnace volume remains constant. This reduces turbulence and lowers the

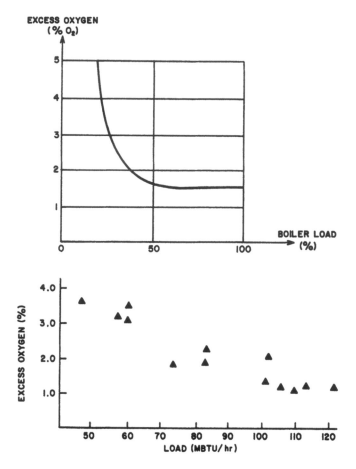

FIGURE 1.26   The top portion of this figure shows theoretical relationships between load and O_2; the bottom portion shows the test-based relationships between load and O_2. (Adapted from Schwartz, J.R., *InTech*, June 1983.)

efficiency of mixing between the fuel and the air. This loss of mixing efficiency is compensated for by the higher percentage of excess oxygen admitted at low loads.

The upper curve shown in Figure 1.26 theoretically illustrates the relationship between the excess O_2 requirement and load, and the lower plot provides actual test data for a specific boiler. Because each boiler has its own unique personality, this relationship must be experimentally determined. Once established, it can be used with a fair degree of confidence, although small shifts are still likely to occur as the equipment ages.

2. Effect of Fuel

In a boiler furnace (where no mechanical work is done), the heat energy evolved from the union of combustible elements with oxygen depends on the ultimate products of combustion.

Because of the weight ratios of oxygen and nitrogen in air (0.2315 and 0.7685, respectively), to supply 1 lb (0.45 kg) of oxygen for combustion, it is necessary to supply $1/.2315 = 4.32$ lb (1.96 kg) of air. In this amount of air, there will be $4.32 \times 0.7685 = 3.32$ lb (1.5 kg) nitrogen, which does not enter directly into the combustion process but which nevertheless remains present.

If theoretical or total air is defined as the amount required on the basis of equation (1), then express air is the percentage over that quantity.

Table 1.2 lists the excess air ranges required to burn various fuels. It can be seen that excess

---

TABLE 1.2
Usual Amount Excess Air Supplied to Fuel-Burning Equipment

Fuel	Type of Furnace or Burners	Excess Air (%)
Pulverized coal	Completely water-cooled furnace for slag-tap or dry ash removal	15–20
	Partially water-cooled furnace for dry ash removal	15–40
Crushed coal	Cyclone furnace–pressure or suction	10–15
Coal	Stoker-fired, forced-draft, B&W chain grate	15–50
	Stoker-fired, forced-draft, underfeed	20–50
	Stoker-fired, natural draft	50–65
Fuel oil	Oil burners, register type	5–10
	Multifuel burners and flat flame	10–20
Acid sludge	Cone and flat flame type burners, steam atomized	10–15
Natural, coke oven and refinery gas	Register-type burners	5–10
	Multifuel burners	7–12
Blast-furnace gas	Intertube nozzle-type burners	15–18
Wood	Dutch oven (10–23% through grates) and Hofft type	20–25
Bagasse	All furnaces	25–35
Black liquor	Recovery furnaces for kraft and soda pulping processes	5–7

Source: From Babcock & Wilcox Co., Reference 10

---

air requirement increases with the difficulty to atomize the fuel for maximum mixing. Figure 1.27 also illustrates that gases require the lowest and solid fuels the highest percentage of excess oxygen for complete combustion. The ranges in Table 1.2 and the curves in Figure 1.27 also illustrate that as the load drops off, the percentage of excess oxygen needs to be increased. The optimum excess percentages for gas, oil, and coal at full load are around 1%, 2%, and 3%, respectively .

3. Detectors

As shown in Figure 1.25, excess air can be correlated to O_2, CO, CO_2, or combustibles present in the flue gas. Combustibles are usually detected either as unburned hydrocarbons or in the form of opacity. These are used for optimization because the goal is not to maintain some optimum concentration but to eliminate combustibles from the flue gas. Therefore, such measurements are usually applied only as limit overrides.

The measurement of CO_2 is not a good basis for optimization either because, as shown in Figure 1.28, its relationship to excess O_2 is very much a function of the type of fuel burned. The CO_2 concentration of the flue gas also varies slightly with the CO_2 content of the ambient air. It can also be noted from Figure 1.25 that CO_2 is not a very sensitive measurement. Its rate of change is rather small at the point of optimum excess air. In fact, the CO_2 curve is at its maximum point when the combustion process is optimized.

Excess O_2 as the basis of boiler optimization can be slow (up to 1 minute dead time if not on-line) and a relatively extensive measurement. Yet, because it is more reliable and less noisy than CO analysis, it is popular. It uses zirconium oxide probes. To minimize duct leakage effects, the probe should be installed close to the combustion zone (Figure 1.29) but still at a point where the gas temperature is below that of the electrically heated zirconium oxide detector (1550°F or

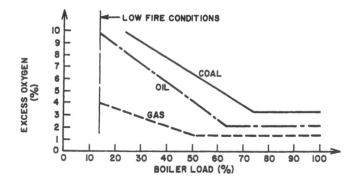

FIGURE 1.27   The ideal amount of excess oxygen provided to a boiler depends on load as well as fuel properties.

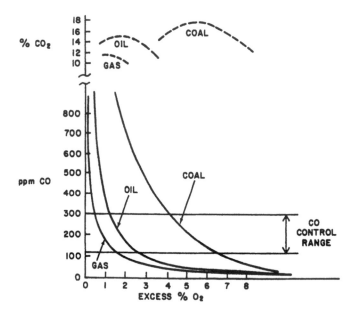

FIGURE 1.28   The relationship between excess O_2 and CO or CO_2 in the flue gas of a boiler operated at a constant load is a function of the type of fuel burned.

850°C). The flow should be turbulent at the sensor location, if possible, to ensure that the sample will be well mixed and representative of flue gas composition. The output signal of these zirconium oxide probes is logarithmic. The correct location of the probe will reduce but not eliminate the bias error caused by air infiltration. Ambient tramp air enters the exhaust ductwork (which is under vacuum) not only through leakage but also to cool unused burners and registers. The O_2 probe cannot distinguish the oxygen that entered through leakage from excess oxygen left over after combustion.

Another limitation of the zirconium-oxide fuel cell sensor is that it measures *net* oxygen. In other words, if there are combustibles in the flue gas, they will be oxidized on the hot surface of the probe and the instrument will register only that oxygen that remains *after* this reaction. This error is not substantial when the total excess oxygen is around 5%, but in optimized boilers, in

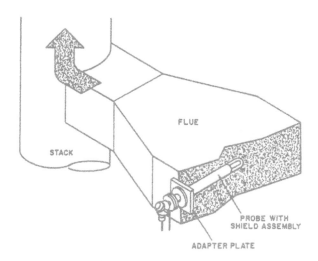

FIGURE 1.29 The probe-type oxygen analyzer should be installed close to the combustion zone but at a point where the temperature is below the limit for the zirconium oxide detector.

which excess oxygen is only 1%, this difference between total and net O_2 can cause a significant error. As infiltration tends to cause an error toward the high side while the fuel-cell effect results in a low reading, the amount of uncertainty is too high to rely on O_2 sensors alone when maximum efficiency is desired.

Other limitations of optimization based on excess oxygen include the precision and accuracy of such excess oxygen curves, as shown in Figure 1.26. This precision is a function not only of the resolution at which the curve was prepared but also of changes in fuel composition and boiler conditions.

4. CO Measurement

As shown in Figure 1.25, the most sensitive measurement of flue-gas composition is the detection of carbon monoxide. As can be seen from Figures 1.28 and 1.30, optimum boiler efficiency can be obtained when the losses due to incomplete combustion *equal* the effects of excess air heat loss. These conditions prevail at the "knee" of each curve. While the excess O_2 corresponding to these knee points varies with the fuel, the corresponding CO concentration is relatively constant.

Theoretically, CO should be zero whenever there is oxygen in the flue gas. In actual practice, maximum boiler efficiency can usually be maintained when the CO is between 100 and 400 ppm. CO is a very sensitive indicator of improperly adjusted burners; if its concentration rises to 1000 ppm, that is a reliable indication of unsafe conditions. Because CO is a direct measure of the completeness of combustion and nothing else, it is also unaffected by air infiltration, other than the dilution effect.

For these reasons, control systems utilizing the measurement of both excess O_2 and carbon monoxide can optimize boiler efficiency, even if load, ambient conditions, or fuel characteristics vary. Also, when these systems detect a shift in the characteristic curve of the boiler, that shift can be used to signal a need for maintenance of the burners, heat transfer units, or air and fuel handling equipment.

Nondispersive infrared (IR) analyzers can be used for simultaneous *in situ* measurement of CO and other gases or vapors such as that of water. This might signal incipient tube leakage. Most IR sensors use a wavelength of 4.7 microns for CO detection because the absorption of CO peaks at this wavelength, whereas that of CO_2 and H_2O does not. CO_2 is also measured and is used to determine the dilution compensation factor for CO.

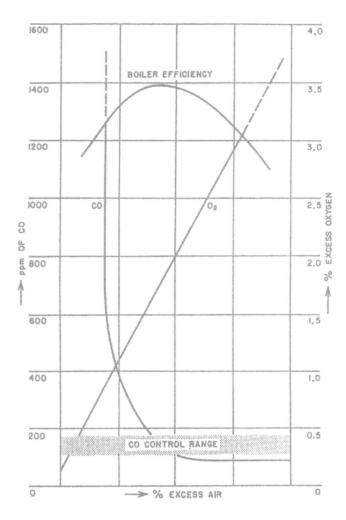

FIGURE 1.30 Gas-burning boiler efficiency is maximum when CO is within control range. (Courtesy of Econics Corporation.)

The CO analyzers are noisy, not very reliable, cannot operate at high temperatures, and therefore are usually located downstream of the last heat exchanger or economizer. At these points, the flue-gas dilution due to infiltration is frequently high enough to require compensation. The measurement of CO_2 is used to calculate this compensation factor.

5. Optimized Air-Fuel Ratio Set Point

One of the ways to optimize boilers is to continuously search for the minimum loss point on Figure 1.23 and operate the boiler at the corresponding air-fuel ratio. In Figure 1.17, the air-fuel ratio (FY-104) is set manually by HIC-104. In Figure 1.9, that same set point is identified as the #1 tie-in point for the optimization controls. That set point can be optimized on the basis of flue gas composition. Because the excess oxygen requirement varies with load (Figure 1.26) and with the type or heating value of the fuel used (Figure 1.27) the optimum set point is a variable. If the heating value of the fuel is the main disturbance (the case with waste incinerators), a sudden change in heating value will cause a change in both the steam pressure in the drum and in the excess oxygen content of the flue gas. Therefore, both of these loops will respond. The pressure controller (PIC-101 in Figure 1.9 is usually the faster and) will correct the firing rate, while the

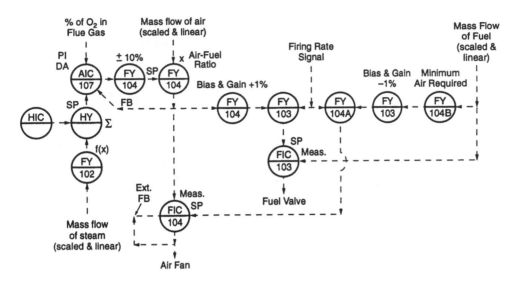

FIGURE 1.31 Optimizing the air-fuel ratio set point.

oxygen controller (usually the slower) will trim the air-fuel ratio setting. If properly tuned, these loops will assist each other.

Figure 1.31 illustrates one such optimized configuration. Here the flue gas analyzer (AIC-107) is a slow integral controller with little or no proportional. This is because the oxygen measurement is noisy and is not desirable to respond to its rapid fluctuations, only to longer-term trends caused by fuel quality variations or by flowmeter errors. antireset wind-up is provided by the external feedback (FB) signal, which is needed due to the adjustable output limits set by the < and > limits on FY-104.

The set point of AIC-107 is modulated in a feedforward manner on the basis of the boiler load, FY-102 detects the actual steam flow and based on the Figure 1.26 function generates the required excess oxygen set point. HIC allows the operator to shift the Figure 1.26 curve up or down to compensate for air infiltration or boiler performance.

6. Multivariable Envelope Control

Multiple measurements of flue-gas composition can be used to eliminate the manual bias in Figure 1.31 and to obtain more accurate and faster control than what is possible with excess O_2 control alone.

Figure 1.32 shows a control system in which the manual bias is replaced by the output signal of a CO controller, trimming the characterized set point of AIC-107. This trimming corrects the characterized excess O_2 controller set point for changes in the characteristic curve. These changes can be caused by local problems at the burners, resulting in incomplete combustion, or by changes in fuel characteristics, equipment, or ambient conditions. The control systems shown in Figure 1.32 could be further improved if the CO measurement signal was corrected for dilution effects due to air infiltration. The control range for CO tends to remain relatively constant. As CO gives an indication of the completeness of combustion, it is *not* a feasible basis for control if the boiler is in poor mechanical condition or if the fuel does not combust cleanly.

Figure 1.33 provides the added feature of an opacity override to meet environmental regulations. In this system, under normal conditions, the cascade master is CO, just as it was in Figure 1.32. Similarly, the cascade slave is excess O_2 but when the set point of the opacity controller is reached, it will start biasing the O_2 set point upward until opacity returns to normal.

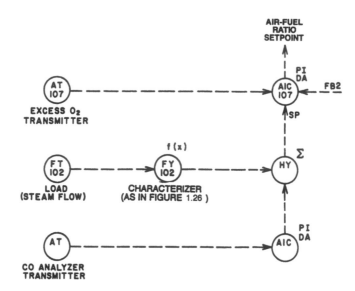

FIGURE 1.32 The need for manual biasing is eliminated by the addition of fine trimming based on CO. (Adapted from McFadden, R.W., *InTech*, May 1984.)

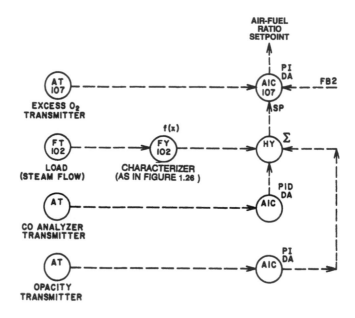

FIGURE 1.33 An opacity override has been added to a characterized CO to O_2 cascade system to meet environmental regulations. (Adapted from American Technical Services, *Boiler Audits*, June 1982.)

With microprocessor-based systems, it is possible to configure a control envelope such as that shown in Figure 1.34. With these control envelopes, several control variables are simultaneously monitored, and control is switched from one to the other, depending on which limit of the envelope is reached. For example, assuming that the boiler is on CO control, the microprocessor will drive the CO set point toward the maximum efficiency (knee point in Figure 1.28), but if in so doing the opacity limit is reached, it will override the CO control and will prevent the opacity limit from being violated. Similarly, if the microprocessor-based envelope is configured for excess oxygen control, it will keep increasing boiler efficiency by lowering excess O_2 until one of the envelope

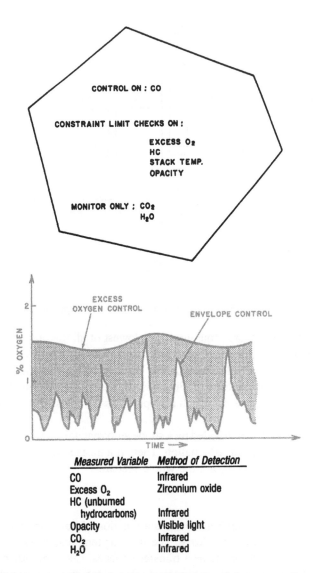

FIGURE 1.34 Multivariable envelope-based constraint control can lower overall excess oxygen. This can be achieved by monitoring carbon dioxide and water and by performing constraint limit checks on excess oxygen, hydrocarbons, stack temperature, and opacity.

limits is reached. When that happens, control is transferred to that constraint parameter (CO, HC, opacity, etc.); through this transfer, the boiler is "herded" to stay within the envelope defined by those constraints. These limits are usually set to keep CO under 400 ppm, opacity below #2 Ringlemann, and HC or NO_x below regulations.

Microprocessor-based envelope control systems usually also include subroutines for correcting the CO readings for dilution effects or for responding to ambient humidity and temperature variations. As a result, these control systems tend to be both more accurate and faster in response than if control was based on a single variable. The performance levels of a gas burning boiler under both excess O_2 and envelope control are shown on the lower part of Figure 1.34.

Envelope control can also be implemented by analog controllers that are configured in a selective manner. This is illustrated in Figure 1.35. Each controller measures a different variable and is set

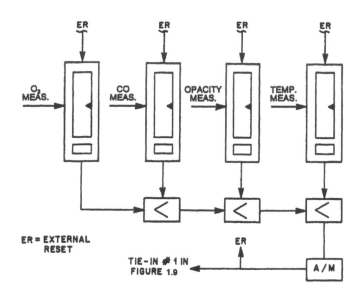

FIGURE 1.35   Controllers configured into a selective loop can also implement envelope control.

to keep that variable under (or over) some limit. The lowest of all the output signals is selected for controlling the air-fuel ratio in Figure 1.9 or 1.31, which ensures that the controller that is most in need of help is selected for control. Through this herding technique, the boiler process is kept within its control envelope. As shown in Figure 1.35, reset wind-up in the idle controllers is prevented by the use of external reset, which also provides bumpless transfer from one controller to the next.

Operator access is shown to be provided by a single auto/manual (A/M) station. A better solution is to provide each controller with an A/M station. Then, if a measurement is lost, only the defective loop needs to be switched to manual, not the whole system.

B. FLUE-GAS TEMPERATURE

The amount of energy wasted through the stack is a function of both the amount of excess air and the temperature at which the flue gases leave. The flue-gas temperature is a consequence of load, air infiltration, and the condition of the heat transfer surfaces. Like any other heat exchanger, the boiler will also give its most efficient performance when clean and well maintained.

In optimized boiler controls, a plot of load versus stack temperature is made when the boiler is in its prime condition, and this plot is used as a reference baseline in evaluating boiler performance. If the stack temperature rises above this reference baseline, this indicates a loss of efficiency. Each 50°F (23°C) increase will lower the boiler efficiency by about 1%. On a 100,000 lb/hr (45,450 kg/hr) boiler, this is a yearly loss of about $20,000.

The reason for rising flue-gas temperatures can be the fouling of heat transfer surfaces in the air preheater, scale buildup on the inside of the boiler tubes, soot buildup on the outside of the boiler tubes, or deteriorated baffles that allow the hot gases to bypass the tubes. Soot is blown by air steam jets using manually or automatically operated, revolving, traversing and/or retractable soot guns or deslaggers.

If the stack temperature drops below the reference baseline, this does *not* necessarily signal an increase in boiler efficiency. More likely, it can signal the loss of heat due to air in leakage. Cold air or cold water can leak into the stack gases if the economizer or the regenerator are damaged.

The consequence of this is loss of efficiency and the danger of corrosive condensation, if the temperature drops down to the dew point. As shown in Figure 1.9, this can be prevented by preheating the air, by steam before it is sent to the air preheater. A drop in flue-gas temperature will also lower the stack effect, thus increasing the load on the induced draft fan.

For the above reasons, the advanced envelope control systems (Figure 1.34) include both high- and low-limit constraints on stack temperature, using the above-described baseline as a reference.

C. Fuel Savings Through Optimization

The overall boiler efficiency is the combined result of its *heat transfer efficiency* and its *combustion efficiency*. Heat transfer efficiency is affected by stack temperature. Combustion efficiency is tied to excess oxygen, which is brought as low as possible without exceeding the limits on CO (usually 400 ppm), opacity (#2 Ringlemann), and unburned carbon and NO_x.

When optimization reduces the flue-gas losses, the resulting savings can be estimated from the amount of reduction in these losses. In the case of a 100,000 lb/hr (45,450 kg/hr) steam boiler, a 1% reduction in fuel consumption (a 1% increase in efficiency) will lower the yearly operating costs by about $20,000 if the fuel cost is $2 per million BTU.

The fuel savings resulting from the lowering of excess O_2 can be estimated from graphs, such as those shown in Figures 1.36 and 1.37. On these graphs, the temperature is the difference between the stack and ambient temperatures. If, in a boiler operating at a 500°F stack temperature difference, optimization lowers the excess O_2 from 5% to 2%, this will result in a fuel savings of 1.5% (2.25% minus 0.75%), according to Figure 1.36. For the same conditions, Figure 1.37 gives a savings of 2% (10.5% minus 8.5%). Such differences are acceptable, as the size and designs of the equipment does influence the results. Although optimization has lowered the excess O_2 in some boilers to around 0.5%, the value of these last increments of savings tends to diminish in proportion to the cost of accomplishing them. Lowering excess O_2 from 1% to 0.5% will increase boiler efficiency by only 0.25%, a savings of about $5,000/yr in a 100,000 lb/hr boiler, and the controls needed to sustain such operation need to be as sophisticated as the ones described in Figure 1.34. For these reasons, an optimization target of 1% excess O_2 on gas fuel is reasonable.

The fuel savings resulting from improved thermal efficiency can be estimated from Figure 1.37 or Figure 1.38. Lowering the stack temperature difference from 500 to 400°F at an excess O_2 of 2% will result in savings of 1.7% (8.5% minus 6.8%), according to Figure 1.37. Assuming an ambient temperature of 60°F, these same conditions are marked in Figure 1.38 at 560 and 460°F, resulting in a savings of about 2%. Therefore, as a crude approximation, a fuel savings of 1% can be estimated for each 50°F reduction in stack gas temperature.

Consequently, the total savings from improved thermal and combustion efficiencies can be estimated on the basis of both stack temperature and excess oxygen being lowered to their limits.

D. Steam Pressure Optimization

In Figure 1.9, the second tie-in point for the optimizer is the set point of the pressure controller PIC-101. In traditional boiler operation, the steam pressure was maintained at a constant value. It is only in recent years that this practice has begun to change.

In cogenerating plants, in which the boiler steam is used to generate electricity and the turbine exhaust steam is used as a heat source, optimization is obtained by maximizing the boiler pressure and minimizing the turbine exhaust pressure, so as to maximize the amount of electricity generated (Figure 1.39).

In plants that do not generate their own electricity, optimization is achieved by minimizing boiler operating pressure. This reduces the pressure drops in turbine governors by opening them,

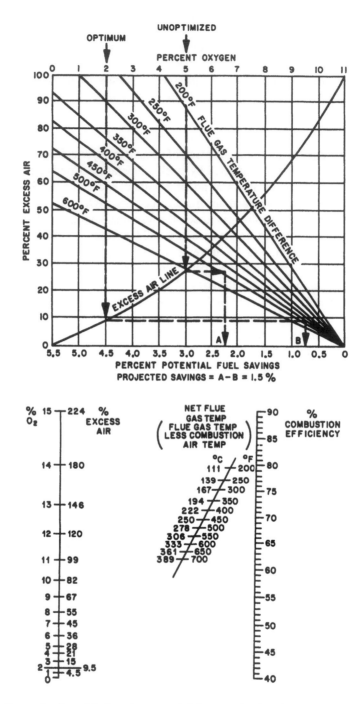

FIGURE 1.36 Determination of fuel-saving potential due to reductions in excess oxygen. Gas combustion efficiency is found in the lower chart at the intersection of a straight line drawn through the applicable excess oxygen and flue-gas temperature points.[13]

lowers the cost of operating feedwater pumps because their discharge pressure is reduced, and generally lowers radiation and wall losses in the boiler and piping.

Pump power is particularly worth saving in high-pressure boilers, because as much as 3% of the gross work produced by a 2400 PSIG (16.56 MPa) boiler is used to pump feedwater. Figure

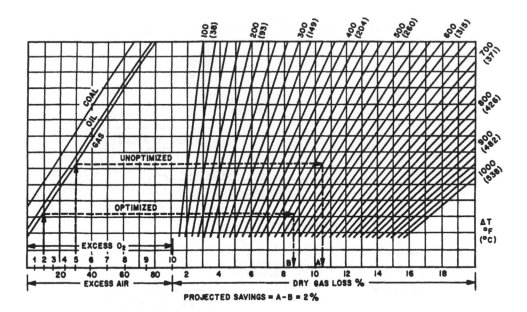

FIGURE 1.37   The fuel-saving potential can also be computed as shown here. (Courtesy of Dynatron, Inc.)

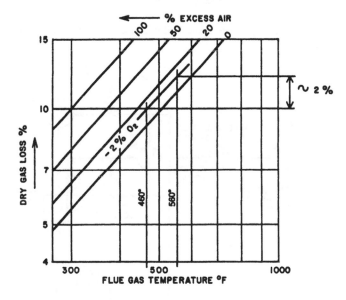

FIGURE 1.38   Temperature of flue gas leaving a medium-size boiler has a direct effect on combustion efficiency. (Adapted from Walsh, T.J., *I&CS*, January 1981.)

1.40 illustrates the method of finding the optimum (minimum) steam pressure, which then becomes the set point for the master controller PIC-101 in Figure 1.9.

As long as all steam user valves (including all turbine throttle valves) are less than fully open, a lowering in the steam pressure will not restrict steam availability because the user valves can open further. The high signal selector (TY-1) selects the most open valve, and the valve position controller (VPC-2) compares that signal with its set point of, for example, 80%.If even the most open valve in the plant is less than 80% open, the pressure controller set point is slowly lowered.

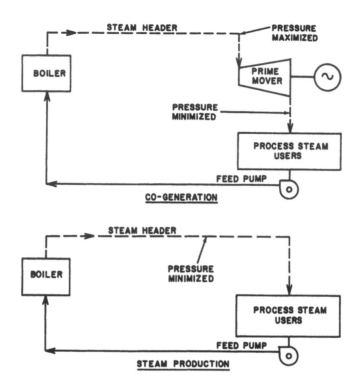

FIGURE 1.39 The optimum steam pressure at the boiler is also a function of what the steam is used for.

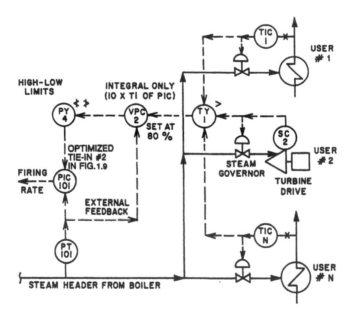

FIGURE 1.40 The optimized floating of steam supply header pressure becomes the set point for the master controller PIC-101.

VPC-2 is an "integral only" controller; its reset time is at least 10 times that of PIC-101. This slow integral action guarantees that only very slow "sliding" of the steam pressure will occur and that noisy valve signals will not upset the system because VPC-2 responds only to the integrated area under the error curve. The output signal from VPC-2 is limited by PY-4, so that the steam pressure set point cannot be moved outside limits set by equipment capability. This necessitates the external feedback to VPC-2, so that when its output is overridden by a limit, its reset will not wind up.

This kind of optimization, in which steam pressure follows the load, not only increases boiler efficiency but also does the following:

1. Prevents any steam valve in the plant from fully opening and thereby losing control
2. Opens all steam valves in the plant, thereby moving them away from the unstable (near-closed) zone of operation
3. Reduces valve maintenance and increases valve life by lowering pressure drop
4. Increases turbine drive efficiencies by opening up all steam governors

The total savings in yearly operating costs resulting from optimizing the steam pressure to follow the load usually justify the cost of implementation.

E. WATER SIDE OPTIMIZATION

The optimization of the water side of a steam generator includes the optimized operation of the feedwater pump, the condensate return system, and the boiler blowdown.

In Figure 1.9, the third tie-in point for the optimizer is the set point of the blowdown flow controller. Blowdown limits the formation of waterside deposits in the tubes, which otherwise could cause caustic corrosion, hydrogen damage and an increase in tube metal temperature. The goal of optimization is to minimize the blowdown as much as possible without causing excessive sludge or scale buildup on the inside surfaces of the boiler tubes. The benefits of such optimization include the reduction in the need for makeup water and treatment chemicals and the reduction in heat loss as hot water is wasted. About 90% of the blowdown should occur continuously, and 10% would result from the periodic blowing down of the mud drum and of the headers.

Blowdown can be optimized by automatically controlling the chloride and conductance of the boiler water. The neutralized conductivity set point is usually around 2500 micromhos. Automatic control maintains this set point within ± 100 micromhos. The required rate of blowdown is a function of the hardness, silica, and total solids of the makeup water and also of the steaming rate and condensate return ratio of the boiler.

The amount of blowdown can be determined as follows:

$$BD = \frac{S-R}{C-1} \qquad (6)$$

where BD = blowdown rate, pounds per hour, R = rate of return condensate, pounds per hour, S = steam load, pounds per hour, and C = cycles of concentration based on makeup.

The value for cycles of concentration is generally determined on the basis of the chloride concentration of the boiler water divided by the chloride content of the makeup water. (If the chloride concentration in the boiler water is 120 ppm and in the feedwater it is 12 ppm, the concentration ratio is 10.) The value is also given by dividing the average blowdown rate into the average rate of makeup water, assuming no mineral contamination in any returned condensate.

Figure 1.41 illustrates that the rate of blowdown accelerates as the boiler water conductivity set point is lowered. A reduction of about 20% can result from converting the blowdown controls from manual to automatic.[14] In the case of a 100,000 lb/hr (45,450 kg/hr) boiler, this can mean a reduction of 1340 lb/hr (600 kg/hr) in the blowdown rate. If the blowdown heat is not recovered, this can lower the yearly operating cost by about $10,000 (depending on the unit cost of the fuel).

Overall boiler efficiency can also be increased if the heat content of the hot condensate is returned to the boiler. Pumping water at high temperatures is difficult; therefore, the best choice is to use pumpless condensate return systems. Figure 1.42 illustrates the operation of such a system,

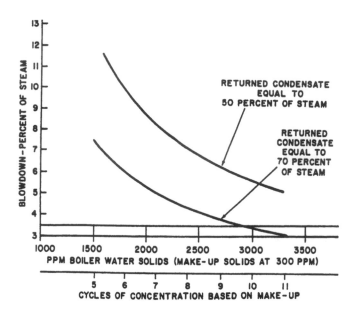

FIGURE 1.41   The rate of blowdown increases as the boiler water conductivity set point is lowered.[14]

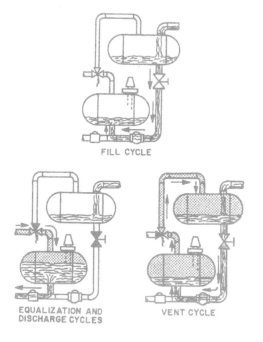

FIGURE 1.42   The pumpless condensate return system uses the steam pressure itself to push the condensate back into the de-aeration tank. (Courtesy of Johnson Corporation.)

which uses the steam pressure itself to push back the condensate into the de-aeration tank. This approach eliminates not only the maintenance and operating cost of the pump but also the flash and heat losses, resulting in the return of more condensate at a higher temperature.

The performance of the steam and condensate piping system in the plant can also be improved if steam flows are metered. Such data are helpful not only in accountability calculations but also in locating problem areas, such as insufficient thermal insulation or leaking traps.

F. LOAD ALLOCATION

The purpose of load allocation between several boilers is to distribute the total plant demand in the most efficient and optimized manner. Such optimization will reduce the steam production cost to a minimum. Such computer-based energy management systems can operate either in an advisory or in a closed-loop mode. The closed-loop systems automatically enforce the load allocation, without the need for operator involvement. The advisory system, on the other hand, provides instructions to the operator but leaves the implementation up to the operator's judgment.

In simple load allocation systems, only the starting and stopping of the boilers is optimized. When the load is increasing, the most efficient idle boiler is started (Figure 1.7); when the load is dropping, the least efficient one is stopped. In more sophisticated systems, the load distribution between operating boilers is also optimized. In such systems, a computer is used to calculate the real-time efficiency of each boiler. This information is used to calculate the incremental steam cost for the next load change for each boiler.

If the load increases, the incremental increase is sent to the set point of the most cost-effective boiler. If the load decreases, the incremental decrease is sent to the least cost-effective boiler (Figure 1.43). The required software packages with proved capabilities for continuous load balancing through the predictions of costs and efficiencies are readily available. With the strategy described in Figure 1.42, the most efficient boiler either will reach its maximum loading or will enter a region of decreasing efficiency and will no longer be the most efficient. When the loading limit is reached on one boiler, or when a boiler is put on manual, the computer will select another as the most efficient unit for future load increases.

The least efficient boiler will accept all decreasing load signals until its minimum limit is reached. Its load will not be increased unless all other boilers are at their maximum load or in manual. As shown in Figure 1.7, some boilers can have high efficiency at normal load while being less efficient than the others at low load. Such units are usually not allowed to be shut down but are given a greater share of the load by a special subroutine.

If all boilers are identical, some will be driven to maximum capacity and others will be shut down by this strategy, and only one boiler will be placed at an intermediate load. Boiler efficiency can be monitored indirectly (by measurement of flue-gas composition, temperature, combustion temperature, and burner firing rate) or directly (through time-averaged steam and fuel flow monitoring). For the direct efficiency measurement, it is important to select flowmeters with acceptable accuracy and rangeability (Table 1.1). In order to arrive at a reliable boiler efficiency reading, the error contribution of the flowmeters, based on actual reading, must not exceed $\pm 1/2$ to $\pm 3/4\%$.

Boiler allocation can be based on actual measured efficiency, on projected efficiency based on past performance, or on some combination of the two. The continuous updating and storing of performance data for each boiler is also a valuable tool in operational diagnostics and maintenance.

IV. CONCLUSIONS

The various goals of boiler optimization include the following:

To minimize excess air and flue-gas temperature
To measure efficiency (use the most efficient boilers, know when to perform maintenance)
To minimize steam pressure (open up turbine governors, reduce feed pump discharge pressures, and reduce heat loss through pipe walls)
To minimize blowdown
To provide accountability (monitor losses, recover condensate heat)
To minimize transportation costs (use variable speed fans, eliminate condensate pumps, and consider variable speed feedwater pumps)

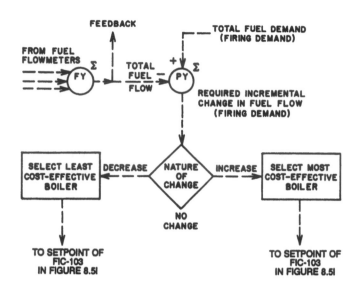

FIGURE 1.43 Computer-based load allocation directs load increases to the most cost-effective boiler and sends load decreases to the least cost-effective boiler.

If the potentials of all of the above optimization strategies are fully exploited, the unit costs of steam generation can usually be lowered by about 10%. In larger boiler houses, this can represent a savings that will pay for the optimization system in a year.[15]

In recent years, nonlinear, neural network-based control systems have also been applied to boilers. These models perform "what if"-type evaluations on fuel fluctuations, soot buildup, and equipment variables, resulting in a matrix of some 45 inputs and 5 or 6 outputs. Such dynamic neural networks periodically update themselves with new data and download new set points based on the patterns evolving in the boiler. From the above, we can only conclude that the potentials for boiling water more efficiently are nearly inexhaustible.

REFERENCES

1. Walsh, T.J., Controlling Boiler Efficiency, *I&CS*, January 1981.
2. Congdon, P., Control Alternatives for Industrial Boilers, *InTech*, December 1981.
3. DeLorenzi, O., Combustion Engineering, Inc., undated publication.
4. Dickey, P.S., A Study of Damper Characteristics, Bailey Meter Co., Reprint No. 8.
5. Lammers, T.F., *Steam-Plant Operation*, 6th ed., McGraw-Hill, New York.
6. Shinskey, F.G., *Process Control Systems*, 4th ed., McGraw-Hill, 1996.
7. Hougen, J.O., *Methods for Solving Process Plant Problems*, ISA, 1996.
8. Lipták, B.G., *Instrument Engineers' Handbook*, Measurement Volume, Chilton Book Co., 1995.
9. Schwartz, J.R., Carbon monoxide monitoring, *InTech*, June 1983.
10. Babcock & Wilcox Co., *Steam, Its Generation and Use*, New York.
11. McFadden, R.W., Multiparameter trim in combustion control, *InTech*, May 1984.
12. American Technical Services, *Boiler Audits*, June 1982.
13. Westinghouse Electric Corp., Oxygen Trim Control, 1985.
14. Schieber, J.R., The Care for Automated Boiler Blowdown, Universal Interlock.
15. Lipták, B.G., Optimizing Boilers, *Chem. Eng.*, May 25, 1987.

BIBLIOGRAPHY

Chattopadhyay, P., *Boiler Operations: Questions and Answers*, McGraw-Hill, New York, 1995.

Hougen, J.O., *Methods for Solving Process Plant Problems*, ISA, 1996.

Kohan, A.L., *Boiler Operator's Guide*, 3rd ed., McGraw-Hill, New York, 1991.

Lindsley, D., *Boiler Control Systems*, McGraw-Hill, New York, 1991.

Lipták, B.G., *Instrument Engineers' Handbook*, Measurement Volume, Chilton Book Co., 1995.

Polonyi, M.J., *Power and Process Control Systems*, McGraw-Hill, New York, 1991.

Prabhakara, F.S. and Smith, R.L., *Industrial and Commercial Power System Handbook,* McGraw-Hill, New York, 1995.

Shinskey, F.G., *Process Control Systems,* 4th ed., McGraw-Hill, New York, 1996.

Woodruff and Lammers, *Steam Plant Operation*, 6th ed., McGraw-Hill, New York, 1992.

2 Chillers

The first section of this chapter describes the thermodynamic and control characteristics of the process and the nature of the equipment and working fluids used in chillers. The second section introduces the reader to the traditional control strategies that are utilized in conventional or packaged chillers. The third section focuses on minimizing the unit cost of chilling by optimizing the flows and temperatures of the heat-transfer fluids as they continuously float the coolant supply to match the load.

I. THE PROCESS

Heat pumps serve to transport heat from a lower to a higher temperature level. They do not make heat; they just transport it. This is similar to a regular pump transporting water from a lower to a higher elevation. The required energy input to a regular pump is a function not so much of the amount of water to be transported but of the elevation to which it is to be lifted. Similarly, the required energy input of a heat pump is a function not only of the amount of cooling it has to do but also of the temperature elevation against which it is pumping. The reduction of this temperature difference is the goal of optimization.

Each °F reduction in the ΔT will lower the yearly operating cost by 1.5% (each °C by 2.7%). There are not many unit operations whose efficiency can be doubled through optimization; chiller optimization is one of these few. It is possible to cut the operating cost of a chiller in half through the reduction of its ΔT by 33°F (18.5°C). Such results from optimization are not unprecedented.[1]

For terminology related to cooling towers and for definitions of terms such as "wet-bulb temperature," refer to Chapter 5.

A. THERMODYNAMICS

Figure 2.1 illustrates how heat pumps can transport heat from a lower to a higher elevation and thereby provide cooling of the low temperature process. The heat pump removes Q_L amount of heat from the chilled water and, at the cost of investing W amount of work, delivers Q_H quantity of heat to the warmer cooling tower (or condenser) water.

In the lower part of this illustration, the idealized temperature-entropy cycle is shown for the refrigerator. The cycle consists of two isothermal and two isentropic (adiabatic) processes:

1→2: Adiabatic process through expansion valve
2→3: Isothermal process through evaporator
3→4: Adiabatic process through compressor
4→1: Isothermal process through condensor

The isothermal processes in this cycle are also isobaric (constant pressure). The efficiency of a refrigerator is defined as the ratio between the heat removed from the process (Q_L) and the work required to achieve this heat removal (W).

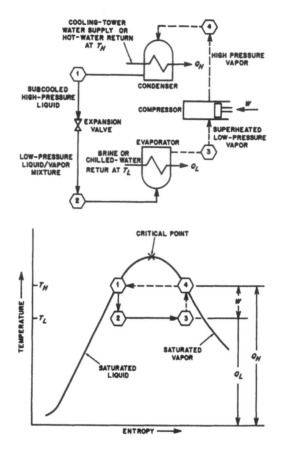

FIGURE 2.1   The refrigeration cycle consists of two isothermal and two isentropic processes. (From Lipták, B.G., *Chem. Eng.*, October 17, 1983. With permission.)

$$\beta = \frac{Q_L}{W} = \frac{T_L}{T_H - T_L} \tag{1}$$

Because the chiller efficiency is much more than 100%, it is usually called the *coefficient of performance* (COP).

If a chiller requires 1.0 kWh (3412 BTU/h) to provide a ton of refrigeration (12,000 BTU/h), its COP is said to be 3.5. This means that each unit of energy introduced at the compressor will pump 3.5 units of heat energy to the cooling water tower. Conventionally controlled chillers operate with COPs in the range of 2.5 to 3.5. Optimization can double the COP by increasing T_L, by decreasing T_H, and through other methods that will be described later.

Figure 2.1 shows the four principle pieces of equipment that make up a refrigeration machine. In path 1→2 (through the expansion valve), the high-pressure subcooled refrigerant liquid becomes a low-pressure, liquid-vapor mixture. In path 2→3 (through the evaporator), this becomes a super-heated low-pressure vapor stream, whereas in the compressor (path 3→4) the pressure of the refrigerant vapor is increased. Finally, in path 4→1, this vapor is condensed at constant pressure. The liquid leaving the condenser is usually subcooled, whereas the vapors leaving the evaporator are usually superheated by controlled amounts.

The unit most frequently used in describing refrigeration loads is the ton. Because several types of tons are referred to in the literature, it is important to distinguish among them:

Standard ton	200 BTU/min (3520 W)
British ton	237.6 BTU/min (4182 W)
European ton (Frigorie)	50 BTU/min (880 W)

B. Refrigerants and Heat Transfer Fluids

The fluid that carries the heat from a low to a high temperature level is referred to as the refrigerant. Table 2.1 provides a summary of the more frequently used refrigerants.

The data are presented on the assumption that the evaporator will operate at 5°F (−15°C) and that the temperature of the cooling water supply for the condenser will allow it to maintain 86°F (30°C). Other temperature levels would have illustrated the relative characteristics of the various refrigerants equally well. It is generally desirable to avoid operation under vacuum in any part of the cycle because of sealing problems. At the same time, very high condensing pressures are also undesirable because of the resulting structural strength requirements. From this point of view, the refrigerants between propane and methyl chloride in Table 2.1 display favorable characteristics. An exception to this reasoning is when very low temperatures are required. For such service, ethane can be the proper selection in spite of the resulting high system design pressure.

Another consideration is the latent heat of the refrigerant. The higher it is, the more heat can be carried by the same amount of working fluid and, therefore, the smaller the corresponding equipment size can be. This feature has caused many users in the past to compromise with the undesirable characteristics of ammonia.

One of the most important considerations is safety. In industrial installations, the desirability of nontoxic, nonirritating, nonflammable refrigerants cannot be overemphasized. It is similarly important that the working fluid be compatible with the compressor lubricating oil. Corrosive refrigerants are undesirable for the obvious reasons of higher initial cost and increased maintenance.

Most of the working fluids listed in Table 2.1 are compatible with reciprocating compressors. Only the last four fluids in the table, which have high volume-to-mass ratios and low compressor discharge pressures, can justify the consideration of rotary or centrifugal machines.

For decades, Freon-12 was the most widely used industrial refrigerant because of its favorable pressure-temperature-latent-heat characteristics. Today, Freon 11, 12, 21, and 22 are (or should be) no longer used, because their release into the atmosphere contributes to skin cancer through the depletion of the ozone layer and because the inhalation of Freon can cause death through causing fibrillation of the heart muscles.

In the majority of industrial installations, the refrigerant evaporator is not used directly to cool the process. More frequently, the evaporator cools a circulated fluid, which is then piped to cool the process.

For temperatures below the point at which water can be used as a coolant, brine is frequently used. It is important to remember that weak brines may freeze and that strong brines, if they are not true solutions, may plug the evaporator tubes. For operation around 0°F (−18°C), the sodium brines (NaCl) are recommended; for services down to −45°F (−43°C) the calcium brines (CaCl) are best.

Care must be exercised in handling brines, because they are corrosive if they are not kept at a pH of 7 or if oxygen is present. In addition, brine will initiate galvanic corrosion between dissimilar metals.

C. System Piping Configuration

The proper control and optimization of chilled water systems can only be accomplished if the plant piping layout allows for it. Figure 2.2 illustrates both a desirable and an undesirable piping layout. There are two problems associates with the undesirable configuration shown at the top of the figure. One problem is that it does not provide chilled water storage, and therefore, if the chiller fails, the

TABLE 2.1
Refrigerant Characteristics

Refrigerant		Applicable Compressor[a]	Boiling Point in °F[b] at Atmospheric Pressure	Evaporator Pressure in PSIA[c] if Operating Temperature is 5°F (-15°C)	Condenser Pressure in PSIA[c] if Operating Temperature is 86°F (30°C)	Latent Heat in BTU/lbm[d] at 18°F (-7.8°C)	Toxic (T), Flammable (F), Irritating (I), Ozone Depleting (O), Interferes with Heartbeat (H)	Mixes and/or Compatible with the Lubricating Oil	Chemically Inert and Noncorrosive	Remarks
Ethane	C_2H_6	R	-127	236	675	148	T&F	No	Yes	For low-temperature service
Carbon dioxide	CO_2	R	-108	334	1039	116	No	Yes	Yes	Low-efficiency refrigerant
Propane	C_3H_8	R	-48	42	155	132	T&F	No	Yes	
Freon-22	$CHClF_2$	R	-41	43	175	92	O&H	(1)	Yes	For low-temperature service
Ammonia	NH_3	R	-28	34	169	555	T&F	No	(2)	High-efficiency refrigerant
Freon-12	CCl_2F_2	R	-22	26	108	67	O&H	Yes	Yes	Most recommended
Methyl chloride	CH_3Cl	R	-11	21	95	178	(3)	Yes	(4)	Expansion valve may freeze if water is present
Sulfur dioxide	SO_2	R	+14	12	66	166	T&I	No	(4)	Common to these refrigerants:
Freon-21	$CHCl_2F$	RO	+48	5	31	108	O&H	Yes	Yes	a. Evaporator under vacuum
Ethyl chloride	C_2H_5Cl	RO	+54	5	27	175	F&I	No	(5)	b. Low compressor discharge pressure
Freon-11	CCl_3F	C	+75	3	18	83	O&H	Yes	Yes	c. High volume-to-mass ratio across compressor
Dichloro methane	CH_2Cl_2	C	+105	1	10	155	No	Yes	Yes	

Note: (1) Oil floats on it at low temperature; (2) Corrosive to copper-bearing alloys; (3) Anesthetic; (4) Corrosive in the presence of water; (5) Attacks rubber compounds.

[a] R = reciprocating, RO = rotary, C = centrifugal

[b] $°C = \dfrac{°F - 32}{1.8}$

[c] PSIA = 6.9 kPa

[d] BTU/lbm = 232.6 J/kg

plant has to shut down. The other, even more serious problem is that if one of the loads (say the reactor) upsets the return water temperature (Tchwr), this in turn will upset the chilled water temperature (Tchws) to the whole plant.

Visualize a situation in which the reactor shown at the top of Figure 2.2 is a batch reactor which has just completed its heat-up phase, and the batch is starting to react. At this point exothermic heat is beginning to be generated, and therefore the jacket is switched to cooling. If the operator is not careful, it is possible that as the chilled water enters the reactor jacket, it will displace its hot water content into the return line back to the storage tank. As this slug of several hundred gallons of hot water travels through the storage tank, it raises Tchwr, but the chiller does not know that until this slug of warm water reaches it. At this point it is too late, because the chiller can only start to increase its rate of cooling when the Tchws has already risen above the set point of the temperature controller (TC). As a result, the chilled water supply temperature to the whole plant will be upset at each such occurrence. Actually, each occurrence will upset the plant supply temperature twice: once when the slug of warm water arrives at the TC and once again when it passes it and therefore the Tchws temperature drops.

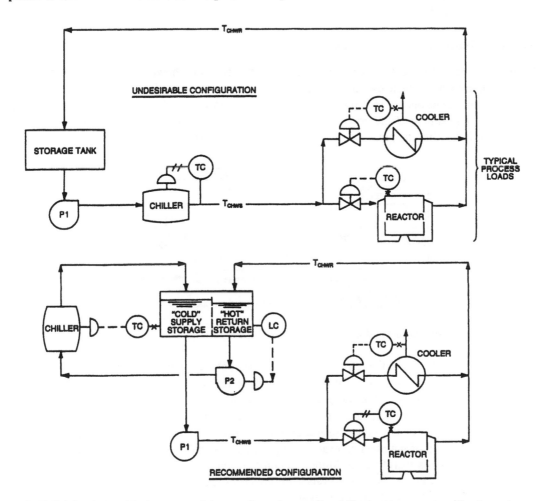

FIGURE 2.2 Only with the proper piping configuration can the chiller system operate without upsets.

The lower part of Figure 2.2 illustrates how one might eliminate both of these problems if the proper tanks and piping configuration are provided. In this recommended piping layout the same slug of warm water will not upset the whole plant, because it is received and blended in the "hot" return storage compartment. Pump P2 takes the water from this tank (or tank section) and sends it through the chiller, which discharges into the "cold" supply storage tank. In this configuration the TC is continuously maintaining the temperature of a fairly large amount of chilled water at the desired Tchws temperature. Therefore, the slug of warm water will not upset the supply to the plant at all. Even more important, the plant will be provided with a reserve storage of ready-made chilled water. This safety storage of coolant can come in handy during emergencies or power failures, or when the chiller is in need of maintenance.

In addition, because the chiller is not in line with P1, but operates on a bypass, the dynamic upsets in the operation of the TC loop will have little or no effect on the chilled water-supply temperature Tchws and therefore, the variations in the dead time or time constant of the loop will not upset the chilled water temperature to the plant.

II. BASIC CONTROLS

On the following pages, the traditional control strategies are described for small-, commercial-, and industrial-sized chillers.

A. SMALL INDUSTRIAL REFRIGERATORS

On the left side of Figure 2.3, the direct expansion-type control is shown. Here a pressure-reducing valve maintains a constant evaporator pressure. The pressure setting is a function of load, and therefore these controls are recommended for constant load installations only. The proper setting is found by adjustment of the pressure-control valve until the frost extends just to the end of the evaporator, indicating the presence of liquid refrigerant up to that point.

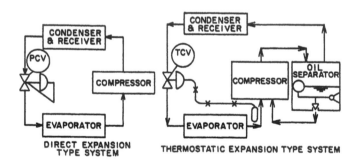

FIGURE 2.3   Small industrial refrigerators with throttling control.

If the load increases, all the refrigerant will vaporize before the end of the evaporator, lowering efficiency as the unit is "starved." This condition will be relieved only by a change in the pressure setting. When the unit is down, the pressure-control valve closes, isolating the high- and low-pressure sides of the system. This guarantees the high start-up torque, which is needed.

On the right side of the same illustration the thermostatic expansion-type control is shown. This system, instead of maintaining evaporator pressure, controls the superheat of the evaporated vapors. This design is therefore not limited to constant loads, because it guarantees the presence of liquid refrigerant at the end of the evaporator under all load conditions.

Figure 2.3 also shows a typical oil separator.

1. Expansion Valves

On the left side of Figure 2.4 a fairly standard superheat control valve is shown. It detects the pressure into and the temperature out of the evaporator. If the evaporator pressure drop is low, these

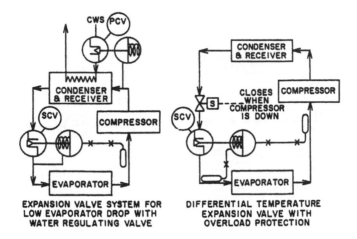

FIGURE 2.4 Expansion valve installation.

measurements (the saturation pressure and the temperature of the refrigerant) are an indication of superheat. The desired superheat is set by the spring in the valve operator, which together with the saturation pressure in the evaporator opposes the opening of the valve. The "superheat feeler bulb" pressure balances these forces when the unit is in equilibrium, operating at the desired superheat (usually 9°F, or 5°C).

If the process load increases, it causes an increase in the evaporator outlet temperature. An increase in this temperature results in a rise in the "feeler bulb" pressure, which in turn further opens the superheat control valve. This greater flow from the condenser to the evaporator increases the saturation pressure and temperature, and the increased saturation pressure balances against the increased feeler bulb pressure at a new (greater) valve opening in a new equilibrium. To adjust to an increased load condition, the evaporator pressure increases, but the amount of the superheat (set by the valve spring) is kept constant.

In this same sketch, the operation of the cooling water regulating valve (PCV) is also illustrated. This valve maintains the condenser pressure constant and at the same time conserves cooling water. At low condenser pressure, such as when the compressor is down, the water valve closes. It starts to open when the compressor is restarted and its discharge pressure reaches the setting of the valve. The water valve opening follows the load, further opening at higher loads to maintain the condenser pressure constant. This feature too can be incorporated into any one of the refrigeration units.

At very low temperatures, a small change in refrigerant vapor pressure is accompanied by a fairly large change in temperature. For example, Freon-12 at the temperature level of −100°F (−73°C) will show a 5°F (2.8°C) temperature change, corresponding to a 0.3 PSIG (21 millibar) variation of saturation pressure. Therefore, the use of a thermal bulb to indirectly detect the saturation pressure in the evaporator will result in a more sensitive measurement. A differential temperature expansion valve, taking advantage of this phenomenon, is illustrated on the right side of Figure 2.4.

2. On-Off Control

Figure 2.5 shows the controls for a fairly simple and small refrigeration unit. This system includes a conventional superheat control valve, a low-pressure-drop evaporator, a reciprocating on-off compressor, and an air-cooled condenser. The purpose of this refrigeration package is to maintain a refrigerated water supply to the plant within some set limits.

The high-temperature switch (TSH) shown in the illustration is the main control device. Whenever the temperature of the refrigerated water drops below a preset value (e.g., 38°F, or 3.3°C),

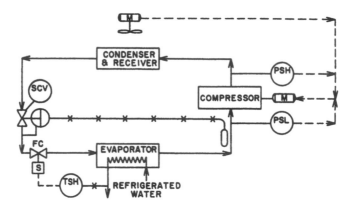

FIGURE 2.5   Small industrial refrigeration unit with electric on-off control.

the refrigeration unit is turned off, and when it rises to some other level (e.g., 42°F, or 5.6°C), it is restarted. This on-off cycling control is accomplished by the closing of the solenoid valve by the temperature switch when the water temperature is low enough. The closing of this valve causes the compressor suction pressure to drop until it reaches the set point of the low-pressure switch, which in turn stops the compressor.

While the unit is running, the expansion valve maintains the refrigerant superheat constant and the safety interlocks protect the equipment. These interlocks include such features as turning off the compressor if the fan motor stops or if the compressor discharge pressure becomes too high for some other reason.

This unit can operate only at full compressor capacity or not at all. This type of machine is referred to as one with two-stage unloading. When varying the cooling capacity of the unit is desired instead of turning it on and off, two possible control techniques are available. One approach involves the multistep unloading of reciprocating compressors. In a three-step system the available operating loads are 100, 50, and 0%, whereas, with five-step unloading, 100, 75, 50, 25, and 0% loads can be handled.

B. COMMERCIAL AIR-COOLED CHILLERS

Figure 2.6 illustrates two methods of controlling an air-cooled commercial chiller. When the control loops are uncoordinated (A), the constant pressure in the evaporator, held by the PC-3, maintains the boiling temperature in the evaporator at an arbitrary constant value (T_L). This is undesirable, because the ΔT in the evaporator ($T_{CHW} - T_L$) should vary with load.

Similarly, by TC-1 modulating the louvers to keep T_H at a constant value, the loop inhibits heat transfer and prevents T_H from being lowered as far as ambient temperature (T_A) would allow. The most efficient chiller operation is obtained when the difference between the condensing and evaporation temperatures ($\Delta T = T_H - T_L$) is the minimum and that will occur when the air flow through the condenser is unrestricted (no louvers) and the heat transfer area of the evaporator is fully utilized (evaporator flooded).

Part B of Figure 2.6 shows coordinated controls, which will respond to changes in both the cooling load (chilled water flow or temperature) or in the ambient temperature (T_A) by adjusting the compressor speed. Heat transfer is maximized in the evaporator by keeping all the tubes immersed in the condensor by maintaining maximum air flow. This way, the suction pressure of the compressor will be as high as it can be, without allowing T_{CHW} to exceed the set point of TC-2, while LC-3 keeps the evaporator tubes flooded. This minimizes the power consumption of the compressor.

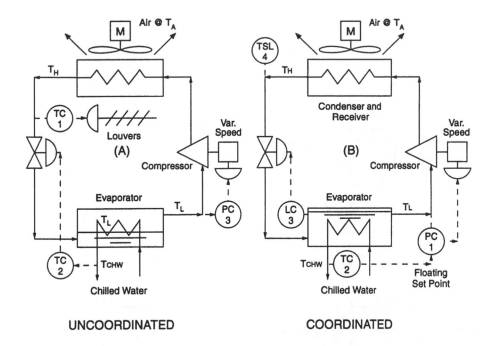

UNCOORDINATED   COORDINATED

FIGURE 2.6 The advantage of coordinated controls (B) is that it maximizes the active heat transfer and thereby reduces operating cost. The disadvantage is increased sensitivity to ambient changes.

Having no controls or limits on T_H makes the operation weather sensitive. A sudden rain storm or low temperatures at night or in the winter might drive T_H down to such low values (very low $\Delta T = T_H - T_L$) that the compressor could not handle the low load and would go into surge. One can protect against this by allowing a low-temperature switch (TSL-4) to turn off some fan units on multiple-fan condensers or to lower the blade pitch on variable-pitch fans.

C. Supermarket and Warehouse Chillers

In supermarkets, warehouses, and other commercial facilities, there usually is a large number of loads (zones, each served by an evaporator coil), and a number of positive-displacement refrigerant compressors, which are turned on or off to follow that load. The control of such a system requires minimizing the cost of operation (meeting any cooling load with a minimum of compressor horsepower) and selecting the operating compressors in such a way that the total run time of each machine is about the same. Figure 2.7 illustrates the control system for such a system, which incorporates some of the ideas contained in the patents listed in Reference 3.

In Figure 2.7, each of the cooling zones can be a separate room, freezer, or other cooled area provided with its own temperature controls (TC). Each of the zones is cooled by an evaporator coil, which is kept cold by boiling liquid refrigerant. The liquid refrigerant is admitted into each coil by an expansion valve that controls the superheat of the evaporated vapors by detecting both the pressure and temperature of those vapors. The expansion valve admits as much refrigerant liquid as is required to keep the evaporator superheat (the temperature above the boiling point at the pressure where the refrigerant is being evaporated) constant.

A balancing valve is also provided for each zone. These are located in the refrigerant vapor line leaving the zone. If all zones are to be kept at the same temperature, all balancing valves can be kept fully open, which will minimize the pressure drop on the suction of the compressors and thereby maximize their efficiency. If some of the zone temperatures are to be controlled (by their TCs) at a higher temperature than that of the others, the balancing valves on these higher temperature

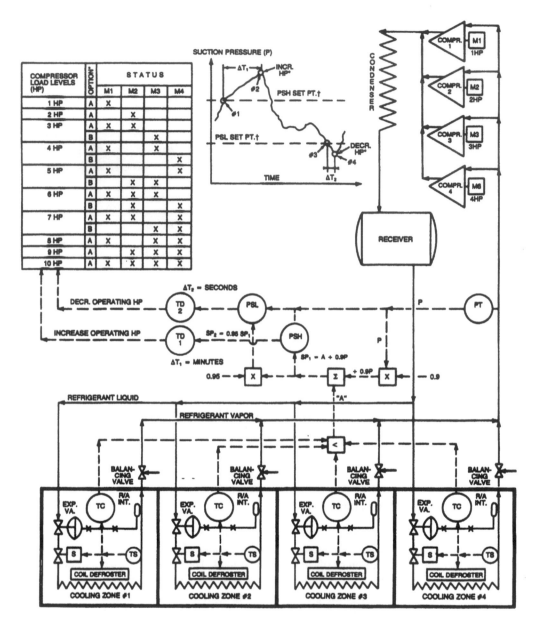

FIGURE 2.7 Minimizing the energy cost and equalizing the operating time of multiple chiller compressors serving an unlimited number of evaporator coils.[3] *When PSH calls for *increasing* operating HP, option A or B is selected to start the *least* used compressor. When PLS calls for *reducing* HP, A or B is selected to turn off the most used compressor. †When temperature in the most critical zone (the one which is the furthest above its set point) increases, both PSL and PSH set points are shifted down; when it drops, both are increased.

zones must be throttled to increase the vaporization pressure and therefore the boiling temperature of the refrigerant.

If the zones are operated below or near to the freezing temperature (below 35°F, or 2°C) it is also necessary to provide defrosting controls because of the ice buildup on the cold evaporator coils. The defrosters are controlled by a coil surface temperature detector switch in each zone (TS), which continuously detects the coil surface temperature. When it has been below some low limit for some preset time period, the switch first shuts off the refrigerant liquid supply by closing a solenoid valve; once all the refrigerant has been vaporized, the switch turns on a coil defroster heater for a short time, to remove the ice. During the defrosting cycle the zone temperature controller is disabled by setting its measurement and set point equal and thereby causing its output signal to stay constant.

The conventional method of turning the refrigerant compressors on and off is to do it in some preset sequence, where each added compressor unit is turned on when the suction pressure to the overall compressor station has dropped to an even lower level. This approach is inefficient because to reach these lower suction pressures requires added compressor horsepower. The control system shown in Figure 2.7 eliminates this waste by turning added compressors on (PSH) or turning unnecessary compressors off (PSL) at the same and higher pressure settings of PSH and PSL. This is achieved by both pressure switches actuating a time delay (TD), and whenever the pressure switch is actuated for the time period set on the time delay, it increases (TD-1) or decreases (TD-2) the horsepower level at which the system is operated. TD-1 is set in minutes while TD-2 is set in seconds. Therefore, a response to a load increase is relatively slow (minutes), while the response to a load decrease is fast (seconds), which serves energy conservation.

Further energy conservation is obtained by pushing the control gap between the pressure switches (PSH and PSL) as high as the loads will allow. This continuous floating of the controlling suction pressure gap is achieved by feedforward action based on zone temperature control. As any one of the zone temperatures rises above the zone thermostat (TC, a reverse-acting, integral-only controller) set point, the output signal of the TC drops. A low signal selector selects the lowest of the TC outputs and thereby identifies the zone having the highest cooling load (the critical zone, which, if satisfied, will result in all others being satisfied).

The set point of PSH is obtained by taking 90% of the suction pressure of the compressor station (P) and adding to it the output signal (A) of the critical TC. The set point of PSL is 95% of the PSH set point. Both the 90% and the 95% are adjustable values, which can be varied from one installation to the next. The energy consumption is minimized and the system is optimized by maximizing these set points, because whenever all the zone temperatures are on set point (or below), the suction pressure of the operating compressor station is increased, and therefore the amount of work the compressors need to do is reduced.

The graph of suction pressure (P) versus time in Figure 2.7 illustrates the operation of the system. As the suction pressure rises, at point #1 the set point of PSH is reached and the time delay TD-1 is actuated. At point #2 the time set on TD-1 has run out, and the operating horsepower level of the compressors is increased by one unit. As a result the suction pressure starts to drop. If at this point the operating compressor capacity exceeds the cooling load, the suction pressure will drop, and when it reaches the set point of PSL (point #3), TD-2 is energized. After the time delay of TD-2 has run out (point #4), the operating horsepower level of the system is reduced by one unit.

In addition to energy optimization, the system is also optimized from a maintenance point of view by keeping all compressor running times approximately the same. The method of achieving this is illustrated by the horsepower table in Figure 2.7. Let us assume that the system is operating at the 4 HP level when PSH calls for an increase to 5 HP. The 5 HP level can be met by either operating compressors 1 and 4 (option A) or by operating compressors 2 and 3 (option B). In this situation the *least used* compressor will be started. If at some time later PSL calls for reducing the operating level to 4 HP, there again are two options. Option A is to run compressors 1 and 3, while option B is to run compressor 4. Here the choice is made in such a manner that the *most used* compressor is turned off. The net result is a continuous balancing of running times between machines. According to Reference 13, the total energy savings obtained from this control system amounts to about 20%.

D. INDUSTRIAL CHILLERS

The refrigeration unit shown in Figure 2.8 contains some of the features typical of conventional industrial units in the 500 ton (1760 kw) and larger sizes. These features include the capability for continuous load adjustment as contrasted with stepwise unloading, the application of the economizer expansion valve system, and the use of a hot gas bypass to increase rangeability.

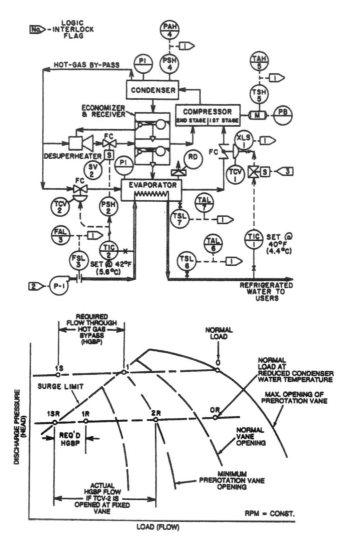

FIGURE 2.8 A conventionally controlled industrial refrigeration system uses a hot gas bypass to increase rangeability.[4]

The unit illustrated provides refrigerated water at 40°F (4.4°C) through the circulating header system of an industrial plant. The flow rate is fairly constant, and therefore process load changes are reflected by the temperature of the returning refrigerated water. Under normal load conditions, this return water temperature is 51°F (10.6°C). As process load decreases, the return water temperature drops correspondingly. With the reduced load on the evaporator, TIC-1 gradually closes the suction damper or the prerotation vane of the compressor. By throttling the suction vane, a 10:1 turndown ratio can be accomplished. If the load drops below this ratio, the hot gas bypass system has to be activated.

The hot gas bypass is automatically controlled by TIC-2. Its purpose is to keep the constant-speed compressor out of surge: when the load drops to levels sufficiently low to approach surge, this bypass valve is opened. If the chilled water flow rate is constant, the difference between chilled water supply and return temperatures is an indication of the load. If full load corresponds to a 15°F (8.3°C) difference on the chilled water side of the evaporator and the chilled water supply temperature is controlled by TIC-1 at 40°F (4.4°C), then the return water temperature detected by TIC-2 is also an indication of load.

If surge occurs at 10% load, this would correspond to a return water temperature of 41.5°F (5.3°C). In order to stay safely away from surge, TIC-2 in Figure 2.8 is set at 42°F (5.6°C), corresponding to approximately 13% load. When the temperature drops to 42°F (5.6°C), this valve starts to open, and its opening can be proportional to the load detected. This means that the valve is fully closed at 42°F (5.6°C), fully open at 40°F (4.4°C), and throttled in between. This throttling action is accomplished by a plain proportional controller that has a 2°F throttling range, which, on a span of 0 to 100°F, corresponds to a proportional band of 2% or a gain of 50.

The hot gas bypass makes it theoretically possible to achieve a very high turndown ratio by temporarily running the machine on close to zero process load. This operation can be visualized as a heat pump, transferring heat energy from the refrigerant itself to the cooling water. In the process, some of the refrigerant vapors are condensed, resulting in an overall lowering of operating pressures on the refrigerant side.

The main advantage of a hot gas bypass, therefore, is that it allows the chiller to operate at low loads without going into surge. The price of this operational flexibility is an increase in operating costs, because the work introduced by the compressor is wasted as friction drop (friction between gas molecules and valve) through the hot gas bypass valve. As will be shown later, optimized control systems eliminate this waste through the use of variable speed compressors, which will respond to a reduction in load by lowering their speed instead of throwing away the unnecessarily introduced energy in TCV-1 and TCV-2.

Instead of controlling the hot gas bypass on the basis of return water temperature (as in the upper section of Figure 2.8), other conventional packages control it on the basis of prerotation vane opening. This is illustrated in the lower portion of Figure 2.8. The problem with opening the hot gas bypass when the prerotation vane has closed to some fixed point is that this control technique disregards condensing temperature. This causes the hot gas bypass to open sooner and to a greater extent than needed. This is illustrated in the lower part of Figure 2.8, where under normal loads the compressor operates at point "O." As the load drops off at constant discharge pressure (constant condenser temperature), the prerotation vane is gradually closed to its minimum allowable opening, until at point "1" it opens the hot gas bypass, to protect the compressor from going into surge. If the actual load drops to "1S," the HGBP will furnish the differential flow between "1S" and "1."

Now, if the cooling water temperature is reduced, normal operation falls to point "OR." As the load drops, the fixed minimum setting on the prerotation vane is reached at point "2R." This is much sooner than necessary. It is possible to obtain microprocessor-based controls that recognize the impact of cooling water temperature variations and will open the HGBP only when load drops to point "1R."

The economizer shown in Figure 2.8 can increase the efficiency of operation by 5 to 10%. This is achieved by the reduction of space requirements, savings on compressor power consumption, reduction of condenser and evaporator surfaces, and other effects. The economizer shown in Figure 2.8 is a two-stage expansion valve with condensate collection chambers. When the load is above 10%, the hot gas bypass system is inactive. Condensate is collected in the upper chamber of the economizer, and it is drained under float level control, driven by the condenser pressure. The pressure in the lower chamber floats off the second stage of the compressor, and it, too, is drained into the evaporator under float level control, driven by the pressure of the compressor second stage. Economy is achieved as a result of the vaporization in the lower chamber by precooling the liquid that enters the evaporator and at the same time desuperheating the vapors that are sent to the compressor second stage.

When the load is below 10%, the hot gas bypass is in operation, and the solenoid valve SV-2, which is actuated by the high pressure switch PSH-2, opens. Some of the hot gas goes through the evaporator and is cooled by contact with the liquid refrigerant, and some of the hot gas flows through the open solenoid. This second portion is desuperheated by the injection of liquid refrigerant upstream of the solenoid, which protects against overheating the compressor.

1. Safety Interlocks

Operating safety in the system illustrated in Figure 2.8 is guaranteed by a number of interlocks. The first interlock system prevents the compressor motor from being started if one or more of the following conditions exist, and it also stops the compressor if any except the first condition listed occurs while the compressor is running.

- Suction vane is open, detected by limit switch XLS-1
- Refrigerated water temperature is dangerously low, approaching freezing, as sensed by TSL-6
- Refrigerated water flow is low, measured by FSL-3
- Evaporator temperature has dropped near the freezing point, as detected by TSL-7
- Compressor discharge pressure (and, therefore, pressure in the condenser) is high, indicated by PSH-4
- Temperature of motor bearing or winding is high, detected by TSH-5
- Lubricating oil pressure is low (not shown in Figure 2.8).

The second interlock system guarantees that the following pieces of equipment are started or are already running upon starting of the compressor:

Refrigerated water pump (P-1)
Lubricating oil pump (not shown)
Water to lubricating oil cooler, if such exists (not shown)

The third interlock usually ensures that the suction vane is completely closed when the compressor is stopped.

E. MULTISTAGE REFRIGERATION UNITS

It is not practical to obtain a compression ratio outside the range of 3:1 to 8:1 with the compressors used in the process industry. This places a limitation on the minimum temperature that a single-stage refrigeration unit can achieve.

For example, in order to maintain the evaporator at −80°F (−62°C) and the condenser at 86°F (30°C)—compatible with standard supplies of cooling water—the required compression ratio would be as follows, if Freon-12 were the refrigerant:

$$\text{Compression ratio} = \frac{\text{refrigerant pressure at } 86°F}{\text{refrigerant pressure at } -80°F} = \frac{108}{2.9} = 3 \qquad (2)$$

Such a compression ratio is obviously not practical; therefore, a multistage system is required.

Figure 2.9 illustrates a cascade refrigeration system, with ethylene as the lower- and propane as the higher-temperature refrigerant. To minimize the total operating cost, the loading of the two stages must be coordinated to balance the total work between the stages. The higher the interstage

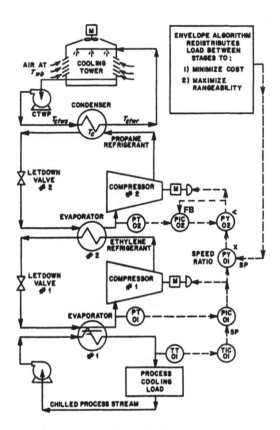

FIGURE 2.9   Multistage refrigeration units can be optimized to balance the work between the stages.

temperature (evaporator #2), the more work is done by stage #1 and the less remains for stage #2. If the compression ratios (and therefore the temperature differences) of the two stages were the same, the work would be nearly equally distributed. If true equality is to be achieved, the differences in the properties of the refrigerants and the added load on the higher stage caused by the work introduced by the lower stage compressor must also be considered.

Another goal is to maximize the rangeability of the multistage unit. The rangeability is maximum when the two stages approach their surge limits together. This, too, can be guaranteed by redistributing the load between the stages so that they maintain equal distance from their respective surge lines as the load drops.

In Figure 2.9, TIC-01, the temperature controller of the chilled process stream, sets the speed of the first stage compressor by modifying the set point of its suction pressure controller. The speed of compressor #2 is set in ratio to #1 by PY-01. This speed ratio set point can be optimized by a multivariable control envelope (to be discussed later), which considers both cost and rangeability in moving the system to its optimum. As long as both stages are far from surge, the algorithm redistributes the load to minimize total operating cost. When one of the stages is nearing surge, the envelope algorithm increases the loading of that stage to keep the system out of surge. Therefore, at high loads the goal of load distribution between the stages is to minimize the total operating cost, and at low loads the goal is to keep the system out of surge.

As ambient temperatures drop, the cascaded stages will operate against a lower total temperature difference; therefore, each of the stages will require less work. As the condenser temperature drops, compressor #2 will lower its suction pressure. This in turn will cause the temperature in evaporator #2 to drop, which will cause compressor #1 to also lower its suction pressure. As the suction

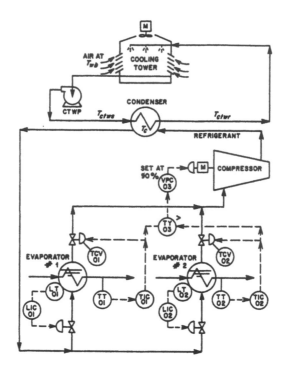

FIGURE 2.10   Load-following control of multiple users can be set up so that the most demanding user sets the speed of the compressor.

pressures drop, PIC-01 and PIC-02 will reduce the speed of their respective compressors, thereby reducing the total work by the heat pumps. Using these procedures, this cascade system responds to ambient variations in a flexible and efficient manner.

F. MULTIPLE USERS

When evaporators operate in parallel, off the same compressor, the load can be followed by varying either the evaporator temperature or the refrigerant level in the evaporators. Figure 2.10 shows the controls when the evaporator levels (their wetted heat transfer areas) are kept constant and load variations result in modifications in the evaporator temperature. TY-03 selects the most open TCV and keeps it from opening to more than 90% by speeding up the compressor when the valve opening reaches 90%. The resulting reduction in compressor suction pressure lowers the evaporator temperature and increases the rate of heat transfer. Thus, no user is ever allowed to run out of coolant, and compressor operating costs are kept at a minimum. This type of load-following optimization system, where compressor speed varies with the requirements of the most heavily loaded user, is very sensitive to dynamic upsets. Therefore, VPC-03 must be tuned for slow adjustment of the compressor speed (wide proportional, mostly integral settings) to avoid upsetting the other users, which are not selected for control.

If one of the user valves is consistently more open than the others, it might be possible to repipe the other loads to the interstage of the compressor. This might provide a better balance between users.

Figure 2.11 illustrates a control system in which the vapor-side control valves have been eliminated. This advantage must be balanced against a number of disadvantages. In this system, heat transfer is modulated by varying the evaporator tube surface areas exposed to boiling. The user with the highest load will have the highest refrigerant level. LY-03 selects the most flooded evaporator for control by LC-03, which increases the speed of the compressor when the level exceeds 90% and lowers it when the level is below 90%.

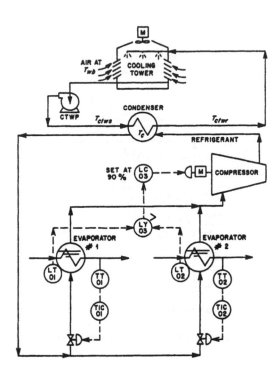

FIGURE 2.11 In this alternative method for controlling multiple users, the vapor-side control valves have been eliminated.

This method of load following has some drawbacks. One problem is that the heat transfer surfaces within the evaporators are not fully utilized. Another difficulty is that the relationship between level and heat transfer area can be confused by foaming, which keeps the tubes wet even when the level has dropped.[5] For these reasons, tuning of such control systems can be difficult, necessitating the sacrifice of responsiveness for stability. Therefore, if responsiveness is critical, it is better to keep the refrigerant level constant and manipulate the evaporating pressure, as shown in Figure 2.10.

G. FEEDFORWARD CONTROL

Feedforward anticipation can increase the responsiveness of the control system while also providing more precise temperature control. Such precision is desired when the goal is to maximize chilling without freezing.

When the heat transfer area is fixed (constant level), the rate of heat transfer (load) is a function of the temperature difference between the process fluid and the boiling refrigerant. Therefore, the boiling temperature (and pressure) must change as the load varies.

Figure 2.12 shows the main components of this control system. The instantaneous load is calculated by multiplying the flow rate of the cooled process fluid (FY-05) by its desired drop in temperature (TY-04). Based on this load and the desired process temperature, FY-06 determines the desired refrigerant temperature. "K" represents heat transfer area and coefficient. The value of K in this summing device is adjustable to match the actual slope of load versus refrigerant temperature. PY-07 serves to convert boiling temperature to the corresponding vapor pressure set point.[5]

Dynamic compensation is provided by FY-03 to match the response of process temperature to refrigerant pressure.[5] A valve position control system, similar to the one shown in Figure 2.10, can also be used here to minimize the operating cost of the compressor.

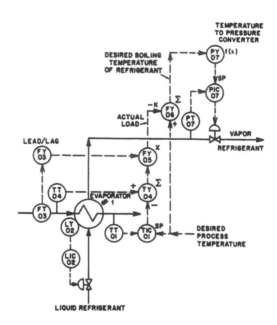

FIGURE 2.12   Increased responsiveness can be achieved through the use of feedforward setting of refrigerant pressure.

III. OPTIMIZED CONTROLS

The piping layout of a conventional cooling system is illustrated in Figure 2.13. To develop a completely generalized method for controlling and optimizing such systems, the duplication of equipment will be disregarded, and all chiller systems will be treated as if they were configured as shown in Figure 2.14. Here any number of cooling towers, pumps, or chillers are represented by single symbols because variations in their numbers will not affect the optimization strategy.

In this generalized cooling system, the cooling load from the process is carried by the chilled water to the evaporator, where it is transferred to the refrigerant. The refrigerant takes the heat to the condenser, where it is passed on to the cooling tower water, so that it might finally be rejected to the ambient air. This heat pump operation involves four heat transfer substances (chilled water, the refrigerant, cooling tower water, and the atmospheric air) and four heat exchanger devices (process heat exchanger, evaporator, condenser, and cooling tower). The total system operating cost is the sum of the costs of circulating the four heat transfer substances (M1, M2, M3, and M4).

In the traditional (unoptimized) control systems, such as the one illustrated in Figure 2.8, each of these four systems was operated independently in an uncoordinated manner. In addition, conventional control systems did not vary the speed of the four transportation devices (M1 to M4). By operating them at constant speeds, they introduced more energy than was needed for the circulation of refrigerant, air, or water and therefore had to waste that excess energy.

Load-following optimization eliminates this waste while operating the aforementioned four systems as a coordinated single process, with the goal of control being to maintain the cost of operation at a minimum. The controlled variables are the supply and return temperatures of chilled and cooling tower waters, and the manipulated variables are the flow rates of chilled water, refrigerant, cooling tower water, and air. If water temperatures are allowed to float in response to load and ambient temperature variations, the waste associated with keeping them at arbitrarily selected fixed values is eliminated and the chiller operating cost is drastically reduced.

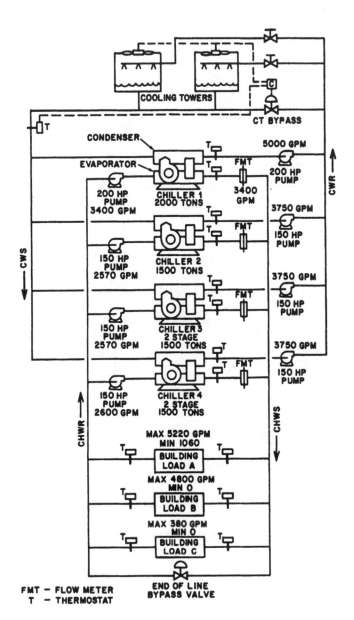

FIGURE 2.13 A typical cooling system layout involving four heat transfer substances and four heat exchange devices.[6]

In order to control the total refrigeration system depicted in Figure 2.14, four control loops must be configured. In these four loops, the controlled (measured) variables are the four water temperatures, and the four manipulated variables are the four motors (M1 to M4) that drive the four transportation devices. Table 2.2 lists the controlled and the manipulated variables in each of these loops and also indicates the optimization criteria for setting the set points of the controllers.

A. Minimizing the Operating Cost

The yearly cost of operating the total cooling system can typically be broken down as follows

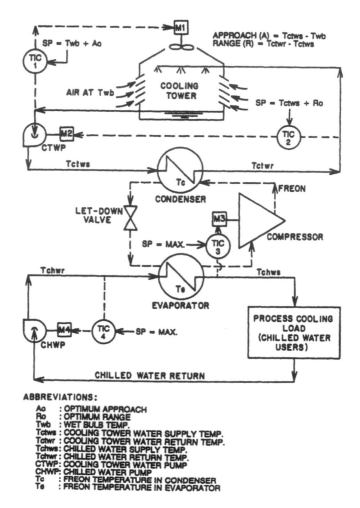

FIGURE 2.14 Optimized control of refrigeration machines can be accomplished by treating the process as an integrated system.[2]

M1 (Fans)	10%
M2 (CTWP)	15
M3 (Compr.)	60
M4 (CHWP)	15
	100%

These cost percentages vary a great deal with geographic region (the proportion of M1 increases in warm weather), with plant layout (M2 and M4 increase when water transport lines are long), and with the nature of the load (the proportion of M3 is lowered as the maximum allowable chilled water temperature rises). Regardless of what these proportions are in a particular installation, the goal of optimization is to find the minimum chilled water and cooling tower water temperatures that will result in meeting the needs of the process at minimum cost. The overall control system can be evaluated in two steps: first the chilled water side (the evaporator side), then the cooling tower water side (condenser side).

The optimization of the lower (evaporator) portion of Figure 2.15 is easily comprehended because the sum of M3 and M4 will be the minimum when both Tchws and Tchwr are as high as the process

TABLE 2.2
Control Configuration and Optimization Criteria for the Four Loops Controlling a Chiller

Loop No.	Controlled Variable	Manipulated Variable	Optimization Criteria (Set point for the Temperature Controller)
1	CT water supply temp. (Tctws)	Level of fan operation (M1)	Optimum approach (Ao) is selected to keep the sum M1 + M2 + M3 to a minimum; therefore, TIC set point becomes Tctws = Twb + Ao
2	CT water return temp. (Tctwr)	Rate of CT water pumping (M2)	The optimum range (Ro) of a CT is found as a function of its Ao; therefore, the TIC set point becomes Tctwr = Tctws + Ro
3	Chilled water supply temp. (Tchws)	Rate of chiller compressor operation (M3)	The optimum TIC set point is the maximum Tchws temperature that will satisfy all the process loads
4	Chilled water return temp. (Tchwr)	Rate of chilled water pumping (M4)	The optimum TIC set point is the maximum Tchwr temperature that will satisfy all the process loads

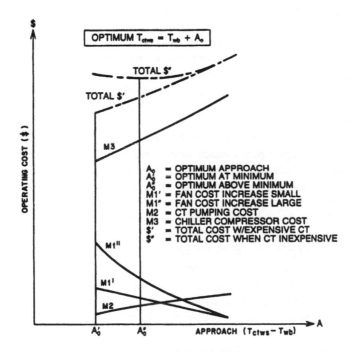

FIGURE 2.15 At each particular load (Twb and Tctws), an empirically determined total cost-versus-approach curve can have a minimum.

will permit. This is true under all load conditions, because the amount of work that the chiller compressor has to do is reduced as the suction pressure rises; this is the case whenever the evaporator temperature (Tchws) rises. (The optimum set point of TIC-3 in Figure 2.14 is the allowable maximum.) Therefore, the minimum cost of operation will be achieved when the chilled water supply temperature has been maximized, but only to the point where it can still provide all the cooling that is required by any part of the process. The method of finding that maximum will be discussed later.

The chilled water pumping cost (M4) depends only on the ΔT across the users. The higher this ΔT, the less water needs to be pumped to transport the amount of heat. As the chilled water supply temperature is set by the load, ΔT can be maximized only by maximizing the chilled water return temperature (Tchwr). Therefore, the optimum set point for TIC-4 in Figure 2.14 is the allowable maximum. The increasing of this temperature therefore has the following benefits:

1. M3 is lowered by about 1.5% per °F rise.
2. The heat transfer efficiency of the evaporator is improved by increasing the ΔT through it.
3. Assuming that on the average the ΔT is 15°F, increasing it by 1°F will lower M4 by about 6%.

As M3 is about 60% and M4 about 15% of the total operating cost, a 1°F decrease in ΔT will lower the yearly cost of operation by about $(1 \times 60)/100 + (6 \times 15)/100 = 1.5\%$.

Finding the optimum cooling tower water temperatures is more complicated because changing the cooling tower water supply and return temperatures will increase some costs while lowering some others. For example, an increase in the cooling tower water supply temperature (Tctws) will increase the cost of chiller operation but will reduce the amount of work the cooling tower fans need to do. Therefore, the optimum temperature is not necessarily the minimum temperature (or approach) that the towers are capable of providing, but rather the temperature which can meet the particular load at the total minimum cost of all operating equipment (M1 + M2 + M3). Figure 2.15 illustrates that this is a function of the load and of the weather. When the conditions are such that a lowering of the approach (or Tctws) would require a large increase in fan operating costs (M1″), the minimum point on the total cost curve (A_o'') will be above the minimum attainable approach (A_o'). Inversely, if the approach can be reduced with a small increase in fan operating costs (M1′), the optimum operating point for the total system corresponds to the minimum Tctws temperature and therefore to the minimum attainable approach (A_o'). Whichever is the applicable total cost curve ($\$'$ or $\$''$), the Tctws temperature which corresponds to the minimum point becomes the set point for TIC-1 in Figure 2.14.

The curves in Figure 2.15 can be based on the measurement of the actual total operating cost, on a projected cost based on past performance, or on some combination of the two. The continuous storing and updating of operating cost history as a function of load, ambient conditions, and equipment configuration not only will provide the curves required for optimization but can also be used to signal the need for maintenance.

In Figure 2.15, the optimum Tctws is found by summing M1, M2, and M3. In most installations the major cost element is M3, which increases by about 1.5% of compressor operating cost (or 1% of total cooling system cost) for each 1°F reduction in approach. The cost-versus-approach curve is not necessarily a smooth one because, if constant-speed machines are stopped and started or if positive-displacement compressors are loaded and unloaded, there will be steps in the M3 curve. This also is true for constant speed fans (M1) or pumps (M2). The pumping cost also increases as approach rises, but M2 is usually only a fraction of M3, unless the piping is very long or undersized.

The fan cost M1 tends to drop off as approach rises. If the load is low relative to the size of the cooling towers, the rate of increase in M1 with a reduction in approach will also be small. In such cases (M1′) the optimum approach (A_o') corresponds to the safe lowest temperature that the tower can generate. If, on the other hand, the load is high, the fan cost of lowering approach will also increase (M1″). In that case, the optimum approach (A_o'') is at some intermediate value.

The optimum return water temperature back to the cooling tower (Tctwr) can be found by determining the optimum range (Ro) for the particular set of operating conditions. The relationship between operating approach and range is a function of the particular cooling tower design. Therefore,

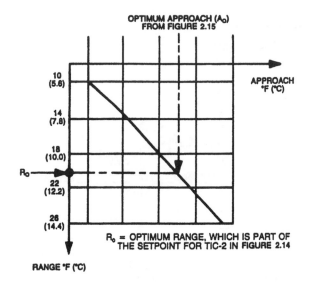

FIGURE 2.16 The optimum range for a particular cooling tower is a function of its design and of the approach at which it is operating. Therefore, R_o can be found on the basis of A_o.

if one has determined the optimum approach for a particular set of conditions A_o (from Figure 2.15), that value can be used to arrive at the optimum range (Ro) in Figure 2.16. The set point for TIC-2 in Figure 2.14 then becomes the sum of the already-determined Tctws and the optimum range:

$$Tctwr = Tctws + Ro \qquad (3)$$

1. Chilled Water Supply Temperature Optimization

The yearly operating cost of a chiller is reduced by 1.5 to 2% for each 1°F (0.6°C) reduction in the temperature difference across this heat pump ($T_H - T_L$ in Figure 2.1). In order to minimize this difference, T_H must be minimized and T_L maximized. Therefore, the optimum value of Tchws is the maximum temperature that will still satisfy all the loads. Figure 2.17 illustrates a method of continuously finding and maintaining this value.

It should be noted that an energy-efficient refrigeration system cannot be guaranteed by instrumentation alone because equipment sizing and selection also play an important part in the overall result. For example, the evaporator heat transfer area should be maximized so that T_L is as close as possible to the average chilled water temperature in the evaporator. Similarly, at the compressor the refrigerant flow should be made to match the load by motor speed control rather than by throttling. TCV-1 and TCV-2 in Figure 2.8 represent sources of energy waste, whereas Figure 2.14 shows the energy-efficient technique of motor speed control. The variable-speed operation can be achieved using variable-speed drives on electric motors or by using steam turbine drives. If several constant-speed motors are used, then all compressors except one should be driven to the load at which they are most efficient (TCV-1 and TCV-2 in Figure 2.8 fully open); the remaining compressor should be used to match the required load by throttling.

Figure 2.17 shows the proper technique of maximizing the chilled water supply temperature in a load-following, floating manner. The optimization control loop guarantees that all users of refrigeration in the plant will always be satisfied while the chilled water temperature is maximized. This is done by selecting (with TY-1) the most open chilled water valve in the plant and comparing that signal with the 90% set point of the valve position controller, VPC-1. If even the most open valve is less than 90% open, the set point of TIC-1 is increased; if the valve opening exceeds 90%, the TIC-1 set point is decreased. This technique allows all users to obtain more cooling (by further

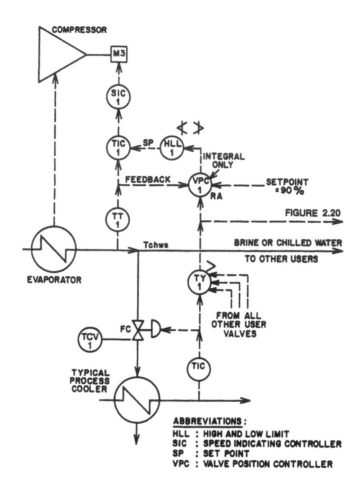

COMPRESSOR

M3

SIC 1

TIC 1 SP HLL 1

INTEGRAL ONLY

FEEDBACK VPC 1 RA SETPOINT = 90%

FIGURE 2.20

TT 1

EVAPORATOR

Tchws

BRINE OR CHILLED WATER
TO OTHER USERS

TY 1

FROM ALL
OTHER USER
VALVES

TCV 1 FC

TYPICAL
PROCESS
COOLER

TIC

ABBREVIATIONS:
HLL : HIGH AND LOW LIMIT
SIC : SPEED INDICATING CONTROLLER
SP : SET POINT
VPC : VALVE POSITION CONTROLLER

FIGURE 2.17 Load-following, floating control of chilled-water supply temperature is one method of continuously finding and maintaining the optimum value of Tchws.

opening their supply valves) if needed, while the header temperature is continuously maximized. The VPC-1 set point of 90% is adjustable. Lowering it gives a wider safety margin, which might be required if some of the cooling processes served are very critical. Increasing the set point maximizes the energy conservation at the expense of the safety margin.

An additional benefit of this load-following optimization strategy is that, because all chilled water valves in the plant are opened as T_{chws} is maximized, valve cycling is reduced and pumping costs are lowered because when all chilled water valves are opened they require less pressure drop. Valve cycling is eliminated when the valve opening is moved away from the unstable region near the closed position.[2]

For the control system in Figure 2.17 to be stable, it is necessary to use an integral-only controller for VPC-1, with an integral time that is tenfold that of the integral setting of TIC-1 (usually several minutes). This mode selection is needed to allow the optimization loop to be stable when the valve opening signal selected by TY-1 is either cycling or noisy. The high/low limit settings on the set point signal (HLL-1) to TIC-1 guarantee that VPC-1 will not drive the chilled water temperature to unsafe or undesirable levels. Because these limits can block the VPC-1 output from affecting T_{chws}, it is necessary to protect against reset wind-up in VPC-1. This is done through the external feedback signal shown in Figure 2.17.

a. Alternative configurations

Figure 2.18 illustrates the ideal water distribution system, in which the individual users are served by two-way valves and the optimum supply temperature is found by keeping the most open valve at 90% opening. Although this is a very effective control configuration, it is not always used in existing installations. If the number of user valves is very high and distributed over a large area or if the loops are not properly tuned and the valves have a tendency of cycling from fully closed to fully open, this control strategy is not practical. In such situations, the controls shown in Figure 2.19 can still be applied by selecting a few representative user valves that are not cycling and by basing the optimization on those.

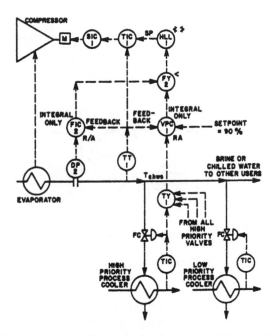

FIGURE 2.18   The chilled water distribution system can be retrofitted for supply temperature optimization. (Adapted from Lipták, B.G., *Chem. Eng.*, October 17, 1983.)

It is also possible to group the user valves into high- and low-priority categories. Figure 2.18 illustrates a control system in which the high-priority users are treated the same way as they were in Figure 2.17. The low-priority users, on the other hand, are grouped together, and their demand is detected through the measurement of total flow. If all low-priority valves are the two-way type, and if supply temperature is constant, flow will vary directly with cooling load. If valve-position-based optimization were applied, it would attempt to maximize valve opening, which in turn would maximize flow by raising T_{chws} to the highest acceptable level. This same goal is achieved by FIC-2 in Figure 2.18. As the flow drops off as a result of a reduction in demand for cooling, FIC-2 will raise the supply temperature, which in turn will increase the total flow, as the valves open up. Therefore, the total flow is kept constant, and load variations result in supply temperature variations. VPC-1 and FIC-2 should be integral-only controllers, with time constants of several minutes.

The control of total flow in Figure 2.18 can satisfy the average user, but not all the users. Therefore, the critical users cannot be treated in this manner, but must be protected by valve position control. Such control is provided by VPC-1, which overrides the flow control signal whenever a high-priority valve reaches 90% opening.

In many existing installations—particularly in air-conditioning applications—three-way control valves are used. This is a highly undesirable practice, because it unnecessarily increases the required

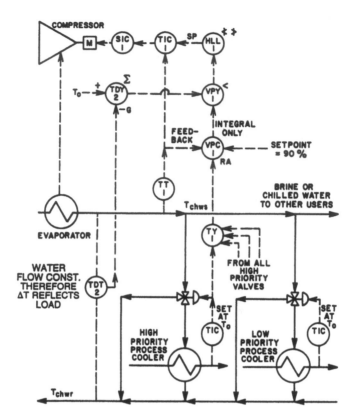

FIGURE 2.19 Chilled water users with three-way valves can also be retrofitted for optimization. (Adapted from Lipták, B.G., *InTech*, September 1977.)

amount of pumping and lowers return water temperature. Yet, if the system is already in operation and if it consists of hundreds of air-conditioning type users, it would be better not to attempt to redesign it but to optimize it in its existing form. A control system that will do that is shown in Figure 2.19.

In this system, the total flow is relatively constant, and it is the return water temperature that reflects the variations in cooling load. The control system in Figure 2.19 determines the set point for TIC-1 by relating the temperature rise (Tchws — Tchwr) and the desired space temperature set points (To) as follows:

$$\text{TIC-1 set point} = \text{To} - \text{G (Tchwr} - \text{Tchws)} \qquad (4)$$

If the thermostats are set at To = 80°F (27°C) and if, at full load, Tchws = 50°F (10°C) and Tchwr = 65°F (18°C), the value of G can be determined as follows:

$$\text{G} = (\text{To} - \text{set point})/(\text{Tchwr} - \text{Tchws}) = (80 - 50)/(65 - 50) = 2 \qquad (5)$$

TDY-2 in Figure 2.19 therefore calculates the desired set point for TIC-1 at any load. For example, if the load drops to 50%, the space temperature To will begin to fall, and the thermostats will divert more coolant into the return line. This will reduce Tchwr from 65°F (18°C) to 57.5°F (14°C). In conventional systems, this then would become the new steady state. With the control system shown in Figure 2.19, TDY-2 will revise the TIC-1 set point as follows:

$$\text{set point} = 80 - 2(57.5 - 50) = 65°F \qquad (6)$$

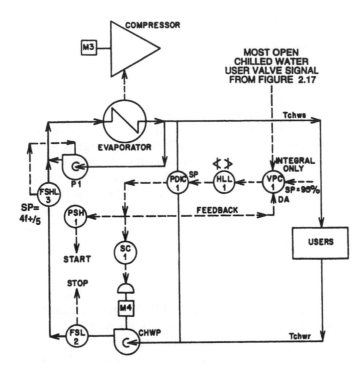

FIGURE 2.20 Floating of chilled water flow rate keeps evaporator ΔT always at maximum (optimum) values.

As Tchws is slowly increasing, the space thermostats will divert less and less water until they come to rest at the same percentage of water diverted as at full load. The benefit of this optimization strategy is in reducing the temperature differential across the chiller, which in turn lowers the operating cost by 15% for each 10°F (5.5°C). The setting of G determines the percentage of coolant diverted by the average valve. If G is set too low, some valves will be 100% open to the chilled water supply and therefore out of control. If G is set too high, the chilled water supply temperature will be lower than needed, and compressor energy will be wasted.

In order to protect the high-priority users from running out of coolant when G is set low, a valve-position-based override is used in Figure 2.19. Whenever a high-priority valve is more than 90% open, VPC-1 is going to override the set point of TIC-1 and lower it until all high-priority valves are less than 90% open.

2. Chilled Water Return Temperature Optimization

The combined cost of operating the chilled water pumps and the chiller compressor (M4 + M3) is a function of the temperature drop across the evaporator (Tchwr – Tchws). Because an increase in this ΔT decreases compressor operating costs (suction pressure rises) while also decreasing pumping costs (the higher the ΔT, the less water needs to be pumped), the aim of this optimization strategy is to maximize this ΔT.

This ΔT will be the maximum when the chilled water flow rate across the chilled water users is the minimum. As Tchws is already controlled, this ΔT will be maximized when Tchwr is maximum. This goal can be reached by looking at the opening of the most open chilled water valve in Figure 2.17. If even the most open chilled water valve is not yet fully open, the chilled water supply temperature (set point of TIC-1 in Figure 2.17) can be increased, or the temperature rise across the users can be increased by lowering the ΔP (set point of PDIC-1 in Figure 2.20); both methods may be used simultaneously. Increasing the chilled water supply temperature reduces the yearly compressor operating cost (M3) by approximately 1.5% for each °F of temperature increase, whereas lowering the ΔP reduces the yearly pump operating cost (M4) by approximately 50 cents/GPM for each PSID.

The set points of the two valve position controllers (VPC-1s in Figures 2.17 and 2.20) will determine if these adjustments are to occur in sequence or simultaneously. If both set points are the same, simultaneous action will result; if one adjustment is economically more advantageous or safer than the other, the set point of the corresponding VPC will be positioned lower than the other.

This will result in sequencing, which means that the more cost-effective or safer correction will be fully exploited before the less effective one is started. In Figures 2.17 and 2.20, it was assumed that increasing Tchws is the more cost-effective step. This is not always the case, and even when it is, there might be process reasons that make it undesirable to float Tchws up or down. If the settings are as shown in Figures 2.17 and 2.20, the system will function as follows: if the most open valve is less than 90% open, VPC-1 in Figure 2.20 will lower the set point of PDIC-1, and VPC-1 in Figure 2.17 will increase the set point of TIC-1. When the opening of the most open valve reaches 90%, VPC-1 in Figure 2.17 will slowly start lowering the TIC set point. If it is not lowered enough (or if the lowering is not fast enough) and the most open valve continues to open further, at 95% the VPC-1 in Figure 2.20 will take fast corrective action, quickly raising the set point of PDIC-1. Thus, no user valve will ever be allowed to open fully and go out of control as long as the pumps and chillers are sized so as to be capable of meeting the load.

VPC-1 in Figure 2.20 is the cascade master of PDIC-1, which guarantees that the pressure difference between the chilled water supply and return is always high enough to motivate water flow through the users but never so high as to exceed their pressure ratings. The high and low limits are set on HLL-1, and VPC-1 is free to float this set point within these limits to keep the operating cost at a minimum.

In order to protect against reset wind-up when the output of VPC-1 reaches one of these limits, an external feedback is provided from the PDIC-1 output signal to the pump speed controller SC-1. The VPC is an integral-only controller, which is tuned to be much more responsive than the VPC in Figure 2.17.

The high speed of response of the VPC in Figure 2.17 is also important from a safety point of view. When a conventional control system is used and the demand for cooling suddenly increases because of some process difficulty, once the valve (TCV-1 in Figure 2.17) is fully open, the amount of cooling provided cannot be increased faster than the rate at which the chiller can lower the chilled water supply temperature. This does take time, and therefore, if the cooling load evolves faster (runaway reaction in a reactor), accidents can occur.

This is not the case when the control system shown in Figure 2.20 is implemented, because as soon as the chilled water valve has opened to 95%, VPC-1 will quickly speed up the chilled water pump (CHWP), and because water is incompressible, this will immediately increase the chilled water supply *pressure* in the main supply header. Consequently, the water supply pressure to TCV-1 in Figure 2.17 will also rise, immediately increasing the water flow and therefore the amount of cooling through that valve. This action, which is practically instantaneous, can be fast enough to arrest a runaway reaction, while utilizing the already-made coolant stored in the distribution headers. Naturally, once the chiller has had time to respond to the increase in load by lowering the chilled water supply temperature, this emergency action will no longer be needed and the water supply pressure will be dropped back to its normal setting.

When the chilled water pump station consists of several pumps, only one of which is variable-speed, additional pump increments are started when PSH-1 signals that the pump speed controller set point is at its maximum. When the load is dropping, the excess pump increments are stopped on the basis of flow, detected by FSL-2. In order to eliminate cycling, the excess pump increment is turned off only when the actual total flow corresponds to less than 90% of the capacity of the *remaining* pumps.

This load-following optimization loop will float the total chilled water flow to achieve maximum overall economy. In order to maintain efficient heat transfer and appropriate turbulence within the evaporator, a small local circulating pump (P1) is provided at the evaporator. This pump is started and stopped by FSHL-3, guaranteeing that the water velocity in the evaporator tubes will never drop below the adjustable limit of about 4 ftps (1.2 m/s).

3. Cooling Tower Supply Temperature Optimization

Minimizing the temperature of the cooling tower water is one of the most effective ways to contribute to chiller optimization. Conventional control systems were operated with constant cooling tower temperatures of 75°F (23.9°C) or higher. A constant utility condition is always in conflict with efficiency and therefore with optimization. Each 10°F (5.6°C) reduction in the cooling tower water temperature will reduce the yearly operating cost of the compressor by approximately 15%. For example, if a compressor is operating at 50°F (10°C) condenser water instead of 85°F (29.4°C), it will meet the same load while consuming half as much power. Operation at condenser water temperatures of less than 50°F (10°C) is quite practical during the winter months. Savings exceeding 50% have been reported.[1]

As shown in Figure 2.21, an optimization control loop is required in order to maintain the cooling tower water supply continuously at an economical minimum temperature. This minimum temperature is a function of the wet-bulb temperature of the atmospheric air. The cooling tower cannot generate a water temperature that is as low as the ambient wet bulb, but it can approach it. The temperature difference between Tctws and Twb is called the *approach*, as was explained in connection with Figure 2.15.

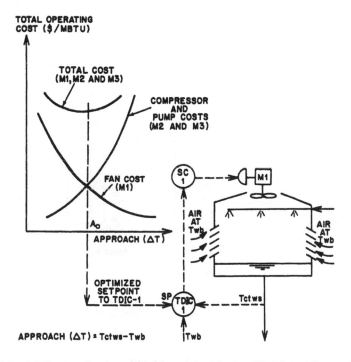

FIGURE 2.21   A load-following floating control loop is required to keep the cooling tower water supply at an economical minimum temperature.

Figure 2.21 illustrates the fact that as the approach increases, the cost of operating the cooling tower fans drops and the costs of pumping and of compressor operation increase. Therefore, the total operating cost curve has a minimum point that identifies the optimum approach that will allow operation at an overall minimum cost. This ΔT automatically becomes the set point of TDIC-1.

This optimum approach increases if the load on the cooling tower increases or if the ambient wet bulb decreases.

If the cooling tower fans are centrifugal units or if the blade pitch is variable, the optimum approach is maintained by continuous throttling. If the tower fans are two-speed or single-speed units, the output of TDIC-1 will start and stop the fan units incrementally in order to maintain the optimum approach. In cases in which a large number of cooling tower cells constitutes the total system, it is also desirable to balance the water flows to the various cells automatically as a function of the operation of the associated fan. In other words, the water flows to all cells whose fans are at high speed should be controlled at equal high rates; cells with fans operating at low speeds should receive water at equal low flow rates, and cells with their fans off should be supplied with water at equal minimum flow rates. Subjects such as these and the cooling towers in general are discussed in more detail in Chapter 5.

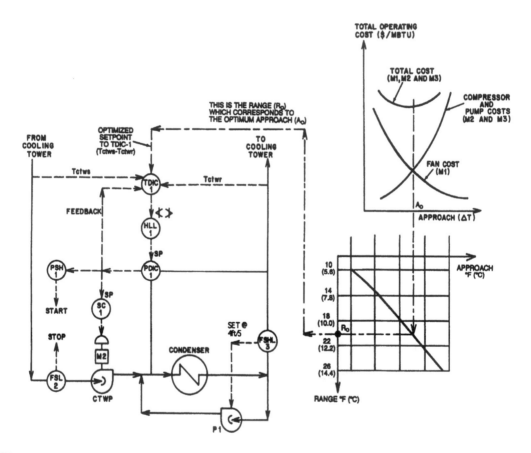

FIGURE 2.22 The cooling tower water flow rate can be floated to keep condenser ΔT always at optimum values.

4. Cooling Tower Return Temperature Optimization

The optimum setting for Tctwr is based on the optimum approach (Ao) obtained from Figure 2.21, using the characteristic curve of the cooling tower depicted in Figure 2.16. This curve relates the approach to the range for the particular tower design, and therefore once Ao is known, Ro can be obtained from it. As shown in Figure 2.22, the ΔT corresponding to this optimum Ro automatically becomes the set point of TDIC-1 in the optimized control loop shown. This controller is the cascade master of PDIC-1, which guarantees that the pressure difference between the supply and return cooling tower water flows is always high enough to provide flow through the users but never so high as to cause damage. The high and low limits are set on HLL-1. TDIC-1 freely floats this set point within these ΔP limits, to keep the operating cost at a minimum.

In order to protect against reset wind-up (when the output of TDIC-1 reaches one of these limits), an external feedback is provided from the PDIC-1 output signal to the pump speed controller SC-1.

When the cooling tower water pump station consists of several pumps, only one of which is variable-speed, additional pump increments are started when PSH-1 signals that the pump speed controller set point is at its maximum. When the load is dropping, the excess pump increments are stopped on the basis of *flow*, detected by FSL-2. In order to eliminate cycling, the excess pump increment is turned off only when the actual total flow has dropped to less than 90% of the capacity of the *remaining* pumps.

This load-following optimization loop will float the total cooling tower water flow to achieve maximum overall economy. In order to maintain efficient heat transfer and appropriate turbulence within the condenser, a small local circulating pump (P1) is provided at the condenser. This pump is started and stopped by FSHL-3, guaranteeing that the water velocity in the condenser tubes will never drop below the adjustable limit of, for example, 4 ftps (1.2 m/s).

5. Heat Recovery Optimization

Figure 2.23 depicts the required optimizing control loop when the heat pumped by the chiller is recovered in the form of hot water.

Like the control system shown for chilled water temperature floating in Figure 2.17, the hot water temperature can also be continuously optimized in a load-following floating manner. If, at a particular load level, it is sufficient to operate with 100°F (37.8°C) instead of 120°F (48.9°C) hot water, this technique will allow the same tonnage of refrigeration to be met at 30% lower cost. The reason is that the compressor discharge pressure is determined by the hot water temperature in the split condenser.

The optimization control loop in Figure 2.23 guarantees that all hot water users in the plant will always obtain enough heat while the water temperature is minimized. TY-1 selects the most open hot water valve in the plant, and VPC-1 compares the transmitter signal with a 90% set point. If even the most open valve is less than 90% open, the set point of TIC-1 is decreased; if the opening exceeds 90%, the set point is increased. This allows the users to obtain more heat (by further opening their supply valves) if needed, while the header temperature is continuously optimized.

Figure 2.23 also shows that an increasing demand for heat will cause the TIC-1 output signal to rise as its measurement drops below its set point. An increase in heat load will cause a decrease in the heat spill to the cooling tower, since the control valve TCV-1A is closed between 3 and 9 PSIG (0.2 and 0.6 bar). At a 9 PSIG (0.6 bar) output signal, all the available cooling load is being recovered and TCV-1A is fully closed. If the heat load continues to rise (TIC-1 output signal rises over 9 PSIG, or 0.6 bar), this will result in the partial opening of the "pay heat" valve, TCV-1B. In this mode of operation, the steam heat is used to supplement the freely available recovered heat to meet the prevailing heat load.

A local circulating pump, P1, is started whenever flow velocity is low. This prevents the formation of deposit in the tubes. P1 is a small 10 to 15 hp pump operating only when the flow is low. The main cooling tower pump (usually larger than 100 hp) is stopped when TCV-1A is closed.

6. Operating Mode Selection

The cost-effectiveness of heat recovery is a function of the outdoor temperature, the unit cost of energy from the alternative heat source, and the percentage of the cooling load that can be used as recovered heat. According to Figure 2.24, if steam is available at $7/MMBTU and only half of the cooling load is needed in the form of hot water, it is more cost-effective to operate the chiller on

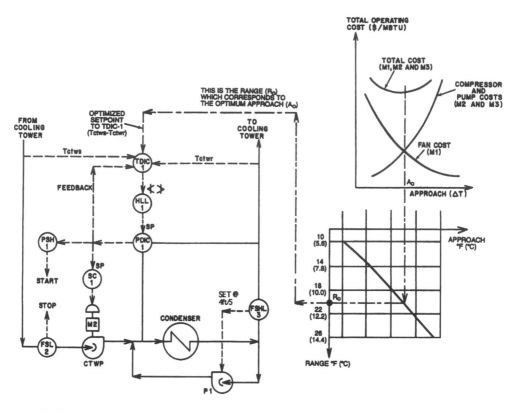

FIGURE 2.23 A load-following control loop is required when the heat pumped by the chiller is recovered as hot water.

cooling tower water and use steam as the heat source when the outside air is below 65°F (18°C). Conversely, when the outdoor temperature is above 75°F (23.9°C), the penalty for operating the split condenser at hot water temperatures is no longer excessive; therefore, the plant should automatically switch back to recovered heat operation. This cost-benefit analysis is a simple and continually used element of the overall optimization scheme.

EXAMPLE 2.1

Temperature of Condenser Water	Cost Components	Mechanical Refrigeration Mode	Heat Recovery Mode
	Cost of Cooling	$4.25	$8.00
65°F (18.3°C)	Cost of Heating $\frac{(0.5)(7.0)}{}=$	$3.50	$0.00
	Total	$7.75	$8.00
	Cost of Cooling	$5.30	$8.00
75°F (23.9°C)	Cost of Heating $\frac{(0.5)(7.0)}{}=$	$3.50	$0.00
	Total	$8.80	$8.00

Note: This example is based on the following assumptions:

[a] The actual cooling load is 50% of chiller capacity. (CL = 0.5 CAP)
[b] The heating load (the demand for hot water) is 50% of cooling load. (HL = 0.5 CL)

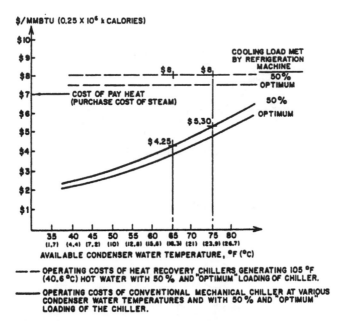

--- OPERATING COSTS OF HEAT RECOVERY CHILLERS GENERATING 105 °F
(40.6 °C) HOT WATER WITH 50 % AND "OPTIMUM" LOADING OF CHILLER.

——— OPERATING COSTS OF CONVENTIONAL MECHANICAL CHILLER AT VARIOUS
CONDENSER WATER TEMPERATURES AND WITH 50 % AND "OPTIMUM"
LOADING OF THE CHILLER.

*Example**

Temperature of Condenser Water	Cost Components	Mechanical Refrigeration Mode		Heat Recovery Mode
	Cost of Cooling		$4.25	$8.00
65°F (18.3°C)	Cost of Heating	$\dfrac{(0.5)(7.0)}{\text{Total}} =$	$\dfrac{\$3.50}{\$7.75}$	$\dfrac{\$0.00}{\$8.00}$
	Cost of Cooling		$5.30	$8.00
75°F (23.9°C)	Cost of Heating	$\dfrac{(0.5)(7.0)}{\text{Total}} =$	$\dfrac{\$3.50}{\$8.80}$	$\dfrac{\$0.00}{\$8.00}$

*This example is based on the following assumptions:
a. The actual cooling load is 50% of chiller capacity. (CL = 0.5 CAP)
b. The heating load (the demand for hot water) is 50% of cooling load.
   (HL = 0.5 CL)

FIGURE 2.24 The most cost-effective mode of refrigeration can be selected automatically.

In plants in locations such as the southern United States, where there is no alternative heat source, another problem can arise because all the heating needs of the plant must be met by recovered heat from the heat pump. It is possible that during cold winter days there might not be enough recovered heat to meet this load. Whenever the heat load exceeds the cooling load and there is no alternative heat source available, an artificial cooling load must be placed on the heat pump.

This artificial heat source can in some cases be the cooling tower water itself. A direct heat exchanger between cooling tower and chilled water streams is also advantageous when there is no heat load but when there is a small cooling load during the winter. At such times, the chiller can be stopped, and the small cooling load can be met by direct cooling from the cooling tower water that is at a winter temperature (Figure 2.25).

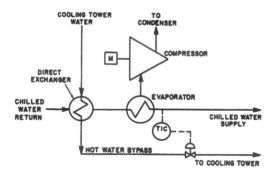

FIGURE 2.25 The cooling tower water can be used either as a heat source (with the compressor on) or as a means of direct cooling (with the compressor off).[7]

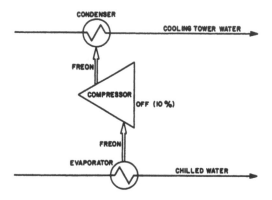

FIGURE 2.26 In the thermosiphon mode of operation, the compressor is off, and refrigerant vaporizes in the evaporator and liquifies in the condenser because the temperature gradient is reversed.[7]

Because coolant can be provided from many sources in a typical plant, another approach to optimization is to reconfigure the system in response to changes in loads, ambient conditions, and utility costs. For example, during some operating and ambient conditions, the cooling tower water may be cold enough to meet the demands of the process directly. Alternatively, if the cooling tower water temperature is below the temperature required by the process, the chillers can be operated in a thermosiphon mode. Refrigerant circulation is then driven by the temperature differential rather than by a compressor (Figure 2.26). In the thermosiphon mode of operation, the chiller capacity drops to about 10% of its rating.

When a chiller or heat pump is operated, the process is said to be cooled by *mechanical refrigeration* (Figure 2.27). When the process is cooled directly by the cooling tower water, without the use of chillers or refrigeration machines, the process is said to be cooled by *free cooling*. Most modern cooling systems are provided with the capability for operating in several different modes as a function of load, ambient conditions, and utility costs.

The decision to switch from one operating mode to another is usually based on economic considerations. The control system will select the mode of operation that will allow the meeting of the load at the lowest total cost. Figure 2.24 illustrates the process of selecting the heat recovery or the mechanical refrigeration mode on the basis of cost-effectiveness. The same kind of logic is applied when the choice is free cooling or mechanical refrigeration.

Once the decision is made to switch modes, the actual reconfiguration of the associated piping and valving is done automatically. In larger, more complex cooling systems, there can be more than 50 different modes of operation (such a system was designed by the author for IBM Corporation

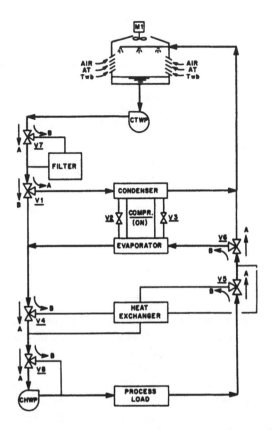

FIGURE 2.27 This mechanical refrigeration system can be automatically reconfigured to operate in any of the three "free cooling" configurations.

Headquarters at 590 Madison Avenue in New York). In this chapter, only four modes will be illustrated.

The cooling system illustrated in Figure 2.27 is configured as a mechanical refrigeration system. The controls of this configuration have been discussed in Figures 2.20 to 2.23. The heavily drawn pipelines show the active flow paths in that mode of operation. When the load drops off or when the ambient temperature decreases, it becomes possible to switch to one of the free-cooling modes after the compressor has been stopped. The mode reconfiguration logic is summarized in Table 2.3, in which the valve and equipment tags are those used in Figure 2.27.

a. Indirect free cooling

Free cooling can be direct or indirect. Indirect free cooling maintains the separation between the cooling tower water and the chilled water circuits. Mode 2 of Table 2.3 describes one of the indirect free-cooling configurations, in which the compressor is off and the heat is transferred from the evaporator to the condenser through the natural migration of refrigerant vapors (Figure 2.28). Opening the refrigerant migration valves equalizes the pressures in the evaporator and condenser. Because the condenser is at a lower temperature, the refrigerant that is vaporized in the evaporator is recondensed there and returns to the evaporator by gravity flow. The cooling capacity of a chiller is about 10% of full load in this mode of operation.

The cooling system is operated in this mode when the load is low and the cooling tower water temperature is about 10°F (5.6°C) below the required chilled water temperature. This usually means 40° to 45°F (4.4° to 7.2°C) cooling tower and 50° to 60°F (10° to 15.6°C) chilled water temperatures. Such high chilled water temperatures are not unrealistic in the winter, because the air tends to be dry and dehumidification is not required.

TABLE 2.3
Mode Reconfiguration Logic

Equipment	Mode 1 (Mechanical Refrigeration)	Mode 2 (Vapor Migration Based Free Cooling)	Mode 3 (Indirect Free Cooling by the Use of Heat Exchanger)	Mode 4 (Direct Free Cooling with Full Flow Filtering)
Compressor	On	Off	Off	Off
Cooling-tower pumps	On	On	On	On
Chilled-water pumps	On	On	On	Off
Valve V1	A	A	B	B
Valves V2 and 3	Closed	Open	Closed	Closed
Valve V4	A	A	B	A
Valve V5	A	A	B	A
Valve V6	A	B	A	A
Valve V7	A	A	A	B
Valve V8	A	A	A	B

Mode 2 is frequently implemented without controls. This means that both sets of pumps are operated and the temperatures are allowed to float as a function of load and ambient conditions. To optimize this operation, the control strategy shown in Figure 2.28 can be implemented. Here TDIC-01 maintains the optimum approach and TDIC-02 maintains the optimum range, unless these controls are overridden by VPC-04 when the refrigerant migration valves approach their full opening, signifying that increased heat transfer is needed at the condenser in order to meet the load. The migration valves V2 and V3 are throttled by TIC-05. The set point of TIC-05 is maximized by VPC-06 by allowing all user valves to open until the most-open valve reaches 90% opening. Chilled water pumping rate is minimized by VPC-07, which provides the set point of PDIC-08. As VPC-06 normally keeps the most open user valve at 90% opening, the set point of VPC-07 is not reached under normal conditions. Therefore, the PDIC-08 set point is kept at its allowable minimum. This in turn results in minimizing the pumping cost as long as all the users are satisfied. If any of the user valves opens to over 95%, VPC-07 will quickly increase the pumping rate to guarantee sufficient coolant to that valve.

At this point the system is using up the cooling capacity stored in the circulated chilled water. Once this is exhausted, control will be lost. To respond to this condition, it is advisable to switch to a different mode of operation: one that can provide more cooling.

b. Plate-type heat exchanger control

Another method of indirect free cooling is to bring the cooling tower water and the chilled water into heat exchange through a plate and frame type heat transfer unit (mode 3). This allows heat to be transferred from the cooling tower to the chilled water system, completely bypassing the refrigeration machine. There is no heat transfer capacity limitation on this system because any cooling load can be handled as long as the plate-type heat exchanger is large enough. Therefore, the main advantage of mode 3 over mode 2 is that it is not restricted to loads of 10% or less. Its main disadvantage is the need for an additional major piece of equipment: the heat exchanger.

Indirect free cooling using the plate-type heat exchanger can be operated manually, without any controls. In that case, the chilled water temperature floats with load and ambient temperature while the pump stations operate at full capacity.

If it is desired to optimize the operation under mode 3, the control system described in Figure 2.29 can be implemented. Here TDIC-01 maintains the optimum approach and TDIC-02 maintains the optimum range, unless overridden by PDIC-03 or by TIC-05. The pumping rate of cooling tower water circulation will be set by the highest of the three controller outputs. The purpose of

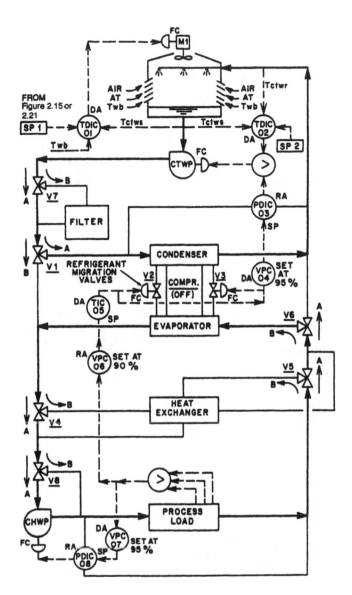

FIGURE 2.28 Free cooling can be achieved by allowing the refrigerant vapor to migrate (mode 2).

PDIC-03 is to guarantee the minimum pressure differential required for the exchanger; TIC-05 can override TDIC-02 and PDIC-03 if more cooling is required.

The set point of TIC-05 is adjusted by VPC-06 to prevent the most-open user valve from exceeding a 90% opening. Under normal conditions, the pumping rate of chilled water is kept at a minimum by PDIC-08, which maintains the minimum ΔP across the load. When VPC-06 is unable to keep the most-open user valve at a 90% opening and it rises to 95%, VPC-07 will start raising the set point of PDIC-08, thereby increasing the pumping rate. This is only a temporary cure, because the added cooling capacity is available only at the expense of heating up the stored chilled water in the pipe distribution system. Therefore, it is advisable to detect this condition; when it occurs for more than a minute or so, the system should be automatically switched to a cooling mode that can handle the increased load. This can be mechanical refrigeration (mode 1) or free cooling through interconnection (mode 4).

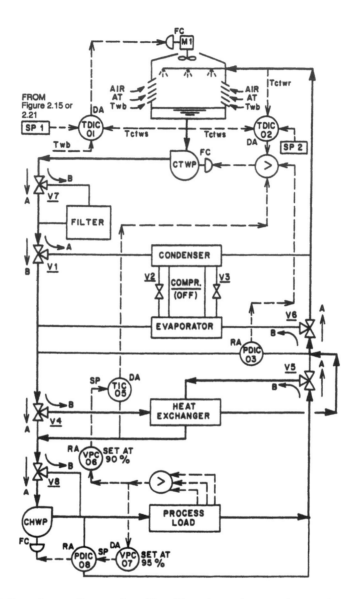

FIGURE 2.29 Indirect free cooling can be achieved by using a plate-type heat exchanger (mode 3).

c. Direct free cooling

In direct free cooling, the cooling tower water is piped directly to the process load, as shown in Figure 2.30. This method (mode 4) is the most cost-effective mode of cooling, because both the compressor and the chilled water pump station are off. This mode of operation can handle high process loads, as it is limited only by the size of the towers and their pumps.

The main disadvantage of direct free cooling is that it brings potentially dirty cooling tower water to the process users, causing plugging and buildup on the heat transfer surfaces. This problem is solved either by full flow filtering (also called strainer cycle), as shown in Figure 2.30, or by the use of closed-circuit, evaporative cooling towers. In such "noncontact" or "closed-loop" cooling towers, the water has no opportunity to pick up contaminants from the air.

The configuration is also frequently operated without automatic controls. In that case, the cooling water temperature floats as the load and ambient conditions vary, and the fan and pumping rates are not optimized. If optimization is desired, the controls shown in Figure 2.30 can be

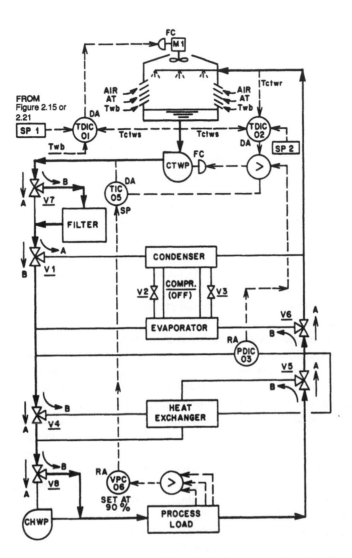

FIGURE 2.30 Direct free cooling can be achieved through the interconnection of cooling circuits (mode 4).

implemented. In this figure, TDIC-01 serves to keep the approach at an optimum value and TDIC-02 optimizes the range. The range controller can be overridden by PDIC-03 or by TIC-05 when either of these controllers require a higher pumping rate than does the range controller. The set point of TIC-05 is optimized to keep the most-open user valve from exceeding a 90% opening.

d. Mode reconfiguration

In cooling systems that also include optional storage and heat recovery systems provided with alternate types of motor drives, the number of possible modes of operation can be rather high. Switching from one mode to another is not as simple as it might appear at first glance, because as a result of reconfiguration, equipment needs to be started or stopped, control loops must be reconfigured, and pump discharge heads are also modified. For example, in the case of mode 4 in Figure 2.30, the water circulation loop served by the cooling tower pumps becomes much longer when the system is switched to direct free cooling. This in turn shifts the operating point of the pump and can lower its efficiency if the system is not carefully designed and evaluated for each mode.

The control loop configuration requirements for modes 1 to 4 are tabulated in Table 2.4.

TABLE 2.4
Control Loop Configuration Requirements

| Controller | Associated Manipulated Variables | | | |
	Mode 1	Mode 2	Mode 3	Mode 4
TDIC-01	Fan	Fan	Fan	Fan
TDIC-02	CTWP	CTWP	CTWP	CTWP
PDIC-03	CTWP	CTWP	CTWP	CTWP
VPC-04	—	PDIC-03 Set point	—	—
TIC-05	Compr.	V2/3	CTWP	CTWP
VPC-06	TIC-05 Set point	TIC-05 Set point	TIC-05 Set point	TIC-05 Set point
VPC-07	PDIC-08 Set point	PDIC-08 Set point	PDIC-08 Set point	—
PDIC-08	CHWP	CHWP	CHWP	—

When the control loops are automatically reconfigured, it is important to revise their tuning constants (as their outputs are directed to different manipulated variables), because the time constants of the loop are also changed. This should preferably be done automatically to minimize the potential for human error.

The switching from one mode to another is done in a stagnant state, while the equipment is turned off, if an interruption of a few minutes can be tolerated. This usually is acceptable for all motors except for the pumps serving the process load. These pumps can be left running, utilizing the coolant storage capacity of the water distribution piping. When transfer to the next mode is initiated, dynamic transfer can be accomplished, because the slowly diverting three-way valves will never block the pump discharge but will only gradually change the destination of the water.

If the mode changes are frequent or if the coolant capacity of the piping headers is very small (insufficient to meet the process load for even a few minutes), all systems must be switched while running. Dynamic switching requires more planning and a higher level of automation because the automatic starting of certain pieces of equipment (such as a chiller driven by a steam turbine) requires a more comprehensive set of safety interlocks than do some other devices.

Automatic operating mode reconfiguration is one of the most powerful tools of optimization, because it makes a previously rigid system flexible. This technique is not limited to cooling systems but is effective in any unit operation in which the system must adapt to changing conditions.

7. Storage Optimization

If storage tanks are available, it is cost-effective to generate the daily brine or chilled water needs of the plant at night, when it is the least expensive to do so, because ambient temperatures are low and night-time electricity is less expensive in some areas.[8]

When demand is low, it may be possible to lower operating costs by operating the chillers part of the time at peak efficiency rather than continuously at partial loading. Efficiency tends to be low at partial loads because of losses caused by friction drop across suction dampers, prerotation vanes, or steam governors. Cycling is practical if the storage capacity of the distribution headers is enough to avoid frequent stops and starts. When operation is to be intermittent, data such as the heat to be removed and the characteristics of the available chillers are needed to determine the most economical operating strategy.

When chiller cycling is used, the thermal capacity of the chilled water distribution system is used to absorb the load while the chiller is off. For example, if the pipe distribution network has a volume of 100,000 gallons (378,000 l), this represents a thermal capacity of approximately 1 million BTUs for each °F of temperature rise (1.9×10^6 J/°C). So, if one can allow the chilled water temperature to float 5°F (2.8°C) (for example, from 40°F to 45°F, or from 4.4°C to 7.2°C) before the chiller is restarted, this represents the equivalent of approximately 400 tons (1405 kW) of thermal capacity. If the load happens to be 200 tons (704 kW), the chiller can be turned off for two hours at a time. If the load is 1000 tons (3,514 kW), the chiller will be off for only 24 minutes. This illustrates the natural load-following, time-proportioning nature of this scheme.

If the chiller needs a longer period of rest than the thermal capacity of the distribution system can provide, three options are available:

1. Tankage can be added to increase the water volume.
2. A second chiller can be started (not the one that was just stopped).
3. The load can be distributed among chillers of different sizes by keeping some in continuous operation while cycling others.

8. Load Allocation

Continuous measurement of the actual efficiency ($/ton) of each chiller can enable all loads to be met through the operation of the most efficient combination of machines for the load. In plants with multiple refrigerant sources, the cost per ton of cooling can be calculated from direct measurements and used to establish the most efficient combination of units to meet present or anticipated loads. As with boilers, this calculation accounts for differences among units as well as for the efficiency-versus-load characteristics of the individual coolant sources, as shown in Figure 2.31.

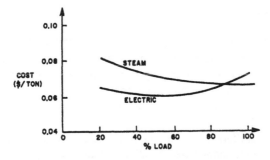

FIGURE 2.31   The unit cost of refrigeration is a function of load but depends on the characteristics of each chiller. The indicated cost figures are for representative installed equipment.[8]

In simple load allocation systems, only the starting and stopping of the chillers is optimized. In such systems, when the load is increasing, the most efficient idle chiller is started; when the load is dropping, the least efficient one is stopped. In more sophisticated systems, the load distribution between operating chillers is also optimized. In such systems, a computer is used to calculate the real-time efficiency of each chiller—that is, to calculate the incremental cost for the next load change for each chiller.

If the load increases, the incremental increase is sent to the set point of the most cost-effective chiller. If the load decreases, the incremental decrease is sent to the least cost-effective chiller (Figure 2.32). The required software packages with proved capabilities for continuous load balancing through the predictions of cost and efficiencies are readily available.[9] With the strategy described in Figure 2.32, the most efficient chiller will either reach its maximum loading or will enter a region of decreasing efficiency and will no longer be the most efficient. When the loading limit is reached on one chiller, or when a chiller is put on manual, the computer will select another as the most efficient unit for future load increases.

The least efficient chiller will accept all decreasing load signals until its minimum limit is reached. Its load will not be increased unless all other chillers are at their maximum load or in

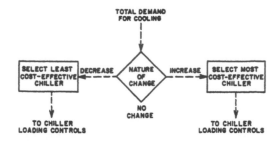

FIGURE 2.32 Computer-based load allocation directs load increases to the most cost-effective chiller and sends load decreases to the least cost-effective chiller.

manual. As shown in Figure 2.31, some chillers can have high efficiency at normal load while being less efficient than the others at low load. Such units are usually not allowed to shut down, but are given a greater share of the load by a special subroutine.

If all chillers are identical, some will be driven to maximum capacity while others will be shut down by this strategy, and only one chiller will be placed at an intermediate load.

In starting up chiller stations, the optimization system "knows" how many BTUs need to be removed before startup and the size and efficiency of the available chillers. Therefore, the length of the "pull-down" period can be minimized and the energy cost of this operation optimized.

B. Retrofit Optimization

In new plants it is easy to install variable-speed pumps, to provide the thermal capacity required for chiller cycling, or to locate chillers and cooling towers at the same elevation and near each other so that the cooling tower water pumping costs will be minimized. In existing plants, the inherent design limitations must be accepted, and the system must be optimized without changes to the equipment. This is quite practical to do and can still produce savings up to 50 percent,[1] but certain precautions are needed.

In optimizing existing chillers, it is important to give careful consideration to the constraints of low evaporator temperature, economizer flooding, steam governor rangeability, surge, and piping/valving limitations.

A detailed discussion of surge controls is given in Chapter 4. Therefore, here it should suffice to just mention that surge occurs at low loads when not enough refrigerant is circulated. A surge condition can cause violent vibration and damage.

Old chillers usually do not have automatic surge controls and have only vibration sensors for shutdown. If the chillers will operate at low loads, it is necessary to add an antisurge control loop. Surge protection is always provided at the expense of efficiency. To bring the machine out of surge, the refrigerant flow must be artificially increased if there is no real load on the machine. The only way to provide this increase in flow is to add artificial and wasteful loads (for example, hot gas or hot water bypasses). Therefore, it is much more economical either to cycle to a large chiller or to operate a small one than to meet low load conditions by running the machine near its surge limit.

Low temperatures can occur in the evaporator when an old chiller is optimized—for example, when one that has been designed for operation at 75°F (23.9°C) condenser water is run at 45°F or 50°F (7.2°C or 10°C) condenser water (in the winter). This phenomenon is exactly the opposite of surge, because it occurs when refrigerant is being vaporized at an excessively high rate. Such vaporization occurs because the chiller is able to pump twice the tonnage for which it was designed as a result of the low compressor discharge pressure. In such a situation, the evaporator heat transfer area becomes the limiting factor; furthermore, the only way to increase heat flow is to increase the temperature differential across the vapor tubes. This shows up as a gradual lowering of refrigerant temperature in the evaporator until it reaches 32°F (0°C), at which point the machine shuts down to protect against ice formation.

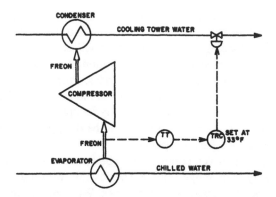

FIGURE 2.33   Protection against evaporator freeze-up can be provided by temperature control.[7]

There are two ways to prevent this phenomenon from occurring in existing chillers. The first is to increase the evaporator heat transfer area (a major equipment modification). The second is to prevent the refrigerant temperature in the evaporator from dropping below 33°F (0.6°C) by not allowing the cooling tower water to cool the condenser to its own temperature. The latter solution requires only the addition of a temperature control loop (Figure 2.33). This prevents the chiller from taking full advantage of the available cold water from the cooling tower by throttling its flow rate, thereby causing its temperature to increase.

1. Economizer and Steam Governor

On existing chillers, the economizer control valves (LCV-1 and LCV-2 in Figure 2.34) are often sized on the assumption that the refrigerant vapor pressure in the condenser (P3) is constant and corresponds to a condenser water temperature of 75°F or 85°F (23.9°C or 29.4°C). Naturally, when such units are operated with 45°F or 50°F (7.2°C or 10°C) condenser water, P3 is much reduced,

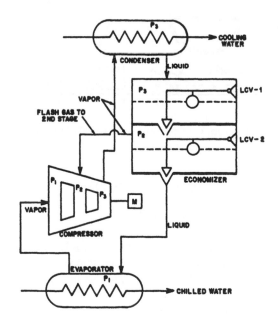

FIGURE 2.34   Economizer flooding can be solved by supplementing LCV-1 and LCV-2 with larger external valves.

as is the pressure differential across LCV-1 and LCV-2. If this occurs—when the refrigerant circulation rate reaches high levels, as it easily can—the control valves will be unable to provide the necessary flow rate, and flooding of the economizer will occur (the flow is higher and the ΔP is lower than was the basis of valve sizing). The solution is to install larger valves, preferably externaland with proportional and integral control modes. Proportional controllers alone cannot maintain the set point as load changes; the addition of the integral mode eliminates this offset. This is important in machines that were not designed originally for optimized, low-temperature condenser water operation, because otherwise the compressor can be damaged by the liquid refrigerant that overflows into it from the flooded flash evaporator.

In order for a steam-turbine-driven compressor to be optimized, its rotational velocity must be modulated over a reasonably wide range. This is not possible with old, existing machines if they are provided with quick-opening steam governor valves. With such governors a slight increase in lift from the fully closed position results in a substantial steam flow and therefore a substantial rotational velocity. If this steam flow is throttled, the valves become unstable and noisy.

The valve characteristics can be changed from quick-opening to linear at minimal cost. The desired wide rangeability can be obtained by welding two rings with V-notches to the seats of the existing steam governor valves, as shown in Figure 2.35.

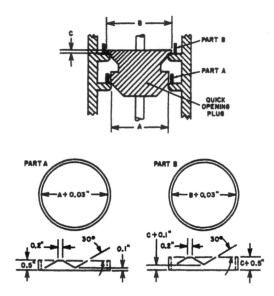

FIGURE 2.35   A double-seated steam governor valve can be rebuilt for optimized variable-speed service. The notched rings provide the necessary rangeability.[7]

IV. CONCLUSIONS

The various goals of chiller optimization include:

- Minimizing the temperature difference across the chiller by minimizing cooling tower water temperature and maximizing chilled water temperature
- Minimizing pumping costs by transporting only as much water as is required to meet the load and by using variable speed pumping
- Minimizing cooling tower operating costs by approach optimization

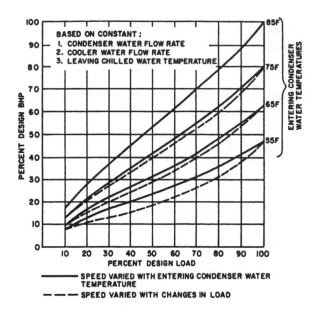

FIGURE 2.36 Combined savings from reducing the condenser water temperature and from using variable-speed compressors.[9]

- Operating chillers at the load at which their efficiency is the highest, by meeting partial loads with part-time operation at maximum efficiency and making coolant when least expensive, if night storage is available.
- Maximizing the overall efficiency by always using the most efficient chiller combination and by also using efficiency information to initiate maintenance

The effects of reducing the condenser water temperature and using variable speed compressors are shown in Figure 2.36. For example, at 50% design load and 55°F condenser water temperature, the power consumption will be less than 20%. This compares to 53% if guide vanes are used on a constant-speed machine operating with 85°F condenser water.

If the potentials of all of the above-described optimization strategies are fully exploited, the operating costs of the chiller stations that operate through the year and are located in the northern states can be cut in half.[1] The corresponding payback period on investment for optimization controls is usually under a couple of years.[8]

With the availability of inexpensive solid-state sensors and microprocessors, it is possible to incorporate all the optimization strategies in a single refrigeration unit controller. Such multivariable controllers make it possible to replace the uncoordinated control of flows and temperatures with load-following floating control, which guarantees the lowest possible cost of operation.

V. TERMINOLOGY

The symbols and abbreviations used in this section are listed below for reference purposes:

CHWP: Chilled water pump
CTWP: Cooling tower water pump
FAH, FAL: Flow alarm; H = high, L = low
FC: Fail closed

FO: Fail open

FSH, FSL, FSHL: Flow switch; H = high, L = low, HL = high-low

HLL: High-low limit switch

HWP: Hot water pump

LCV: Level-control valve

M_{1-5}: Motor that drives equipment or "place" where energy is consumed; 1 = cooling tower fan(s), 2 = cooling water pump(s), 3 = compressor(s), 4 = chilled water pump(s), 5 = hot water pump(s)

P-1: Pump

P, P_{1-3}: Pressures

PAH: Pressure alarm; H = high

PB: Pushbutton

PCV: Pressure-control valve

PDIC: Pressure-differential indicating controller

PI: Pressure indicator

PSH, PSL: Pressure switch; H = high, L = low

Q_H: Amount of heat delivered to cooling tower water

Q_L: Amount of heat removed from chilled water

RD: Rupture disk

S: Solenoid

SC: Speed controller

SIC: Speed-indicating controller

SP: Set point

SV: Solenoid valve

T_c: Temperature of refrigerant in condenser inlet

T_e: Temperature of refrigerant in evaporator inlet

T_H: Temperature of cooling water at condenser exit, absolute

T_L: Temperature of chilled water at evaporator exit, absolute

T_{chwr}: Temperature of chilled water return

T_{chws}: Temperature of chilled water supply

T_{ctwr}: Temperature of cooling tower water return

T_{ctws}: Temperature of cooling tower water supply

T_{hwr}: Temperature of hot water return

T_{hws}: Temperature of hot water supply

T_{wb}: Temperature of wet bulb

TAH, TAL: Temperature alarm; H = high, L = low

TCV: Temperature control valve

TIC: Temperature-indicating controller

TSH, TSL: Temperature switch; H = high, L = low

TT: Temperature transmitter

TY: Temperature relay

VPC: Valve position controller

W: Work

XLS: Limit switch

XSCV: Superheat control valve

REFERENCES

1. Romita, E., "A Direct Digital Control for Refrigeration Plant Optimization," *ASHRAE Trans.*, Vol. 83, Part 1, 1977.

2. Lipták, B.G., Optimizing controls for chillers and heat pumps, *Chem. Eng.*, October 17, 1983.

3. U.S. Patent Numbers 4,612,776 (dated September 23, 1986) and 4,628,700 (dated December 16, 1986). Both patents assigned to R.H. Alsenz.
4. Cooper, K.W., Saving energy with refrigeration, *ASHRAE J.* December 1978.
5. Shinskey, F.G., *Energy Conservation Through Control*, Academic Press, 1978.
6. Hallanger, E.C., Operating and controlling chillers, *ASHRAE J.*, September 1981.
7. Lipták, B.G., Optimizing plant chiller systems, *InTech*, September 1977.
8. Lipták, B.G., Save energy by optimizing your boilers, chillers, and pumps, *InTech*, March 1981.
9. Carrier Corporation, Centrifugal liquid chillers, *Appl. Data,* 17-IXA, 1975.

BIBLIOGRAPHY

Anderson, E.P., *Air Conditioning: Home and Commercial*, 5th ed., Macmillan, New York.

ASHRAE Handbook and Product Directory, Latest Edition, ASHRAE, Inc.

Brumbaugh, J.E., *Heating, Ventilating and Air Conditioning Library,* rev. ed., Macmillan, New York.

Chopey, N.P., *Handbook of Chemical Engineering Calculations*, 2nd ed., McGraw-Hill, New York, 1994.

Lipták, B.G., Optimizing controls for chillers and heat-pumps, *Chem. Eng.*, October 17, 1983.

McMillan, G.K., *Tuning and Control Loop Performance*, 3rd ed., Instrument Society of America, 1994.

O'Callaghan, P.W., *Energy Management*, McGraw-Hill, New York, 1992.

Shinskey, F.G., *Process Control Systems,* 4th ed., McGraw-Hill, New York, 1996.

Wang, S.K., *Handbook of Air Conditioning and Refrigeration*, McGraw-Hill, New York, 1994.

Whitman, W.C., *Refrigeration and Air Conditioning Technology,* Delmar Publishing, 1991.

3 Clean Rooms

Clean rooms are used for experimental studies of a wide variety. Testing and analysis laboratories, used in the medical, military, and processing industries, are all designed to operate as clean rooms. The optimization of clean rooms, therefore, is an important subject to many industries. The production of semiconductors is one of several processes done in a clean-room environment. Optimization of the clean-room control system will drastically reduce both the energy cost of operation and the cost of manufacturing of off-spec products.

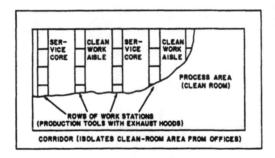

FIGURE 3.1 A semiconductor manufacturing clean room has rows of workstations.

I. THE PROCESS

The process, or production, area of a typical semiconductor manufacturing laboratory is between 100,000 and 200,000 square feet (9,290 and 18,580 square meters). The market value of the daily production from this relatively small area is over $1 million. Plant productivity is increased if the following conditions are maintained in the process area:

1. No drafts (+0.02 in. H_2O ± 0.005 in. H_2O, or 0.05 ± 0.012 millibars)
2. No temperature gradients (72°F ± 1°F, or 22°C ± 0.6°C)
3. No humidity gradients (35% RH ± 3%)
4. No air flow variations (60 air changes/hr ± 5%)

Therefore, the primary goal of optimization is to maximize plant productivity continuously through the accurate control of these parameters. The second goal is to conserve energy. The main control elements and control loop configurations required for arriving at a high-productivity semiconductor manufacturing plant will be described in this chapter.

In order to prevent contamination through air infiltration from the surrounding spaces, the clean-room pressure must be higher than that of the rest of the building. As shown in Figure 3.1, the clean room is surrounded by a perimeter corridor. The pressure within the clean room is higher than that in the surrounding corridor to protect against the leakage of dirty air. The clean room is

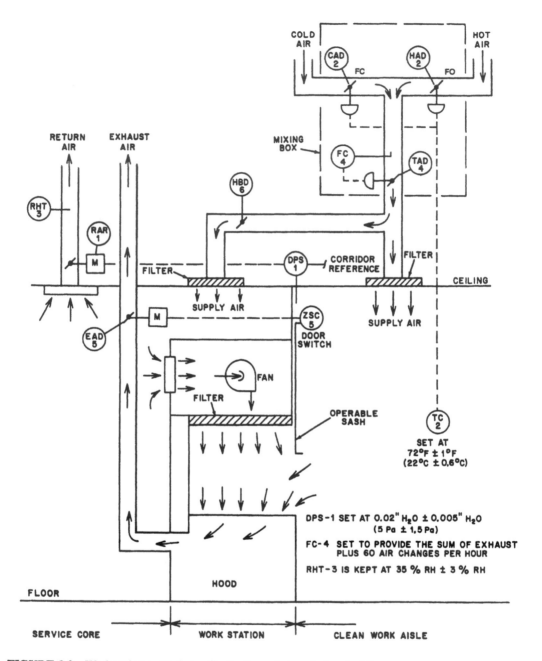

FIGURE 3.2   Workstation controls handle the flow of air and chemical fumes.

made up of rows of workstations. These rows are also referred to as zones. A typical semiconductor producing plant has approximately 200 workstations, also called subzones. Each subzone faces the clean work aisle, and behind it is the service core.

II. BASIC CONTROLS OF SUBZONES

As shown in Figure 3.2, air is supplied to the subzone through filters located in the ceiling in front of the workstation. The total flow and the temperature of this air supply are both controlled. The evacuated exhaust air header is connected to the lower hood section of the workstation. It pulls in all the air and chemical fumes that are generated in the workstation and safely exhausts them into the atmosphere. In order to make sure that none of the toxic fumes will spill into the work aisle, clean air must enter the workstation at a velocity of about 75 fpm (0.38 m/s). The air that is not pulled in by the exhaust system enters the service core, where some of it is recirculated back into the workstation by a local fan. The rest of the air is returned from the service core by the return air header. A damper in this header (RAR-1) is modulated to control the pressure (DPS-1) in the work aisle.

A. Pressure Controls

The pressure in the corridor is the reference for the clean-room pressure controller DPS-1 in Figure 3.2. This controller is set to maintain a few hundredths of an inch of positive pressure relative to the corridor. The better the quality of the building construction, the higher this set point can be, but even with the lowest quality buildings, a setting of approximately 0.02 in. H_2O (0.05 millibars) can be maintained easily. Because at such near-atmospheric pressures, air behaves as if it were incompressible, the pressure control loop shown in Figure 3.2 is both fast and stable. When the loop is energized, DPS-1 quickly rotates the return air control damper (RAR-1) until the preset differential is reached. At that point, the electric motor stops rotating the damper, and it stays at its last opening. This position will remain unaltered as long as the air balance in the area remains the same:

$$\text{return airflow} = \text{supply airflow} - (\text{exhaust airflow} + \text{pressurization loss}) \qquad (1)$$

When this airflow balance is altered (for example, as a result of a change in exhaust airflow), it will cause a change in the space pressure, and DPS-1 will respond by modifying the opening of RAR-1.

B. Elimination of Drafts

Plant productivity is maximized if drafts are eliminated in the clean work aisle. Drafts would stir up the dust in this area, which in turn would settle on the product and cause production losses. In order to eliminate drafts, the pressure at each workstation must be controlled at the same value. Doing so eliminates the pressure differentials between stations and therefore prevents drafts. When a DPS-1 unit is provided to control the pressure at each workstation, the result is a uniform pressure profile throughout the clean room.

DPS-1 usually controls the pressure in the clean work aisle on the "process" side of the workstation. Yet, it is important to make sure that *all* points, including the service aisle, are under positive pressure. Because the local circulating fan within the workstation draws the air in from the service core and discharges it into the clean work aisle, the pressure in the service core will always be lower than that on the process side (see Figure 3.2). Therefore, it is possible for localized vacuum zones to evolve in the service core, which could cause contamination by allowing air infiltration. To prevent this from happening, several solutions have been proposed.

One possibility is to have DPS-1 to control the service core pressure. This is not recommended, because a draft-free process are a can be guaranteed only if there are no pressure gradients on the work aisle side; this can be achieved only by locating the DPS-1 units on the work aisle side of the workstations.

Another possibility is to leave the pressure controls on the work aisle side but raise the set point of DPS-1 until all the service cores in the plant are also at a positive pressure. This solution cannot be universally recommended either, because the quality of building construction might not be high enough to allow operation at elevated space pressures. The pressurization loss in badly sealed buildings can make it impossible to reach the elevated space pressure. Yet another solution is to install a second DPS controller, which would maintain the service core pressure by throttling the damper HBD-6. This solution will give satisfactory performance but will also increase the cost of the control system by the addition of a few hundred control loops (one per workstation).

A more economical solution is shown in Figure 3.2. Here, a hand-operated bypass damper (HBD-6) is manually set during initial balancing. This solution is reasonable for most applications, because the effect of the workstation fan is *not* a variable, and therefore a constant setting of HBD-6 should compensate for it.

C. TEMPERATURE CONTROLS

The temperature at each workstation is controlled by a separate thermostat, TC-2 in Figure 3.2. This temperature controller adjusts the ratio of cold air to hot air within the supply air mixing box to maintain the space temperature.

Unfortunately, conventional thermostats cannot be used if the temperature gradients within the clean room are to be kept within ± 1°F of 72°F (–0.6°C of 22°C). Conventional thermostats cannot meet this requirement with regard to measurement accuracy or control quality. Even after individual calibration, it is unreasonable to expect less than a ± 2°F or ± 3°F (± 1°C or ± 1.7°C) error in overall loop performance if HVAC-quality thermostats are used. Part of the reason for this is the fact that the "offset" cannot be eliminated in plain proportional controllers, such as thermostats.

Thus, in order for the thermostat to move its output from the midscale value (50%-50% mixing of cold and hot air) an error in room temperature must first develop. This error is the permanent offset. The size of this offset error for TC-2 in Figure 3.2 for the condition of maximum cooling can be determined as follows:

$$\text{offset error} = \frac{\text{spring range of CAD}}{2 \ (\text{thermostat gain})} \tag{2}$$

Assuming that CAD-2 has an 8 to 13 PSIG (0.54 to 0.87 bar) spring (a spring range of 5 PSI, or 34.5 kPa) and that TC-2 is provided with a maximum gain of 2.5 PSI/°F (0.17 bar/°C), the offset error is 1°F (0.6°C). Under these conditions, the space temperature must permanently rise to 73°C (22.8°C) before CAD-2 can be fully opened. The offset error will increase as the spring range increases or as the thermostat gain is reduced. Sensor and set point dial error are always additional to the offset error.

Therefore, in order to control the clean room temperature within ± 1°F (± 0.6°C), it is necessary to use an RTD-type or a semiconductor transistor-type temperature sensor and a proportional-plus-integral controller, which will eliminate the offset error. This can be most economically accomplished through the use of microprocessor-based shared controllers that communicate with the sensors over a pair of telephone wires that serve as a data highway.

D. HUMIDITY CONTROLS

The relative humidity sensors are located in the return air stream (RHT-3). In order to keep the relative humidity in the clean room within 35% RH ± 3% RH, it is important to select a sensor with a lower error than ± 3% RH. The repeatability of most human hair element sensors is approximately ± 1% RH. These units can be used for clean room control applications if they are *individually calibrated* for operation at or around 35% RH. Without such individual calibration, they will not perform satisfactorily because their off-the-shelf inaccuracy, or error, is approximately ± 5% RH. The controller associated with RHT-3 is not shown in Figure 3.2 because relative humidity is not controlled at the workstation (subzone) level but at the zone level (as mentioned earlier, a zone is a row of workstations). The control action is based on the relative humidity reading in the combined return air stream from all workstations within that zone.

E. FLOW CONTROLS

The proper selection of the mixing box serving each workstation is of critical importance. Each mixing box serves the dual purpose of providing accurate control of the total air supply to the subzone and modulating the ratio of "cold" and "hot" air to satisfy the requirements of the space thermostat TC-2.

The total air supply flow to the subzone should equal 60 air changes per hour *plus* the exhaust rate from that subzone. This total air supply rate must be controlled within ± 5% of actual flow by FC-4, over a flow range of 3:1. The rangeability of 3:1 is required because as processes change, their associated exhaust requirements will also change substantially. FC-4 in Figure 3.2 can be set manually, but this setting must change every time a new tool is added to or removed from the subzone. The setting of FC-4 must be done by individual in-place calibration against a portable hot wire anemometer reference. Settings based on the adjustments of the mixing box alone (without an anemometer reference) will not provide the required accuracy.

Some of the mixing box designs available on the market are not acceptable for this application. Unacceptable designs include the following:

- "Pressure-dependent" designs, in which the total flow will change if air supply pressures vary. Only "pressure-independent" designs can be considered because both the cold and the hot air supply pressures to the mixing box will vary over some controlled minimum.
- Low rangeability designs. A 3:1 rangeability with an accuracy of ± 5% of *actual* flow is required.
- Selector or override designs. In these designs either flow or temperature is controlling on a selective basis. Such override designs will periodically disregard the requirements of TC-2 and will thus induce upsets, cycling, or both.

If the mixing box is selected to meet the aforementioned criteria, both airflow and space temperature can be accurately controlled.

III. OPTIMIZING CONTROLS

Each row of workstations shown in Figure 3.1 is called a zone, and each zone is served by a cold deck (CD), a hot deck (HD), and a return air (RA) subheader. These subheaders are frequently referred to as *fingers*. The control devices serving the individual workstations (subzones) will be able to perform their assigned control tasks if the "zone finger" conditions make it possible for them to do so.

For example, RAR-1 in Figure 3.2 will be able to control the subzone pressure as long as the ΔP across the damper is high enough to remove all the return air without requiring the damper to open fully. As long as the dampers are throttling (neither fully open nor completely closed), DPS-1 in Figure 3.2 is in control.

PIC-7 in Figure 3.3 is provided to control the vacuum in the RA finger and thereby to maintain the required ΔP across RAR-1 in Figure 3.2. PIC-7 is a nonlinear controller with a fairly wide neutral band. This protects the CD finger temperature (TIC-6 set point) from being changed until a sustained and substantial change takes place in the detected RA pressure.

Similarly, the mixing box in Figure 3.2 will be able to control subzone supply flow and temperature as long as its dampers are not forced to take up extreme positions. Once a damper is fully open, the associated control loop is out of control. Therefore, the purpose of the damper position controllers (DPCs) in Figure 3.3 is to prevent the corresponding dampers from having to open fully.

Lastly, the relative humidity in the return air must also be controlled within acceptable limits.

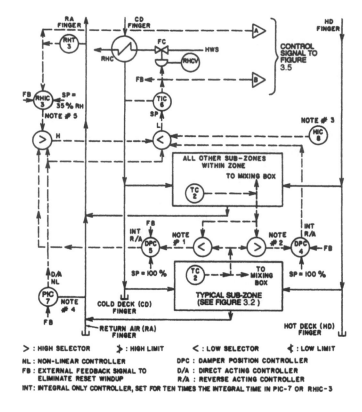

FIGURE 3.3 Typical zone control system with envelope control. Note 1: This is the opening of the most open hot air damper, HAD-2 (shown in Figure 3.2) is fully open when the TC-2 output is 3 PSIG (0.2 bar). Note 2: This is the opening of the most open cold air damper. CAD-2 (shown in Figure 3.2) is fully open when the TC-2 output is 15 PSIG (1.0 bar). Note 3: HIC-8 sets the maximum limit for the CD temperature at approximately 70°F (21°C). Note 4: As vacuum increases (and pressure decreases), the output of PIC-7 also increases. Note 5: D/A in summer, R/A in winter.

Therefore, at the zone level there are five controlled, or limit, variables but only one manipulated variable:

Limited or Controlled Variables	Manipulated Variable
RA pressure	TIC-6 set point
RA relative humidity	
Max. CAD opening	
Max. HAD opening	
Max. TIC-6 set point	

Figure 3.3 shows the control loop configuration required to accomplish the aforementioned goals. It should be noted that dynamic lead/lag elements are not shown and that this loop can be implemented in either hardware or software.

A. ENVELOPE OPTIMIZATION

Whenever the number of control variables exceeds the number of available manipulated variables, it is necessary to apply multivariable envelope control. This means that the available manipulated

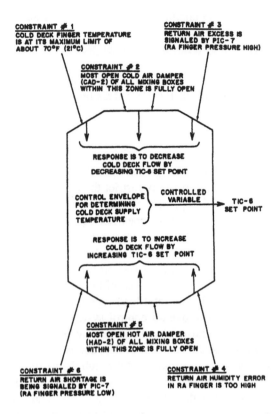

FIGURE 3.4 A constraint control envelope may be used for cold deck temperature optimization.

variable (TIC-6 set point) is not assigned to serve a single task but is selectively controlled to keep many variables within acceptable limits—within the "control envelope" shown in Figure 3.4.

By adjusting the set point of TIC-6, it is possible to change the cooling capacity represented by each unit of CD air. Because the same cooling can be accomplished by using less air at lower temperatures or by using more air at higher temperatures, the overall material balance can be maintained by manipulating the set point of the heat balance controls. Return air humidity can similarly be affected by modulating the TIC-6 set point, because, when this set point is increased, more CD air will be needed to accomplish the required cooling. Increasing the ratio of humidity-controlled CD air in the zone supply (HD humidity is uncontrolled) also brings the zone closer to the desired 35% RH set point.

In this envelope control system, the following conditions will cause an increase in the TIC-6 set point:

- Return air shortage detected by a drop in the RA finger pressure (increase in the detected vacuum) measured by PIC-7. An increase in the TIC-6 set point will increase the CD demand (opens all CAD-2 dampers in Figure 3.2), which in turn lowers the HD and, therefore, the RA demand.
- Hot air damper in mixing box (HAD-2 in Figure 3.2) is near to being fully open. This, too, necessitates the aforementioned action to reduce HD demand.
- RA humidity does not match the RHIC-3 set point. This condition also necessitates an increase in the proportion of the CD air in the zone supply. Because the moisture content of the CD air is controlled, an increase in its proportion in the total supply will help control the RH in the zone.

On the other hand, the following conditions will require a decrease in the TIC-6 set point:

- Return air excess detected by a rise in RA finger pressure (drop in the detected vacuum) measured by PIC-7.
- Cold air damper in mixing box (CAD-2 in Figure 3.2) is fully open.
- CD finger temperature exceeds 70°F (21°C). This limit is needed to keep the CD always cooler than the HD.

B. Plantwide Optimization

In conventionally controlled semiconductor plants, both the temperature and the humidity of the main CD supply air are fixed. This can severely restrict the performance of such systems, because as soon as a damper or a valve is fully open (or closed), the conventional system is out of control. The optimized control system described here does not suffer from such limitations, because it automatically adjusts both the main CD supply temperature and the humidity so as to follow the load smoothly and continuously, never allowing the valves or dampers to lose control by fully opening or closing. The net result is not only increased productivity but also reduced operating costs.

A semiconductor production plant might consist of two dozen zones. Each of these zones can be controlled as shown in Figure 3.3. The total load represented by all the zones is followed by the controls shown in Figure 3.5. The overall control system is hierarchical in its structure: the subzone controls in Figure 3.2 are assisted by the zone controls in Figure 3.3, and the plantwide controls in Figure 3.5 guarantee that the zone controllers can perform their tasks. The interconnection among the levels of the hierarchy is established through the various valve and damper position controllers (VPCs and DPCs). These guarantee that no throttling device, such as mixing boxes or RHC valves, anywhere in this plant will ever be allowed to reach an extreme position. Whenever a control valve or a damper is approaching the point of losing control (nearing full opening), the load-following control system at the next higher level modifies the air or water supply conditions so that the valve or damper in question will not need to open fully. This hierarchical control scheme is depicted in Figure 3.6.

The overall plantwide control system shown in Figure 3.5 can be viewed as a flexible combination of material balance and heat/humidity balance controls. Through the load-following optimization of the set points of TIC-1 and TIC-11, the heat balance controls are used to assist in maintaining the material balance around the plant. For example, if the material balance requires an increase in airflow and the heat balance requires a reduction in the heat input to the space, both requirements will be met by admitting more air at a lower temperature.

1. Material Balance Controls

The plantwide material balance is based on pressure control. PC-9 and PC-10 modulate the variable-volume fans so as to maintain a minimum supply pressure in each of the CD and HD fingers. The suction side of the CD supply fan station is open to the outside and will draw as much outside air as the load demands (Figure 3.5).

The suction pressure of the HD supply fan station is an indication of the balance between RA availability and HD demand. This balance is maintained by PIC-7 in Figure 3.3, at the zone level. Because this controller manipulates a heat transfer system (RHCV), it must be tuned for slow, gradual action. This being the case, it will not be capable of responding to sudden upsets or to emergency conditions, such as the need to purge smoke or chemicals. Such sudden upsets in material balance are corrected by PC-6 and PC-7 in Figure 3.5. PC-6 will open a relief damper if the suction pressure is high (vacuum is low), and PC-7 will open a makeup damper if it is low (vacuum is high). In between these limits, both dampers will remain closed and the suction pressure will be allowed to float.

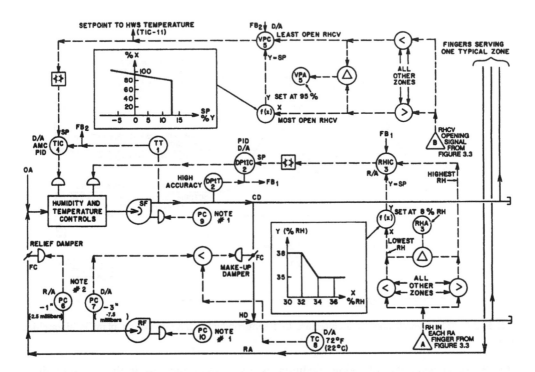

FIGURE 3.5 The total load represented by all zones is handled by the plantwide optimization control system. DPtIC = Dew point controller, PID = controller with proportional, integral, and derivative control modes, AMC = controller with positions for automatic-manual-cascade modes of operation, VPC = valve position controller, INT = controller with integral control action only, which is set at 10 times the integral time setting of the associated process controller(s). Note 1: This pressure controller keeps the lowest of all finger pressures at a value above some minimum. It modulates the fan volume or speed and starts extra fan units to meet the load. When the load drops, the unnecessary fans are stopped by flow (not pressure) control. Note 2: If the pressure rises above −1 in. H_2O (−2.5 millbars), the relief damper starts to open, and if the pressure drops below −3 in. H_2O (−7.5 millbars), the makeup damper opens. This opening is limited by TC-8, which prevents the HD temperature from dropping below 72°F (22°C). Between the settings of PC-6 and PC-7, both dampers are tightly closed and the suction pressure floats.

2. Heat Balance Controls

Heat balance also requires a multivariable envelope control system (similar to that shown in Figure 3.4), because the number of controlled variables exceeds the number of available manipulated variables. Therefore, the plantwide air and water supply temperature set points (TIC-1 and TIC-11) are not adjusted as a function of a single consideration but are selectively modulated to keep several variables within acceptable limits inside the control envelope (Figure 3.5).

By adjusting the set points of TIC-1 and TIC-11, it is possible to adjust the cooling capacity represented by each unit of CD supply air and the heating capacity of each unit of hot water. This then provides not only for load-following control but also for minimization of the operating costs. Cost reductions are accomplished by minimizing the HWS temperature and by minimizing the amount of simultaneous cooling and reheating of CD air.

This control envelope is so configured that the following conditions will decrease the TIC set points (both TIC-1 and TIC-11):

1. The least open RHCV is approaching full closure.
2. CD supply temperature is at its maximum limit.

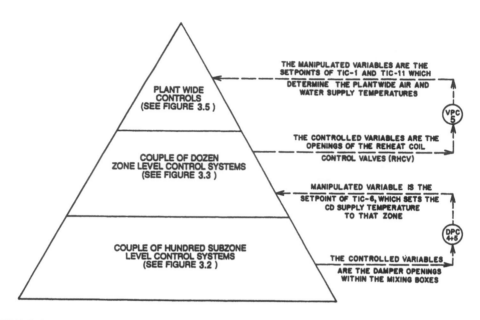

FIGURE 3.6 The overall control architecture in a semiconductor manufacturing plant is hierarchical in structure.

On the other hand, the following conditions will require an increase in the TIC set points:

1. The most open RHCV is approaching full opening.
2. The CD supply temperature is at its minimum limit.

The set point of VPC-5 is produced by a function generator, f(x) in Figure 3.5. The dual purposes of this loop are to prevent any of the reheat coil control valves (RHCVs) from fully opening and thereby losing control and to force the least open RHCV toward a minimum opening (to minimize wasteful overlap between cooling and reheat) but to keep it from full closure as long as possible. These purposes are accomplished by f(x), which keeps the least open valve at approximately 12% opening as long as the most open valve is less than 90% open. If this 90% opening is exceeded, f(x) prevents the most open valve from fully opening. It does so by lowering the set point of VPC-5, which in turn will increase the set points of the TICs. Increasing the set point of TIC-1 reduces the load on the most open RHCV, whereas increasing of the TIC-11 set point increases the heating capacity at the reheat coil. Through this method of load-following and through the modulation of main supply air and water temperatures, all zones in the plant will be kept under stable control if the loads are similar in each zone (Figure 3.6).

3. Mechanical Design Limitations

All control systems—including this one—will lose control when the design limits of the associated mechanical equipment are reached or if the dissimilarity in load distribution is greater than was anticipated by the mechanical design. Therefore, if a condition ever arises in which one zone requires large amounts of cooling (associated RHCV closed) while some other zone at the same time requires its finger reheat coil valve (RHCV) to be fully open, then the control system must decide which condition it is to correct, because it cannot correct both. The control system in Figure 3.5 is so configured that it will give first priority to preventing the RHCV valve from fully opening. Therefore, if as a result of mechanical design errors or misoperation of the plant there is no CD supply temperature that can keep all valves from fully closing or fully opening, then the control system will allow some RHCV valves to close while keeping all of them less than 100% open. Whenever such a condition is approaching (that is, when the difference between the opening of the most open and the least open valves reaches 95%), a valve position alarm (VPA-5) is actuated. This allows the operator to check the causes of such excessively dissimilar loads between zones and to take corrective action by revising processes, relocating tools, modifying air supply ducts, and/or adding or removing mixing boxes.

By keeping all RHCV valves from being nearly closed, this control system simultaneously accomplishes the following goals:

1. Eliminates unstable (cycling) valve operation by not allowing valves to operate nearly closed
2. Minimizes pumping costs by minimizing pressure losses through throttling valves
3. Minimizes heat pump operating costs by minimizing the required hot water temperature
4. Provides a means of detecting and thereby smoothly following the variations in the plantwide load

4. Humidity Controls

At the zone level, the RA humidity is controlled by RHIC-3 in Figure 3.3. In addition, the dew point of the main CD air supply must be modulated. The main CD supply temperature is already being modulated to follow the load. This is accomplished by measuring the relative humidity in all RA fingers (RHT-3 in Figure 3.3) and selecting the one finger that is farthest away from the control target of 35% RH.

The control system is similar to the TIC set point optimization loop discussed earlier. The first elements in the loop are the selectors shown in Figure 3.5. They pick out the return air fingers with the highest and lowest relative humidities. The purpose of the loop is to "herd" all RH transmitter readings in such a direction that *both* the highest and the lowest readings will fall within the acceptable control gap limits of 35% ± 3% RH. This is accomplished by sending the highest reading to RHIC-3 in Figure 3.5 as its measurement and also sending the lowest reading to this RHIC-3 as its modified set point.

A humidity change in either direction can be recognized and corrected through this herding technique. The set point of RHIC-3 is produced by the function generator, $f(x)$. Its dual purposes are to prevent the most humid RA finger from exceeding 38% RH and to keep the driest RA finger humidity from dropping below 32% RH, as long as this can be accomplished without violating the first goal.

These purposes are accomplished by keeping the set point of RHIC-3 at 35% as long as the driest finger reads 34% RH or more. If it drops below that value, the set point is raised to the limit of 38% RH in order to overcome this low humidity condition, without allowing excessive humidity in the return air of some other zone.

This control strategy and the RHIC-3 controller in Figure 3.3 will automatically respond to seasonal changes and will give good RH control as long as the loads in the various zones are similar. If the humidity loads are substantially dissimilar, this control system is also subject to mechanical equipment limitations. In other words, the mechanical equipment is so configured that addition or removal of moisture is possible only at the main air supply to the CD. Consequently, if some zones are moisture-generating and others require humidification, this control system can respond to only one of these needs. Therefore, if the lowest RH reading is below 32% while the highest is already at 38%, the low humidity condition will be left uncontrolled, and the high humidity zone will be controlled to prevent it from exceeding the 38% RH upper limit. Allowing the minimum limit to be temporarily violated while maintaining control on the upper limit is a logical and safe response to humidity load dissimilarities between zones. It is safe because the intermixing of the return airs will allow self-control of the building by transferring moisture from zones with excess humidity to ones with humidity deficiency.

Whenever the difference between minimum and maximum humidity readings reaches 8% RH, the RHA-3 alarm is actuated. This will alert the operator that substantial moisture load dissimilarities exist so that he or she can seek out and eliminate the causes.

5. Exhaust Air Controls

The exhaust air controls at the subzone level are shown in Figure 3.2. At this level, the control element is a two-position damper, EAD-5. When the workstation is functioning, its operable sash is open. This condition is detected by the door position limit switch, ZSC-5, which in that case fully opens the two-position damper. When the workstation is out of service, its operable sash is closed, and therefore the door switch closes EAD-5 to its minimum position. This minimum damper position still provides sufficient air exhaust flow from the workstation to guarantee the face velocity needed for operator safety. In other words, when the sash is closed, the air will enter the workstation over a smaller area; therefore, less exhaust airflow is required to provide the air inflow velocity that is needed to keep the chemical fumes from leaking out. Through this technique, the safety of operation is unaffected, and operating costs are lowered because less outside air needs to be conditioned if the exhaust airflow is lowered.

In order for EAD-5 to maintain the required exhaust air flows accurately, it is necessary to keep the vacuum in the exhaust air collection ductwork at a constant value. This pressure control loop is depicted in Figure 3.7. In each zone exhaust finger, the PC-1 shown in Figure 3.7 keeps the vacuum constant by throttling the EAD-1 damper as required. In order for these dampers to stay in control while the exhaust airflow varies, it is important to keep them from fully opening. This is accomplished first by identifying the most open finger dampers (low selector) and then by comparing its opening with the set point of the damper position controller, DPC-2. The task of this

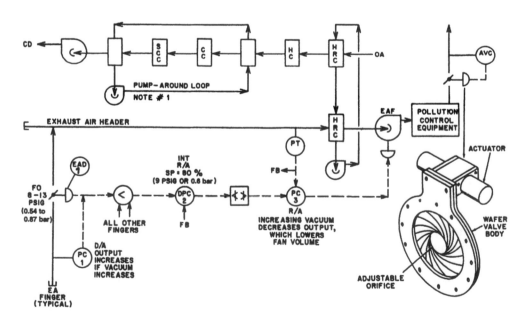

FIGURE 3.7 Exhaust air optimization controls are used to keep the vacuum in the exhaust air collection ductwork at a constant value and to maximize heat recovery. Note #1: When the need for dehumidification in the summer results in excessive overcooling of the CD air supply, and therefore in increased demand for reheat at the CD fingers, the pump-around economizer loop is started to lower operating cost and increase efficiency by reducing the degree of overcooling and thereby lowering the need for reheat.

controller is to limit the opening of the most open EAD-1 to 80% and to increase the vacuum in the main EA header if this opening would otherwise exceed 80%. Therefore, if the measurement of DPC-2 exceeds 80% (drops below 9 PSIG, or 0.6 bar), the vacuum set point of PC-3 is increased (pressure setting lowered). This in turn will increase the operating level of the exhaust fan station (EAF).

The limits on the set point of DPC-2 will prevent mechanical damage, such as collapsing of the ducts as a result of excessive vacuum or manual misoperation. The reasons for integral action and external feedback are the same as in the case of all other damper and valve position controllers discussed earlier.

Figure 3.7 also shows a glycol-circulating heat recovery loop that can be used as shown to preheat the entering outside air or that can be used as a heat source to a heat pump in the winter (not shown). In either case, the plant operating cost is lowered by recovering the heat content of the air before it is exhausted in the winter.

The discharging of chemical vapors into the atmosphere is regulated by pollution considerations. The usual approach is to remove most of the chemicals by adsorption and scrubbing prior to exhaustion of the air. An added measure of safety is provided by exhausting the air at high velocity, so as to obtain good dispersion in the atmosphere. Because the volume of air being exhausted varies, an air velocity controller (AVC in Figure 3.7) is used to maintain the velocity of discharge constant. This is accomplished by modulation of a variable orifice iris damper, also illustrated in Figure 3.7.

IV. CONCLUSIONS

The productivity of semiconductor manufacturing plants can be greatly increased and the operating costs can be lowered through the instrumentation and control methods described previously and by such added control system features as:

Low leakage dampers (0.5 CFM/ft^2 at 4 in. H$_2$O ΔP, or 2.5 l/s/m^2 at 10 millibars ΔP)
Accurate airflow metering for material balance and pressurization loss control
Pump-around economizers (see Figure 3.7)

The initial cost of the previously described control system is not higher than the cost of conventional systems, because the added expense of more accurate sensors is balanced by the much-reduced installation cost of distributed shared controls. Therefore, the benefit of increased productivity is a result not of higher initial investment but of better control system design.

It should be emphasized that the described load-following optimization envelope control strategy is far from being typical of today's practices in the semiconductor manufacturing industry. In fact, it has never been fully implemented, and many plants are still operated under conventional HVAC controls. Therefore, the conclusion that yields will increase and operating costs will drop when such improved control systems are installed is an assumption. This appears to be supported by the results of partial system implementations and by the experiences in other industries but will remain an assumption until the first semiconductor manufacturing plant with all the features of such modern controls is operating.

BIBLIOGRAPHY

Anderson, E.P., *Air Conditioning: Home and Commercial*, 5th ed., Macmillan, New York.
Brumbaugh, J.E., *Heating, Ventilating and Air Conditioning Library*, rev. ed., Macmillan, New York.
Cooper, F.G., Low pressure sensing and control, *InTech*, September 1992.
Demster, C.S., *Variable Air Volume Systems for Environmental Quality*, McGraw-Hill, New York, 1995.

DHO/Atlanta Corporation, *Conserving Energy*, Power Induction Unit, Data Sheet.

Grimm, H.R. and Roasler, R.C., *Handbook of HVAC Design*, McGraw-Hill, New York, 1990.

Haines, R.W. and Wilson, C.L., *HVAC Systems Design Handbook*, McGraw-Hill, New York, 1994.

Hartman, T.B. *Direct Digital Control for HVAC Systems,* McGraw-Hill, New York, 1993.

Huntington, W.C., *Building Construction*, 4th ed., John Wiley & Sons, New York.

Kerin, T.W. and Katz, E.M., Temperature measurement in the 1990s, *InTech*, August 1990.

Lipták, B.G., Applying the techniques of process control to the HVAC process, *ASHRAE Trans.*, Vol. 89, Part 2A, 1983.

Lipták, B.G., Reducing the operating costs of buildings by the use of computers, *ASHRAE Trans.*, 83:1, 1977.

O'Callaghan, P.W., *Energy Management,* McGraw-Hill, New York, 1993.

Parmley, R.O., *HVAC Design Data Sourcebook*, McGraw-Hill, New York, 1994.

Sabatini, J.N. and Smith, R.D., *Building and Safety Codes for Industrial Facilities,* McGraw-Hill, New York, 1993.

Spielvogel, L.G., Exploding some myths about building energy use, *Arch. Rec.*, February 1976.

Stein/Reynolds/McGuinness, *Mechanical and Electrical Equipment for Buildings*, 8th ed., John Wiley & Sons, New York.

Strother, E.F., *Thermal Insulation Building Guide*, R.E. Krieger, 1990.

Sun, T.-Y., *Air Handling Systems Design*, McGraw-Hill, New York, 1994.

Wang, S.K., *Handbook of Air Conditioning and Refrigeration*, McGraw-Hill, New York, 1994.

Wendes, H.C., *HVAC Retrofits*, Fairmont Press, 1994.

Wendes, H.C., *HVAC Energy Audits*, Fairmont Press, 1996.

Whitman, W.C. and Johnson, W.M., *Refrigeration and Air Conditioning Technology*, Delmar, 1991.

4 Compressors

The transportation of vapors and gases is a major element in the operating cost of processing plants. The optimization of this unit operation can substantially lower the total operating cost of the plant. The goal of this chapter is to describe the strategies available to increase the safety and energy efficiency in these systems. Prior to the discussion of state-of-the-art advanced controls, the basic equipment and the conventional strategies will be described.

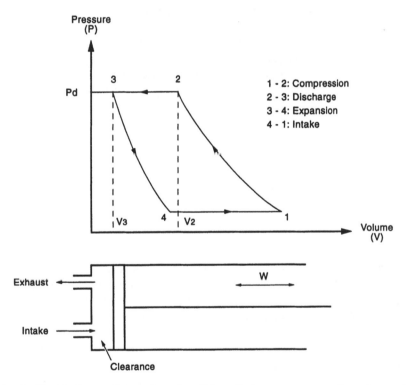

FIGURE 4.1 An ideal indicator diagram is a plot of the cylinder pressure vs. volume in a reciprocating air compressor.

I. THE PROCESS

In a reciprocating compressor, an indicator diagram describes the idealized process (Figure 4.1). This cycle consists of two isentropic (1-2 and 3-4) and two constant-pressure (4-1 and 2-3) processes. When serving a chiller, these constant-pressure processes are also isothermal. An actual indicator diagram will differ significantly from the ideal one (Figure 4.1) because of pressure drops across intake and discharge valves, leakages around the piston and through the valves, the time required to open valves, and the inertia effects in the connected pipes.

The volumetric efficiency of a compressor is the ratio between the mass of the gas discharged and the mass of the gas filling the piston prior to compression. This efficiency is reduced as clearance volume, piston and valve leakage, or valve-pressure drops increase, while it could theoretically be increased by cooling during the suction stroke. The work of compression is the area under the curve in the P-V diagram (Figure 4.1). Between points 2 and 3, the pressure is constant and therefore the work is $_2W_3 = P_d(V_2 - V_3)$.

Compressors are gas-handling machines that perform the function of increasing gas pressure by confinement or by kinetic energy conversion. Methods of capacity control for the principal types of compressors are listed in Table 4.1. The method of control to be used is determined by process requirements, the type of driver, and the cost.

TABLE 4.1
Capacity Control Methods of Compressors

Compressor Type	Capacity Control Method
Centrifugal	Suction throttling
	Discharge throttling
	Variable inlet guide vanes
	Speed control
Rotary	Bypassing
	Speed control
Reciprocating	On-off control
	Constant-speed unloading
	Speed control
	Speed control and unloading

Since the driver constitutes half the compressor installation, careful selection must be made in order to ensure trouble-free performance. When the control involves variable speed, this can be accomplished by the use of a steam turbine, a gas turbine, or gasoline, or diesel engines. For constant speed control, electric motors are well suited; however, variable speed can also be obtained from electric motors.

The operating ranges of various compressor designs are shown in Figure 4.2., while those of the most popular compressor drives are shown in Figure 4.3. Also given in that figure is the classification of two-speed governors. Regulation error is related to the difference in speed at zero power output and at the rated power output. Variation error is related to the effect of whether the speed change occurred above or below set point.

A. Centrifugal Compressors and their Throttling

The centrifugal compressor is a machine that converts the momentum of gas into a pressure head.

$$H = \frac{\tau\omega}{w} = \frac{n}{n-1}ZRT_1\left[\left(P_D/P_1\right)^{\frac{n}{n-1}} - 1\right]$$

$$(1)$$

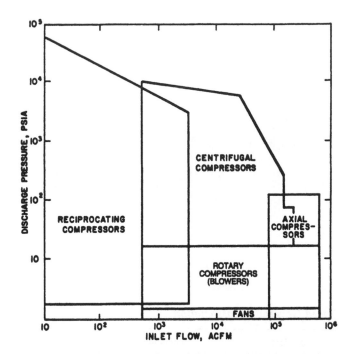

FIGURE 4.2 Operating ranges vary for different compressors and fans.[1]

Equation 1 is the basis for plotting the compressor curves and for understanding the operation of capacity controls.

In Equation 1, the pressure ratio (P_D/P_I) varies inversely with mass flow (W). For a compressor running at constant speed (ω), constant inlet temperature (T_I), constant molecular weight (implicit in R), and constant n, τ, and Z, the discharge pressure may be plotted against weight flow as in Figure 4.4 (curve I). The design point (1) is located in the maximum efficiency range at design flow and pressure.

Positive-displacement compressors pressurize gases through confinement. Dynamic compressors pressurize them by acceleration. The axial compressor drives the gas parallel to the shaft; in the centrifugal compressor, the gas receives a radial thrust toward the wall of the casing where it is discharged. The axial compressor is better suited for constant flow applications, whereas the centrifugal design is more applicable for constant pressure applications. This is because the characteristic curve of the axial design is steep, and that of the centrifugal design is flat (Figure 4.5). The characteristic curve of a compressor plots its discharge pressure as a function of flow, and the load curve relates the system pressure to the system flow. The operating points (L1 or L2 in Figure 4.5) are the intersections of these curves. Axial compressors are more efficient; centrifugal ones are better suited for dirty or corrosive services.

The normal operating region falls between the low and the high demand load curves in Figure 4.5.

Compressor loading can be reduced by throttling a discharge or a suction valve by modulating a prerotation vane or reducing the speed. As was shown in Figure 1.4, discharge throttling is the least energy efficient and speed modulation is the most energy efficient method of turndown. Suction throttling is a little more efficient and gives a little better turndown than discharge throttling, but it is still a means of wasting that transportation energy, which should not have been introduced in the first place.

Guide vane positioning, which provides prerotation or counter-rotation to the gas, is not as efficient as speed modulation, but it does provide the greatest turndown. As was shown in Figure

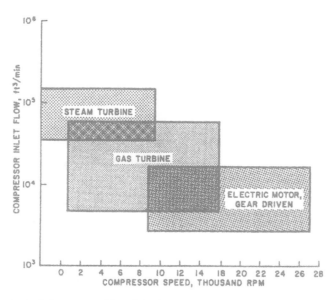

Speed Governor Classficiation

Governor class	Regulation	Variation
A	10%	¾%
B	6%	½%
C	4%	¼%
D	½%	¼%

FIGURE 4.3   Each drive has its own throughput and speed range.[1]

1.4, speed control is the most efficient, as small speed reductions result in large power savings because of the cubic relationship between speed and power. If the discharge pressure is constant, flow tends to vary linearly with speed. If the discharge head is allowed to vary, it will change with the square of flow and, therefore, with the square of speed as well. This square relationship between speed and pressure tends to limit the speed range of compressors to the upper 30% of their range.

Constant-speed steam turbine governors can be converted into variable-speed governors by revising the quick-opening characteristics of the steam valve into a linear one. The efficiency of variable-speed drives varies substantially with their design (Figure 4.6). Electronic governors tend to eliminate the dead bands that are present in mechanical designs. They also require less maintenance because of the elimination of mechanical parts. Electronic governors also give better turndowns and are quicker and simpler to interface with surge or computer controls.

The error in following the load increases as the speed of process disturbances increases or as the speed control loop speed is reduced. Therefore, it is desirable to make the speed loop response as fast as possible. On turbines, this goal is served by the use of hydraulic actuators, and the motor response is usually increased by the use of tachometer feedback.

For the purposes of the control systems shown in this chapter, it will be assumed that the compressor throughput is controlled through speed modulation with tachometer feedback.

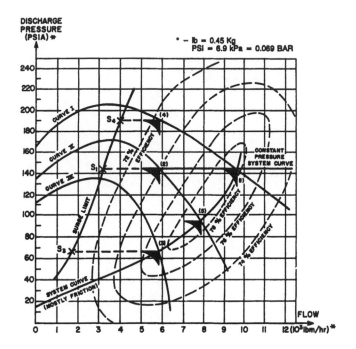

FIGURE 4.4 Centrifugal compressor curves.

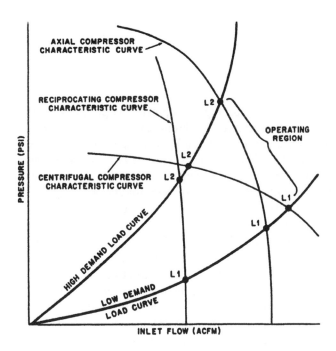

FIGURE 4.5 The characteristic curve of the axial compressor is steep; that of the centrifugal compressor is flat.[1]

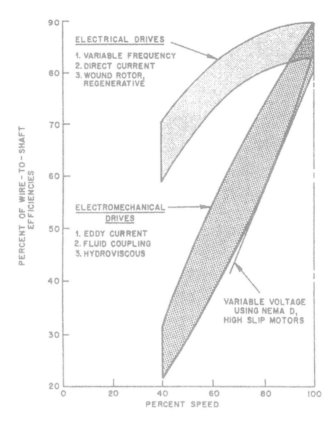

FIGURE 4.6   Wire-to-shaft efficiencies differ for various types of variable-speed drives.[2]

1. Suction Throttling

One can control the capacity of a centrifugal compressor by placing a control valve in the suction line, thereby altering the inlet pressure (P_I). From Equation 1 it can be seen that the discharge pressure will be altered for a given flow when P_I is change, and a new compressor curve will be generated. This is illustrated in Figure 4.4 (curves II and III). Consider first that the compressor is operating at its normal inlet pressure (following curve I) and is intersecting the "constant pressure system" curve at point (1) with a design flow of 9600 lbm/hr (4320 kg/hr) at a discharge pressure of 144 PSIA (1 MPa) and 78% efficiency. If it is desired to change the flow to 5900 lbm/hr (2655 kg/hr) while maintaining the same discharge pressure, it would be necessary to shift the compressor from curve I to curve II. The new intersection with the "constant pressure system" curve is at the new operating point (2), at 74% efficiency. In order to shift from curve I to curve II, one must change the discharge pressure of 190 PSIA (1.3 MPa) at the 5900 lbm/hr (2655 kg/hr) flow on curve I, to 144 PSIA (1 MPa) on curve II. If the pressure ratio is ten ($P_D/P_I = 10$), then it would be necessary to throttle the suction by only $\Delta P_1 = 46/10 = 4.6$ PSI (32 kPa) to achieve this shift.

It is also important to consider how close the operating point (2) is to the surge line. The surge line represents the low-flow limit for the compressor, below which its operation is unstable as a result of momentary flow reversals. Methods of surge control will be discussed later. At point (2) the flow is 5900 lbm/hr (2655 kg/hr), and at the surge limit (S_1) it is 3200 lbm/hr (1440 kg/hr). Thus, the compressor is operating at 5900/3200 = 184% of surge flow. This may be compared with curve I at point (1), where prior to suction throttling the machine is operating at 9600/3200 = 300% of surge flow.

The same method of suction throttling may be applied in a "mostly friction system" also shown in Figure 4.4. In order to reduce the flow from 9600 lbm/hr (4320 kg/hr) to 5900 lbm/hr (2655 kg/hr), it is necessary to alter the compressor curve to curve III, so that the intersection with the "mostly friction system curve" is at the new operating point (3), at 77% efficiency. In order to do this, one must change the discharge pressure from 190 PSIA (13 bars–on curve I) to 68 PSIA (5 bars–on curve III). Thus, $\Delta P_D = 190 - 68 = 122$ PSI (8 bars), and the amount of inlet pressure throttling for a machine with a compression ratio of 10 is $\Delta P_1 = 122/10 = 12.2$ PSI (0.84 bars). The corresponding surge flow is at 1700 lbm/hr (765 kg/hr), which means that the compressor is operating at 5900/1700 = 347% of surge flow. Therefore, surge is less likely in a "mostly friction system" than in a "constant-pressure system" under suction throttling control.

2. Discharge Throttling

A control valve on the discharge of the centrifugal compressor may also be used to control its capacity. In Figure 4.4, if the flow is to be reduced from 9600 lbm/hr (4320 kg/hr) at point (1) to 5900 lbm/hr (2655 kg/hr), the compressor must follow curve I and therefore operate at point (4), at 190 PSIA (1.3 MPa) discharge pressure and 72% efficiency. However, the "mostly friction system" curve at this capacity requires only 68 PSIA (5 bars) discharge pressure. The surge flow (at S_4) is 4000 lbm/hr (1800 kg/hr), and the compressor is therefore operating at 5900/4000 = 148% of surge. Thus, surge is more likely to occur in a mostly friction system when discharge throttling is used.

The various parameters involved in suction and discharge throttling of this sample compressor are compared in Table 4.2.

TABLE 4.2
Compressor Parameters as a Function of Throttling

	Control Valve ΔP(PSI)	Compressor Efficiency	Operation Above Surge By
Suction throttling "Constant pressure system"	4.6	74%	184%
Suction throttling "Mostly friction system"	12.2	77%	347%
Discharge throttling "Mostly friction system"	122	72%	148%

3. Inlet Guide Vanes

This method of control uses a set of adjustable guide vanes on the inlet to one or more of the compressor stages. By prerotation or counter rotation of the gas stream relative to the impeller rotation, the stage is unloaded or loaded, thus lowering or raising the discharge head. The effect is similar to suction throttling as illustrated in Figure 4.4 (curves II and III), but less power is wasted because pressure is not throttled directly. Also, the control is two-directional, since it may be used to raise as well as to lower the band. It is more complex and expensive than throttling valves but may save 10 to 15% on power and is well suited for use on constant-speed machines in applications involving wide flow variations.

The guide vane effect on flow is more pronounced in constant discharge pressure systems. This can be seen in Figure 4.4 (curve II), in which the intersection with the "constant pressure system" at point (2) represents a flow change from the normal design point (1) of 9600 — 5900 = 3700 lbm/hr (1665 kg/hr), whereas the intersection with the "mostly friction system" at point (5) represents a flow change of only 9600 — 7800 = 1800 lbm/hr (810 kg/hr).

4. Variable Speed

The pressure ratio developed by a centrifugal compressor is related to tip speed by the following equation:

$$\psi\, u^2/2g = \frac{ZRT_1}{(n-1/n)}\left[\left(P_D/P_1\right)^{(n-1/n)} - 1\right] \tag{2}$$

From this relation the variation of discharge pressure with speed may be plotted for various percentages of design speed, as shown in Figure 4.7. The obvious advantage of speed control from a process viewpoint is that both suction and discharge pressures can be specified independently of the flow. The normal flow is shown at point (1) for 9700 lbm/hr (4365 kg/hr) at 142 PSIA (9.8 bars). If the same flow is desired at a discharge pressure of 25 PSIA (1.73 bars), the speed is reduced to 70% of design, shown at point (2). In order to achieve the same result through suction throttling with a pressure ratio of 10 to 1, the pressure drop across the valve would have to be (142 —25)/10 = 11.7 PSI (0.81 bars), with the attendant waste of power, as a result of throttling. This is in contrast with a power saving accomplished with speed control, since *power input is reduced as the square of the speed.*

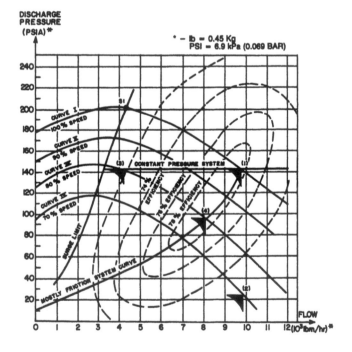

FIGURE 4.7   Control of centrifugal compressor capacity by speed variation.

One disadvantage of speed control is apparent in constant-pressure systems, in which the change in capacity may be overly sensitive to relatively small speed changes. This is shown at point (3), where a 20% speed change gives a flow change of (9600 — 4300)/9600 = 55%. The effect is less pronounced in a "mostly friction system," in which the flow change at point (4) is (9600 — 8100)/9600 = 16%.

II. BASIC CONTROLS

In this part of the chapter, the basic controls for rotary, reciprocating, and centrifugal compressors are described, together with installation and safety considerations, such as surge protection.

A. ROTARY COMPRESSORS AND BLOWERS

The rotary compressor is essentially a constant-displacement, variable-discharge pressure machine. Common designs include the helical-screw, the lobe, the sliding vane, and the liquid ring types. The characteristic curves for a lobe-type unit are shown in Figure 4.8. As shown by curves I and II, the inlet flow varies linearly with the speed of this positive-displacement machine. The small decrease in capacity (at constant speed) with an increase in pressure is a result of gas slippage through impeller clearances. It is necessary to compensate for this by small speed adjustments as the discharge pressure varies. For example, when the compressor is operating at point 1, it delivers the design volume of 66 ACFM (1.9 m³/m) at 3¹/₂ PSIG (0.24 bar). In order for the same flow to be maintained when the discharge pressure is 7 PSIG (0.48 bar), the speed must be increased from 1420 rpm to 1550 rpm at point 2 by the flow controller in the discharge line.

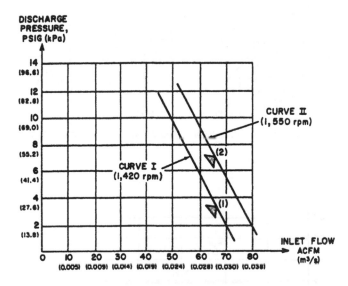

FIGURE 4.8 As shown by the rotary compressor curves, the inlet flow varies linearly with the speed of this positive-displacement machine.

In addition to speed variation, the discharge flow can also be varied by throttling the suction, the bypass, or the vent line from the rotary blower. In Figure 4.9 excess gas is vented to the atmosphere as the temperature control valve closes. The temperature of the outlet gas is controlled to prevent product degradation and to provide the proper product dryness. In systems in which the gas is not vented, it may be returned to the suction of the blower on pressure control.

An important application of the liquid ring rotary compressor is in vacuum service. The suction pressure is often controlled by bleeding gas into the suction on pressure control. This is shown in Figure 4.10, where suction pressure control is used on a rotary filter maintaining the proper drainage of liquor from the cake on the drum.

B. RECIPROCATING COMPRESSORS

The reciprocating compressor is a constant-volume, variable-discharge pressure machine. A typical compressor curve is shown in Figure 4.11, for constant-speed operation. The curve shows no

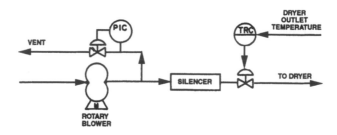

FIGURE 4.9 A rotary blower with bypass capacity control vents excess gas as the temperature control valve closes.

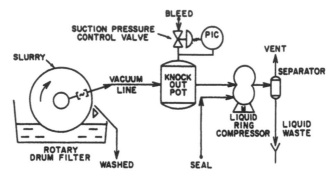

FIGURE 4.10 The liquid ring rotary compressor is shown with suction pressure control.

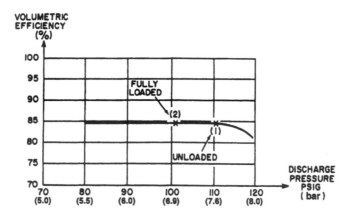

FIGURE 4.11 Reciprocating compressor curve.

variation in volumetric efficiency in the design pressure range, which may vary by 8 PSIG (0.54 barg) from unloaded to fully loaded

The volumetric inefficiency is a result of the clearance between piston end and cylinder end on the discharge stroke. The gas that is not discharged re-expands on the suction stroke, thus reducing the intake volume.

The relationship of speed to capacity is a direct ratio, since the compressor is a displacement-type machine. The typical normal turndown with gasoline or diesel engine drivers is 50% of maximum speed, in order to maintain the torque within acceptable limits.

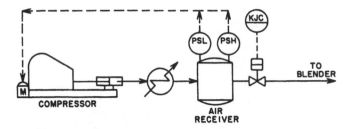

FIGURE 4.12   On-off control on reciprocating compressor.

1. On-Off Control

For intermittent demand, where the compressor would waste power if run continuously, the capacity can be controlled by starting and stopping the motor. This can be done manually or by the use of pressure switches. Typical switch settings are on at 140 PSIG (9.6 barg), off at 175 PSIG (12 barg). This type of control would suffice for processes in which the continuous usage is less than 50% of capacity, as shown in Figure 4.12, where an air mix blender uses a rapid series of high-pressure air blasts when the mixer becomes full. The high-pressure air for this purpose is stored in the receiver.

2. Constant-Speed Unloading

In this type of control, the driver operates continuously, at constant speed, and one varies the capacity in discrete steps by holding suction valves open on the discharge stroke or opening clearance pockets in the cylinder. The most common schemes are three- and five-step unloading techniques. The larger number of steps saves horsepower because it more closely matches the compressor output to the demand.

In three-step unloading, capacity increments are 100%, 50%, or 0% of maximum flow. This method of unloading is accomplished by the use of suction valve unloading in the double-acting piston. At 100%, both suction valves are closed during the discharge stroke. At 50%, one suction valve is open on the discharge stroke, wasting half the capacity of the machine. At 0%, both suction valves are held open on the discharge stroke, wasting total machine capacity.

For five-step unloading, a clearance pocket is used in addition to suction valve control. The capacity can be 100, 75, 50, 25 or 0% of maximum flow. This is shown in Figure 4.13. At 100%, both suction valves and the clearance pocket are closed; at 75%, only the clearance pocket is open; at 50%, only one suction valve is open on the discharge stroke; at 25%, one suction valve and the clearance pocket are open; at 0%, both suction valves are opened during the discharge stroke.

The use of step unloading is most common when the driver is inherently a constant-speed machine, such as an electric motor.

The pressure controller signal from the air receiver operates a solenoid valve in the unloader mechanism. The action of the solenoid valve directs the power air to lift the suction valves or to open the clearance port, or both.

For three-step unloading, two pressure switches are needed. The first switch loads the compressor to 50% if the pressure falls slightly below its design level, and the second switch loads the compressor to 100% if the pressure falls below its setting.

For five-step unloading, a pressure controller is usually substituted for the four pressure switches (Figure 4.14) otherwise required, and the range between unloading steps is reduced to not more than 2 PSI (0.14 bar) deviation from the design level, keeping the minimum pressure within 8 PSI (0.55 bar) of design. In cases in which exact pressure conditions must be met, a throttling valve is installed, which bypasses the gas from the discharge to the suction of the compressor. This device

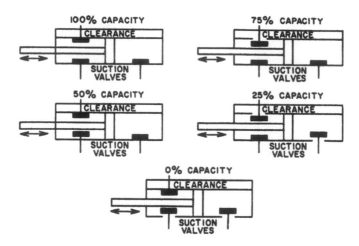

FIGURE 4.13   Constant-speed, five-step unloading with valves and clearance pockets.

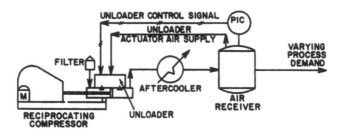

FIGURE 4.14   Constant-speed capacity control of reciprocating compressor.

smoothes out pressure fluctuations and, in some cases, eliminates the need for a gas receiver in this service. This can prove economical in high-pressure services above 500 PSIG (34.5 barg), in which vessel costs become significant.

3. Stand-Alone Air Compressors

A typical air compressor, together with the controls normally provided by its manufacturer, is shown in Figure 4.15. Such a stand-alone compressor usually operates as follows:

When the operator turns the control switch to "On," this will energize the time delay (TR) and will open the water solenoid valve (WSV) if neither the air temperature (ATSH) nor the cooling water temperature (WTSH) is high.

After seven seconds, TR-1 opens. Assuming that this time delay was sufficient for the oil pressure to build up, the opening of TR-1 will have no effect, because OPSL will have closed in the meantime.

If the opening of WSV results in a cooling water flow greater than the minimum setting of the low water flow switch, then WFSL closes and the compressor motor (M) is started. Whenever the motor is on, the associated "Run" pilot light is energized. This signals to the operator that oil (OPSL), water (WFSL), and air (ATSH) conditions are all acceptable and therefore the unit can be loaded.

When the operator turns the load/unload control switch to "Load," the TR-2 contact is already closed because the seven second delay has passed. Therefore, the APSL contact will determine the status of the machine. If the demand for air at the users is high, the pressure at APSL will drop, which in turn will cause the APSL contact to close and the compressor to load. As a result of

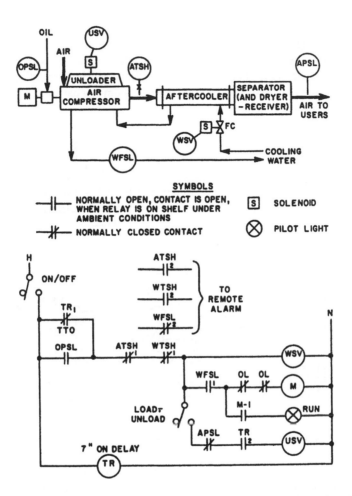

FIGURE 4.15 A typical stand-alone air compressor is provided with this type of equipment layout and control logic.[2]

loading the compressor, the pressure will rise until it exceeds the control gap of APSL, causing its contact to reopen and the machine to unload.

The compressor continues to load and unload automatically as a function of plant demand. As the demand rises, the loaded portion of the cycle will also rise. Once the load reaches the full capacity of the compressor, the unit stops cycling and stays in the loaded state continuously. If the demand rises beyond the capacity of this compressor, it is necessary to start another one. The following paragraphs describe how this is done automatically, requiring no operator participation.

If at any time during operation the cooling water flow (WFSL) drops too low, the motor will stop. If either water or air temperatures rise to a high value, this condition not only will stop the motor but will also close the water solenoid valve (WSV). The above three conditions (WFSL, WTSH, ATSH) will also initiate remote alarms, as shown in Figure 4.15, to advise the operators of the possible need for maintenance. If the oil pressure drops below the setting of OPSL, it will also cause the stoppage of the motor and the closure of WSV, but after such an occurrence, the compressor will not be allowed to restart automatically when the oil pressure returns to normal. In order to restart the machine, the operator will have to go out to the unit and reset the system by turning the control switch to "Off" and then to "On" again to repeat the complete start-up procedure.

a. Local/remote switch

The first step in integrating several compressors into a single system is the addition of a local/remote switch at each machine. As shown in Figure 4.16, when this switch is turned to "Local," the compressor operates in the stand-alone mode, as was described in connection with Figure 4.15. When this switch is turned to "Remote" the compressor becomes a part of the integrated plantwide system, consisting of several compressors.

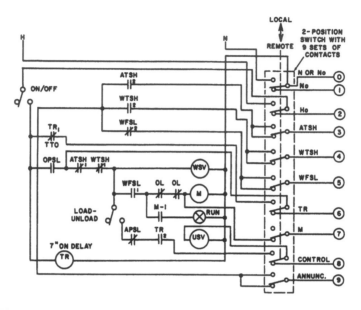

FIGURE 4.16   The local/remote switch is wired to allow stand-alone or integrated compressor operation.[2]

As shown in Figure 4.16, this two-position switch has nine sets of contacts and is mounted near the compressor. It can be installed in a few hours, and once installed, the compressor can again be operated in the "Local" mode.

Only 10 wires need to be run from each compressor to the remote controls. These 10 wires serve the following functions:

#0   The working neutral, N in local, No in the remote mode of operation
#1   The common neutral (No) of the integrated controls
#2   The common hot (Ho) of the integrated controls
#3   The high air temperature (ATSH) alarm
#4   The high water temperature (WTSH) alarm
#5   The low water flow (WFSL) alarm
#6   The 7 second time delay (TR) of the integrated controls
#7   The motor (M) status indication
#8   The load/unload control signal from the remote system
#9   The common hot for the remote annunciator (Annunc.)

When integrating several compressors into a single system, it is advisable to number these ten wires in a consistent manner, such as

Compressor #1 Wire #10 to #19
Compressor #2 Wire #20 to #29
Compressor #3 Wire #30 to #39
etc.

In this system, the first digit of the wire number indicates the compressor and the second digit describes the function of the wire. Immediately knowing, for example, that wire number 45 comes from compressor #4 and serves to signal a low water flow condition on that machine simplifies checkout and start-up.

Of the ten wires from each of the compressors, six are used for remote alarming:

"Run" light
High air temperature light
High water temperature light
Low water flow light
Audible alarm bell with silencer
Alarm reset buttons

The only time these circuits are deactivated is when the associated compressor is off.

b. Lead-lag selector

As plants grow, their compressed air requirements also tend to increase. As a result of such evolutionary growth, many existing plants are served by several uncoordinated compressor stations. When, because of space limitations, the new compressors are installed in different locations, the manual operation of such systems becomes not only inefficient but unsafe.

The steps involved in integrating such stand-alone compressor stations into an automatically operating, load-following single system are described here. In such integrated systems, the identity of the "lead" and "lag" compressors, or the ones requiring maintenance ("Off"), can all be quickly and conveniently altered, while the system continues to efficiently meet the total demand for air. Thus, air supply shortage or interruption is eliminated, together with the need for continuous operator's attention.

When it is desirable to combine two compressors into an integrated lead-lag station, all that needs to be added are two pressure switches, as shown in Figure 4.17. APSL-A is the lead and APSL-B is the lag control pressure switch. To maintain a 75 PSIG (5.2 barg) air supply at the individual users, these normally closed pressure switches can be set as follows:

Pressure Control Switch	Pressure Below Which Switch Will Close (PSIG)	Differential Gap (PSI)
APSL-A	90	10
APSL-B	85	10

With the above settings, the system performance will be as shown in Table 4.3. When two stand-alone compressors are integrated into such a lead-lag system, only five wires need to be brought from each compressor to the lead-lag switch, shown in Figure 4.18. Turning this single switch automatically reverses the lead-lag relationship between the two compressors. In addition to the two-position lead-lag switch, running lights are also provided; they show whether either or both compressors are unloaded (standby) or loaded.

As the demand for air increases, the lead machine will load and unload to meet that demand. If the lead machine is already continuously loaded and the demand is still rising, the lag compressor will be started automatically.

The interlocks provided are as follows:

The on-time delays 2TR and 3TR in Figure 4.18 are provided for stabilizing purposes only. They guarantee that a system configuration (or reconfiguration) will be recognized only if it is maintained for at least three seconds. Responses to quick changes are thus eliminated.

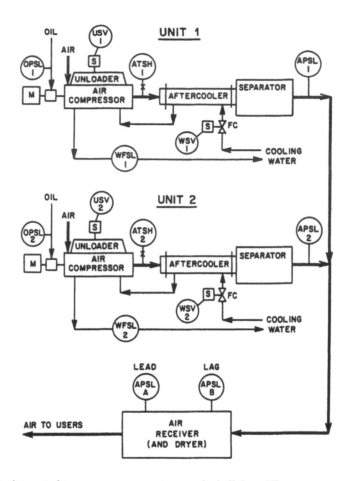

FIGURE 4.17   An integrated two-compressor system can be built by adding two pressure switches.[3]

TABLE 4.3
System Performance with APSL-A Set to Close at 90 PSIG,
APSL-B Set to Close at 85 PSIG, and Differential Gap Set
at 10 PSI

Pressure in Air Receiver (PSIG)	APSL-A Contact	APSL-B Contact	Lead Compressor	Lag Compressor
Over 100	Open	Open	Off	Off
100	Open	Open	Off	Off
95	Open	Open	Off	Off
90	Closed	Open	On	Off
85	Closed	Closed	On	On
80	Closed	Closed	On	On
85	Closed	Closed	On	On
90	Closed	Closed	On	On
95	Closed	Open	On	Off
100	Open	Open	Off	Off

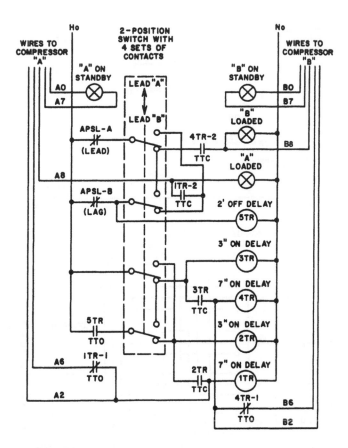

FIGURE 4.18   The lead-lag selector switch automatically reversed the lead-lag relationship between the two compressors.[2]

The on-time delays 1TR and 4TR are provided to give time for the oil pressure to build up in the system. If after seven seconds the oil pressure is not yet established, and therefore OPSL in Figure 4.16 is still open, the contacts 1TR-1 and 4TR-1 in Figure 4.18 will open and the corresponding compressor will be stopped.

The off-time delay 5TR guarantees that the lag machine will not be cycled on and off too frequently. Once started, the lag compressor will not be turned off (but will be kept on standby) until the off delay of two minutes has passed.

c. Large systems

From the building blocks discussed in the previous paragraphs (and in Reference 2), an integrated remote controller can be configured for every compressor combination. Figure 4.19 illustrates how six stand-alone compressors might be integrated into a single load-following system. As the system pressure drops below 9 PSIG, APSL-A closes, starting the first compressor. As the load rises, causing the system pressure to drop further, more switches will close, until at 80 PSIG all six switches will be closed and all six compressors will be running.

The "priority selector" shown in Figure 4.19 is provided for the convenient reconfiguration of compressor sequencing. When the selector is in position #1, as the load increases, the compressors will be started in the following order: A, B, C, D, E, F. When switched to position #2, the sequence becomes C, D, E, F, A, B; in position #3, the sequence is E, F, A, B, C, D.

If a system consists of nine compressors, and full flexibility in integrated remote control is desired, the control cabinet might look like Figure 4.20. With this system, the operator can set the

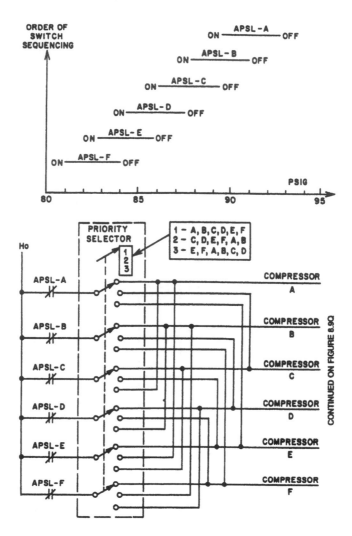

FIGURE 4.19 A priority selector can be used to change the order of priorities in large compressor systems.[2]

lead, lag, or off status for all nine compressors and can also select the order in which they are to come on.

d. Flexibilities

From the previous discussion it can be seen that the building-block approach to compressor system design is very flexible. Any number of stand-alone compressors can be integrated into an automatic load-following control system, with complete flexibility for priority, lead-lag, or off selection. The logic blocks described can be implemented either in hardware or software. All wires and terminals are prenumbered, which minimized installation errors. The potential benefits are listed below:

1. Unattended, automatic operation relieves operators for other tasks.
2. Automatic load-following eliminates the possibility of accidents caused by the loss of the air supply.
3. Because the supply and the demand are continuously and automatically matched, the energy cost of operation is minimized.

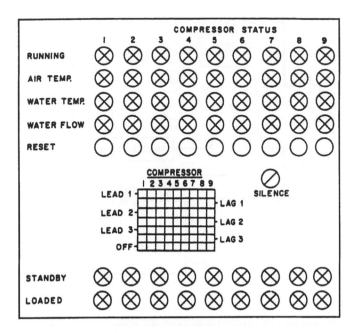

FIGURE 4.20 A control cabinet can provide full flexibility in remote control for an integrated nine-compressor system. The mechanically interlocked pushbutton station is so designed that only one button can be pressed in each column or row, except in the bottom one. In the bottom row any or all buttons can be simultaneously pressed.[2]

C. Centrifugal Compressors

Figures 4.4 to 4.7 described the characteristics of the centrifugal compressor. Here, the associated lube, seal, surge, and safety systems and their associated basic controls will be discussed in connection with both single and multiple compressor installations.

1. Lube and Seal Systems

A typical lube oil system is shown in Figure 4.21. Dual pumps are provided to ensure the uninterrupted flow of oil to the compressor bearing and seals. Panel alarms on low oil level, low oil pressure (or flow), and high oil temperature are provided. A head tank provides oil for coasting down in case of a power failure. The design of these systems is critical, because failure of the oil supply could mean a shutdown of the entire process. In cases in which oil cannot be tolerated in contact with the process gas, an inert gas seal system may be used. This is shown in Figure 4.22 for a centrifugal compressor with balanced seals.

2. Surge Controls

The design of compressor control systems is not complete without consideration of surge control, because this affects stability of the machine. Surging begins at the positively sloped section of the compressor curve. In Figure 4.7 this occurs at S_1 on the 100% speed curve at 4400 lbm/hr (1980 kg/hr). This flow will ensure safe operation for all speeds, but some power will be wasted at speeds below 100% because the surge limit decreases at reduced speed. Even for a compressor running at a constant speed, surge point changes as the thermodynamic properties vary at the inlet. This is shown in Figure 4.23. Although an inaccurate surge control point can put the compressor into a deep surge, a conservative surge point results in useless recycling and wasted horsepower. Various schemes to control surge are outlined in the following paragraphs. These include

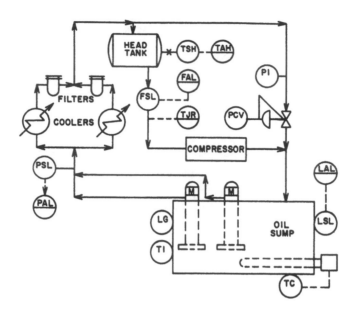

FIGURE 4.21 Compressor lube oil system for bearings and seals.

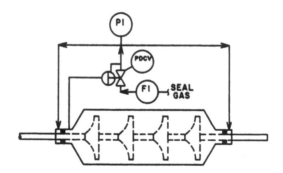

FIGURE 4.22 Compressor seal system.

1. Compressor pressure rise ($\Delta P = P_D - P_I$) versus differential across suction flow meter (h)
2. Pressure ratio (P_D/P_I) versus actual volumetric flow (Q)
3. Break horsepower versus mass flow (m)
4. Pressure ratio (P_D/P_I) versus Mach number squared
5. Incipient surge
6. Minimum flow
7. Surge spike detection

a. *The phenomenon of surge*

In axial or centrifugal compressors, the phenomenon of momentary flow reversal is called surge. The surge region of the compressor is characterized by negative resistance to flow. During surging, the compressor discharge pressure drops off and then is reestablished on a fast cycle. This cycling, or surging, can vary in intensity from an audible rattle to a violent shock. Intense surges are capable of causing complete destruction of compressor parts, such as blades and seals.

The characteristic curves of compressors are such that, as the flow drops, at each speed they reach a maximum discharge pressure (Figure 4.24). A line connecting these points (A to F) is the

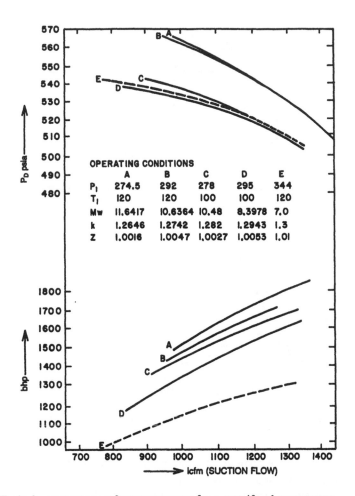

OPERATING CONDITIONS

	A	B	C	D	E
P_I	274.5	292	278	295	344
T_I	120	120	100	100	120
Mw	11.6417	10.6364	10.48	8.3978	7.0
k	1.2646	1.2742	1.282	1.2943	1.3
z	1.0016	1.0047	1.0027	1.0053	1.01

FIGURE 4.23 Typical compressor performance curves for a centrifugal compressor.

surge line. If flow is further reduced, the pressure generated by the compressor drops below that which is already existing in the pipe, and momentary flow reversals occur. The frequency of these oscillations is between 0.5 and 10 Hz. The surge frequency of most compressor installations in the processing industries is slightly less than 1 Hz.[5] Surge is usually preceded by a stall condition, which is caused by localized flow oscillations around the rotor at frequencies of 50 to 100 Hz.

At the beginning of surge, the total flow drops off within 0.05 seconds, and then it starts cycling rapidly at a period of less than 2 seconds.[1] This period is usually shorter than that of the flow control loop that controls the capacity of the compressor. If the flow cycles occur faster than the control loop can respond to them, this cycling will pass through undetected as uncontrollable noise. Therefore, fast sensors and instruments are essential for this loop.

As is shown in Figure 4.24, at low compression ratios, the surge line is a parabolic curve on a plot of pressure rise (discharge pressure minus suction pressure) versus flow. This function shows an increasing nonlinearity as the compression ratio increases. If the Figure 4.24 surge line is plotted as P_D-P_I versus the square of flow (orifice differential = h), it becomes a straight line (Figure 4.25).

On a plot of ΔP versus Q (Figure 4.24), the following changes will increase the probability of surge by reducing the distance between the operating point and the surge line: a decrease in suction pressure; an increase in suction temperature; a decrease in molecular weight; a decrease in specific heat ratio. On a plot of ΔP versus h (Figure 4.25), these effects are more favorable. A decrease in suction pressure moves the surge line in the safe direction, temperature has no effect, and the effect of the other variables is also less pronounced. Therefore, although the ΔP versus h plot is accurate

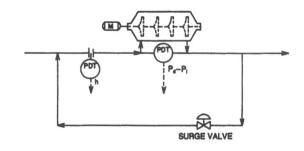

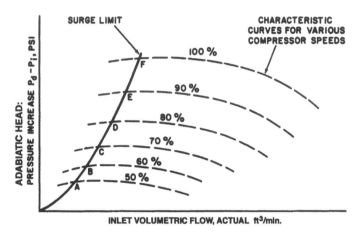

FIGURE 4.24 The surge line of a speed-controlled centrifugal compressor is parabolic.[4]

only at low compression ratios, it does have the advantage of being independent from the effects of composition and temperature changes. However, suction pressure should be included in the model in order to be exact; yet most ΔP versus h plots disregard it.

The adiabatic head (P_d-P_i) varies not only with compression ratio, but also with process properties (molecular weight, heat-capacity ratio, temperature, and supercompressibility) and with compressor speed or vane settings. As the compression ratio rises, (P_d-P_i) no longer relates linearly to volumetric flow.

If only the inlet pressure (P_i) varies, instead of using the orifice differential at the inlet (hi) as the surge measurement, the ratio hi/P_i should be used. If all process properties vary, hi should be replaced by the volumetric inlet flow (Fi), measured by a vortex or similar flowmeter, and the ratio Fi/P_I should be held constant. If compressor speed (N) also varies, the ratio to be held constant becomes Fi/(N.P_i). When adjustable guide vanes are used, there is a separate curve associated with each vane position, and the surge curve can no longer be linear.

b. Surge curve variations

Figure 4.24 shows the surge and speed characteristic curves of a centrifugal compressor with low compression ratio; Figure 4.26 shows these curves for an axial compressor. The characteristic curves of the axial compressor are steeper, which makes it better for constant-flow services. The centrifugal design is better for constant pressure control.

The effect of guide vane throttling is also shown for both centrifugal and axial compressors (Figure 4.27). As can be seen, the shape of the surge curve varies with the type of equipment used. The surge curve of speed controlled centrifugal compressors bends up (Figure 4.24), whereas for axial (Figure 4.26) and vane-controlled machines (Figure 4.27), the surge line bends over. It is this negative slope of the axial compressor's surge curve that makes it sensitive to speed variations, because an increase in speed at constant flow can quickly bring the unit into surge.

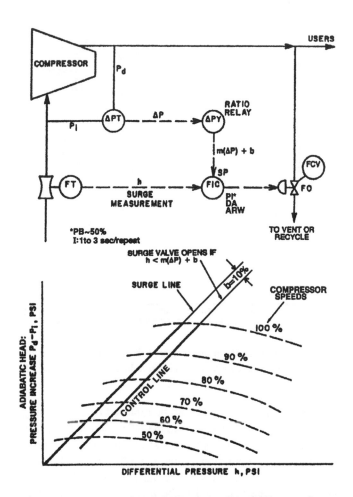

FIGURE 4.25 The surge curve becomes a straight line on a plot of ΔP versus h.

As can be seen from the above information, the shape of the surge curve varies with compression ratio and with equipment design. It should also be noted that in the case of multistage compressors, the surge line is discontinuous. If the compressor characteristics are as shown in Figure 4.28, the addition of a compressor stage causes a break point in the surge curve. With more stages, more break points would also be added, and the resulting net effect is a surge curve that bends over instead of bending up, as does the surge curve for the single-stage compressor in Figure 4.24.

Figure 4.28 also shows the choke curve. This curve connects the points at which the compressor characteristic lines become vertical. Below this curve, flow will stay constant even if pressure varies, as long as the compressor is at a constant speed.[1] As speed is reduced, the surge and choke lines intersect. Below this intersection, the traditional methods of surge protection (venting, recycling) are ineffective; only a quick raising of the compressor speed can bring the machine out of surge.

This is similar to the situation when the compressor is being started up. As the operating point is moving on the load curve (Figure 4.29), it must pass through the unstable region on the left of the surge curve as fast as possible in order to avoid damage from vibration.

The shape of the control line shown in Figure 4.25 is applicable only when the surge curve is parabolic. As shown in Figures 4.25, 4.26, and 4.27, surge curves can also be linear or have negative slopes and discontinuities. If the surge line is not parabolic, the following techniques can be considered for generating the surge control lines.

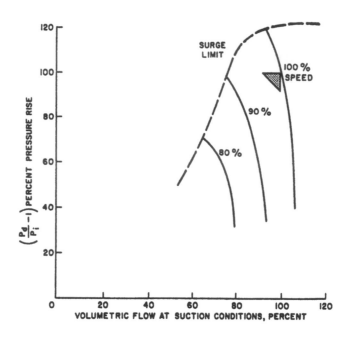

FIGURE 4.26 The surge line for an axial compressor goes through a slope reversal at higher flows. (From McMillan, G.K., *Centrifugal and Axial Compressor Control*, Instrument Society of America, 1983. With permission.)

If the surge curve is linear, the signal h should be replaced by a signal representing flow. This can be obtained by adding a square root extractor to the FT in Figure 4.25. If the surge curve bends down, the use of two square root extractors in series[1] can be used to approximate its shape. On multistage machines with high compression ratios, it might be necessary to substitute a signal characterizer for ΔPY in Figure 4.25. On variable vane designs, it has been proposed that the flowmeter differential (h) could be corrected to be a function of actual vane position to better approximate the surge curve[3] (see Figure 4.30). If a flowmeter differential cannot be obtained from either the suction or the discharge side of the compressor, the surge curve can also be approximated on the basis of speed, power, or vane position measurements.[3]

c. Flow measurement

The two critical components of a surge control loop are the flow sensor and the surge valve. Both must possess speed and reliability.

Flow oscillations under surge conditions occur on a cycle of little more than a second. The cycle starts with a sudden drop of pressure. The flow transmitter should be fast enough to detect this. The time constants of various transmitter designs are as follows:[1]

Pneumatic with damping: Up to 16 seconds
Electronic d/p: 0.2 to 1.7 seconds
Diffused silicone d/p: Down to 0.005 seconds

Only the diffused silicone design is fast enough to follow the precipitous flow drop at the start of the surge cycle or the oscillations during surge.

Measurement noise is another serious concern, because it necessitates a greater margin between the surge and the control lines. Noise can be minimized by the use of 20 pipe diameters

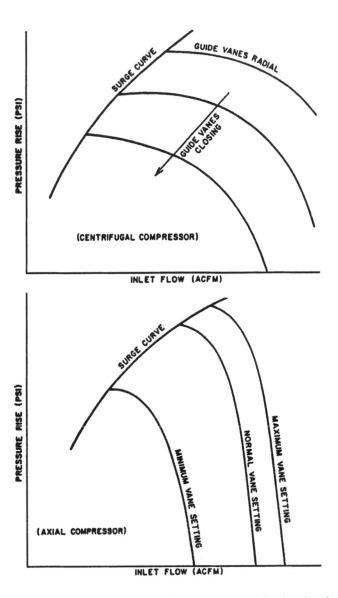

FIGURE 4.27 The surge curve bends over when guide vanes are used for throttling.[1]

of upstream and 5 diameters of downstream straight runs around the streamlined flow tube type sensor. Noise will also be reduced if the low-pressure tap of the d/p cell is connected to a piezometric ring in the flow tube. The addition of straightening vanes will also contribute to the reduction of noise.

Flow for anti-surge control is usually detected on the suction side of the compressor. If good, noise-free flow measurement cannot be obtained on that side, a corrected discharge side differential pressure reading (h_d) can be substituted. The suction side differential pressure (h_s) reading can be obtained from a reading of (h_d) with the following corrections for the suction and discharge pressures (P_s and P_d) and temperatures (T_s and T_d):

$$h_s = h_d(P_d/P_s)(T_s/T_d) \tag{3}$$

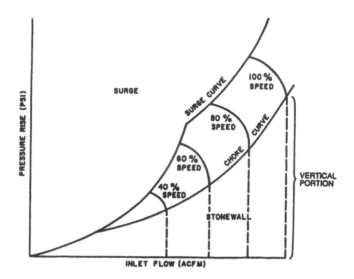

FIGURE 4.28 With multistage compressors, operation must be confined between the surge and choke curves. (From McMillan, G.K., *Centrifugal and Axial Compressor Control*, Instrument Society of America, 1983. With permission.)

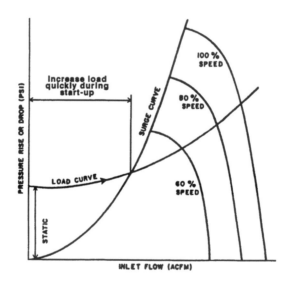

FIGURE 4.29 The speed and surge curves are characteristics of the compressor, while the load curve is a function of the load.[1]

d. Surge control valves

Surge valves should be located in the compressor bypass (Figure 4.31), should fail open, have linear characteristics, and, whenever possible, should be provided with volume booster (rather than positioner) for fast and precise control. In a surge application, the time constant of the valve is usually the largest time constant in the loop, and therefore, the surge valve must be made to throttle quickly in *either direction*. It is desirable to cut the sum of prestroke dead time and full stroking time of the valve to under 1 second. Valves, as large as 18 in. (46 cm) have been made to throttle to any position in under a half second.[4] Even faster performance (0.1 second stroke without

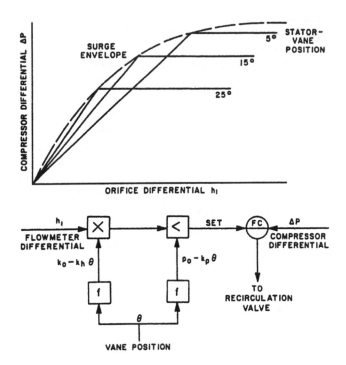

FIGURE 4.30   When the surge curve bends over, the control model should include vane position.[3]

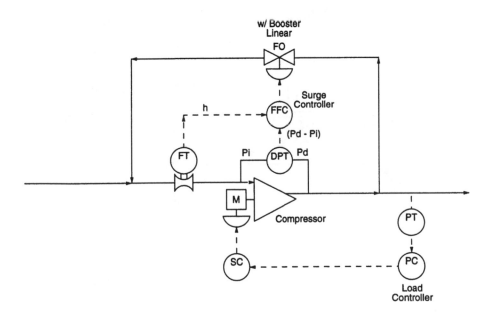

FIGURE 4.31   Surge valves must throttle quickly in either direction.

overshoot) can be obtained from digital valves, but their high cost, difficult maintenance, and ease of plugging limits their use.

The use of butterfly or eccentric disc valves is not recommended because of their high breakaway torque, which drops drastically as the valve opens. If the valve does not have a positioner, the opening is likely to be first delayed and then be followed by an overshoot as the torque drops. This makes the loop unstable.

The use of quick-exhaust relays (to speed the opening of the surge valves) is not recommended either, because they result in a similarly large overshoot as the valve starts to open, which is followed by slow throttling and erratic stroking.[4]

The surge valve should be sized to pass the full compressor capacity at 70% of the full compressor discharge pressure. The reduction in sizing pressure is recommended because, during surge episodes, the flow reversals cause a drop in pressure. Since the valve is the weak link in the loop, and because testing or maintenance can seldom be performed on-line, for critical installations, total redundancy (two surge valves in parallel) is recommended.

Surge Valves without Positioners — The positioner is a position controller. As such, it is the cascade slave of the controller that throttles the valve. Because of its small output air-flow capacity, the positioner slows the valve response. If the positioner is slower than the surge controller, the positioner will deteriorate the overall control quality of the loop. Therefore, in general, one would try to avoid the use of positioners on surge valves.

Figure 4.32 illustrates the design of a surge valve without a positioner. In such a design, the booster air supply regulator capacity should exceed the capacity of the booster. The booster should have high inlet and low outlet sensitivity for fast throttling. Even with an outlet sensitivity of 1 in. H_2O (25 mm H_2O) or less, the booster dead band (minimum input change that will have an effect on the output) can be 0.25 PSI (17 millibar) or more. To minimize booster dead band, a 6 to 30 PSIG (0.4 to 2.0 bar) I/P air signal is preferred.

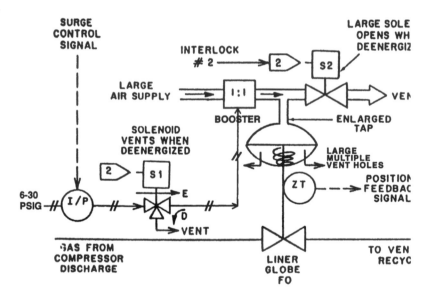

FIGURE 4.32 Surge valve without positioner.

The booster capacity is to be selected to fully stroke the valve in 0.5 seconds and it should be close-coupled to the valve to minimize the length of the large diameter pipe between it and the enlarged tap. The spring side of the actuator should be provided with multiple vent holes to minimize pressure buildup as the spring is compressed.

Surge Valves with Positioners — In the following applications the use of positioners is unavoidable: (1) to reduce valve hysteresis caused by packing (such as Grafoil) friction, (2) to reduce the dead band of the air booster used, (3) to reduce disc flutter in butterfly valves, (4) to reduce overshoot with butterfly or eccentric disc valves, which are not provided with torque compensation, and (5) when piston actuators are used.

Under these conditions, it is necessary to use positioners. This will require the detuning (lowering the gain or increasing the proportional band) of the surge controller. Because the air capacity of the positioner is too low to quickly change the pressure in the valve actuator, a volume booster needs to be added to its output. Unfortunately, a series pairing of a positioner and booster combination will be unstable (go into a fast limit cycle of 20% amplitude and 1 Hz frequency[4]) because the positioner is able to change the pressure in the booster inlet chamber faster than the booster can change the air pressure in the large actuator. To eliminate this limit cycle, a bypass around the booster with an adjustable restriction must be added (Figure 4.33).

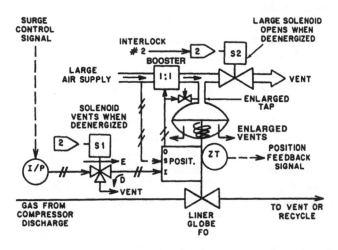

FIGURE 4.33 A well-designed surge control valve will have these features and accessories.[1]

Figure 4.33 shows some of the main components of a surge control valve.[1] The electronic control signal is converted to a pneumatic one by the I/P converter, and this pneumatic signal is sent to the positioner through a three-way solenoid (S1). The output signal from the positioner is sent simultaneously into a booster relay (1:1) and an adjustable bypass. The output from the booster relay is connected by large-diameter, short tubes to the valve actuator with an enlarged pressure tap and to the large venting solenoid (S2). Some of the operational features of the system are as follows:

When interlock #2 is de-energized, requiring instantaneous full opening of the surge valve, S1 vents the positioner inlet and S2 vents the actuator topwork. Because S2 is large, air removal is fast and the valve opens quickly. On return to normal, both solenoids are energized. This closes S2 and opens S1 to the control signal.

All surge valves should be throttle-tested before shipment. It is also desirable to monitor the surge valve opening through the use of a position transmitter.

e. Surge flow controller

The surge controller (FIC) in Figure 4.25 is a direct-acting controller with proportional and integral action and with antireset wind-up (ARW) features. It is a flow (or flow ratio) controller with a narrower proportional band than would be used for flow control alone. In a well-tuned surge controller, both the proportional band and the integral settings must be minimized (PB ~ 50%, 1 to 3 sec/repeat). With this level of responsiveness, the FIC set point can be 10% from the surge curve, as shown in Figure 4.25.

Under normal conditions, the actual flow (h) is much higher than the surge control limit (SP), and therefore the integral mode of the FIC has a tendency to wind up on its positive error toward a saturated maximum output (exceeding the 100% output at which the FO valve is fully closed). If this was allowed to happen, the controller could not respond to a surge condition (a quick drop in h) because it would first need to develop a negative area of error equal to that which caused the saturation.

The aim of an antireset wind-up (ARW) feature in this controller is to hold the FIC output under normal conditions at around 105%, so that the valve is closed but the signal is just lingering above the 100% mark without saturation. As soon as h starts dropping toward the set point, the valve should start to open, and it should reach full opening *before* the set point is reached. As soon as h starts dropping, the proportional contribution to the FIC output decreases. If the approach is slow, the increase in reset contribution will be greater than the decrease caused by the proportional contribution. Therefore, if h does not drop to SP, the valve slowly returns to the closed position. The effect of the internal feedback in the FIC is to activate the integral action only when the valve is *not* closed, while keeping the proportional action operational all the time.

The main goal of the feedback surge controller is to protect the machine from going into surge. Once the compressor is in surge, the FIC is not likely to be able to bring it out of it, because the surge oscillations are too fast for the controller to keep up with them. This is why backup systems are needed.

The surge controller is usually electronic, and the more recent installations tend to be microprocessor based. Such digital units can memorize complex, nonlinear surge curves and can also provide adaptive gain, which is a means of increasing controller gain as the operating point approaches the surge curve. In digital controllers, it is desirable to set the sample time at about one tenth of the period of surge oscillation, or at 0.1 seconds, whichever is shorter. If a flow-derivative backup control is used, the sample time must be 0.05 seconds or less, because if it is slower, the backup system would miss the precipitous drop in flow as surge is beginning.

ARW or Batch Switch — In surge applications, the antireset wind-up (ARW) feature is provided by a batch switch (Figure 4.34). It switches the feedback (F) from the controller output (O) to the preload (PL) whenever the output to the valve is more than what it takes to close it. In the case of a fail-open (FO) valve, a 100% output signal will keep the valve closed; therefore, HL is set at 100%, and the batch unit switches the feedback at that point to equal the preload (F = PL = 105%).

This way, as long as the compressor capacity is well above the surge point (E > PB(PL-HL)/100), the batch unit holds the controller output at or near the trip point (HL = 100%). When the compressor capacity drops to the controller set point (usually set at 10% above the surge point), and therefore the error drops to zero, the output will immediately start to open the valve (by dropping its value below 100%), as the batch unit transfers the feedback to the controller output (O).

Tuning the Controller — The *closed loop ultimate oscillation* method cannot be used for tuning the surge loop because there is a danger of this oscillation triggering the start of surge. The *open loop reaction curve* technique of tuning is used with both the surge and the throughput controller in manual. The output of each controller is changed by 5 or 10%, and the resulting measurement response is analyzed on a high-speed recorder, yielding the time constant and the dead time of the process. The tuning settings of the surge controller can be tested by first increasing the bias (b in Figure 4.25) until the surge valve opens and then analyzing the loop response. The FIC must be so tuned that its overshoot is *less* than half the distance between the control and surge lines in Figure 4.25. The throughput controller (PC) should be kept in automatic while the surge controller is tested (Figure 4.31) so that interaction problems will be noted. Once the surge controller is tuned, it should not be switched to manual but should always remain in automatic.

The ARW feature is not effective on slow disturbances, because the integral contribution that increases the FIC output outweighs the proportional contribution that lowers it; thus, the valve stays closed until the surge line is crossed. In order for the ARW to be effective, the time for the controller error to drop from its initial value to zero (T) must fall within two limits. It must be slower than the stroking time of the surge valve (Tv), but it must be faster than two integral times of the controller:[1]

$$Tv < T < 2Ti \qquad\qquad (4)$$

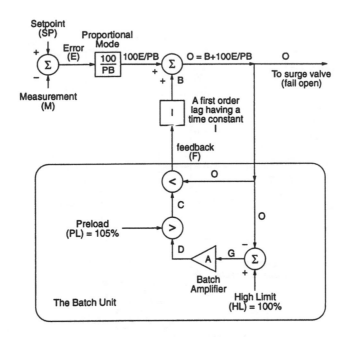

FIGURE 4.34 Batch switch holds valve output at O = 105% as long as the compressor is safely away from surge.

If the above requirement cannot be satisfied—because, for example, one of the potential disturbances is a slow-closing discharge valve—an "anticipator" control loop should be added, which detects the drop in or disappearance of the load and opens the surge valve in a feedforward manner.

Surge Protection Backup — As was shown in Figure 4.25, a parabolic surge curve with a positive slope will appear as a straight line on a ΔP versus h plot. The purpose of surge control is to establish a surge control line to the right of the actual surge line, so that corrective action can be taken *before* the machine goes into surge. Such a control system is shown in Figure 4.35.

The biased Surge control line is implemented through a biased ratio relay (ΔPY), which generates the set point for FIC as follows:

$$SP = m(\Delta P) + b \tag{5}$$

where SP = desired value of h in inches H_2O, m = slope of the surge line at the operating point, ΔP = compressor pressure rise in PSI, and b = bias of the surge set point in inches H_2O.

The offset between the surge and the control lines should be as small as possible for maximum efficiency, but it must be large enough to give time to correct upsets without violating the surge line.[6] The slower the upsets and the faster the control loop, the less offset is required for safe operation. In a good design, the bias b is about 10%, but as disturbances get faster or instruments slower, it can grow to 20%.

A second line of defense is the backup interlock FSL. It is normally inactive, as the value of h is normally above the FIC set point SP. When the surge controller FIC is not fast enough to correct a disturbance, h will drop below SP. FSL is set to actuate when h has fallen 5% below SP, and at that point it fully opens the surge valve through interlock #1.

The purpose of FIC in Figure 4.35 is to protect the compressor from going into surge. The feedback surge controller is usually not fast enough to bring the machine out of surge once it has developed because the surge oscillations are too fast for this controller. Yet even with the best

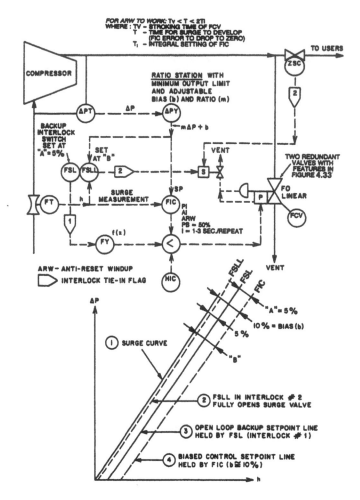

FIGURE 4.35 The illustrated surge control loops and lines are used to protect centrifugal machines at low compression ratios.[1]

feedback surge controller design, the machine will go into surge. This can be intentionally initiated to verify the surge curve, it can be caused by errors or shifts in the surge curve as the unit ages, and it can occur because of misoperation or failure of either the controller or downstream block and check valves. For these reasons, a second line of defense—a backup system—is required in addition to the surge FIC.

The backup system usually consists of two subsystems, identified as interlocks #1 and #2 in Figure 4.35. In interlock #2, FSLL detects conditions that will cause the complete stopping of all forward flow from the compressor; when such a condition is detected, FSLL instantaneously and fully opens the surge valve. One such condition in Figure 4.35 is signaled by a closed position limit switch (ZSC) on the block valve.

Interlock #1 in Figure 4.35 provides open-loop backup as follows: the FSL detects an approach to surge that is closer than the FIC set point; when it detects such a condition, it takes corrective action. Under normal conditions, FSL senses an h value that is higher than SP, meaning that the operating point is to the right of curve (4). As surge approaches, the operating point crosses the control line (curve (4)) as h drops below SP. The FSL is set to actuate interlock #1 when h has crossed curve (3). This occurs when the flow has dropped below the FIC set point by the amount "A" (usually 5%).

When interlock #1 is actuated, it triggers the signal generator FY to drop its output signal to zero and then, when the operating point has returned to the right of curve (3), gradually increase it back to 100%. When the FY output drops to zero, the low signal selector (in the output of the FIC) will select it and send it to the valve. This causes the surge valve to open fully in less than a second, followed by a slow closure according to the program in the signal generator FY. As the FY output rises, it will reach the output of FIC, at which point control is returned from the backup system to the feedback controller.

This type of backup is called the *operating point method*. Its advantage is that it takes corrective action *before* the surge curve (curve (1)) is reached. Its disadvantages are that it is not usable on multistage machines or on systems with recycle and that the protection it provides is lost if ΔPY in Figure 4.35 fails by dropping its output to zero. For this reason, it is desirable to provide a minimum limit on the output of ΔPY.

The manual loader input (HIC) to the low signal selector is used during start-up. Because its output passes through the low selector, it can increase but cannot decrease the valve opening.

Flow-Derivative Backup — The difference between the operating point technique of backup (Figure 4.35) and the flow-derivative method (Figure 4.36) is that the first method acts before the surge curve is reached, whereas the second is activated by the beginning of surge. Because flow drops off very quickly at the beginning of surge, the instruments that measure the rate of this drop must be very fast. If implemented digitally, the sample time of the flow derivative loop must be 0.05 seconds or less in order not to miss the initial drop in flow. The lead-lag station (L/L) that detects the rate of flow reduction is adjusted so that the lag setting will filter out noise and the lead setting will give a large output when the flow drops.

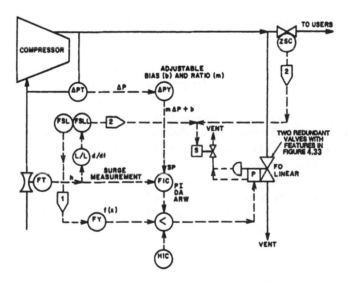

FIGURE 4.36   In the flow-derivative method of backup, the sample time must be 0.05 seconds or less.[1]

The function of interlock #1 is the same as was described in connection with Figure 4.35. When FSL detects the drop in flow, it causes the signal generator (FY) output to drop to zero, fully opening the valve within a second. If this action succeeds in arresting the surge, the signal generator output slowly rises and control is returned to the feedback FIC.

If interlock #1 does not succeed in bringing the machine out of surge, then after a preset number of oscillations FSLL is actuated. This triggers interlock #2, which keeps the surge valve fully open until surge oscillations stop.

The flow derivation method of backup has the advantage of providing protection even if the compressor pressure-rise instruments (ΔPT, ΔPY) fail, but it also has the disadvantage of not being usable when the flow signal is noisy.

f. Override controls

Figures 4.35 and 4.36 show backup and overrides implemented by low signal selectors on the outlet of the feedback FIC. Figure 4.37 shows some additional overrides, which might be desirable in some installations. Under normal conditions all override signals are above 100%, and they leave the feedback FIC in control. Sending the override signals through the low signal selector guarantees that they can only increase the opening of the surge valve, not decrease it.

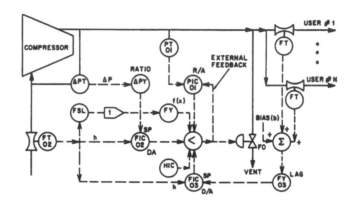

FIGURE 4.37 This control system includes high pressure and user shutdown overrides.[1]

The high-pressure override controller (PIC-01) protects downstream equipment from overpressure damage. Normally its measurement is far below its set point and therefore its output signal is high. In order to prevent reset wind-up, external feedback is used.

The other override loop (FIC-03) guarantees that the shutdown of a user or sudden change in user demand will not upset the mass balance (balance between inlet and outlet flow) around the compressor. This is an important feature, because without it the resulting domino effects could shut down other user reactors or the whole plant.

In Figure 4.37, when a user is shut down, FT-02 detects the drop in total flow within milliseconds, if a fast flow transmitter is used. This drop in flow reduces the measurement signal (h) to FIC-03, while its set point is still unaltered because of the lag (FY-03). Therefore, the surge valve is opened to maintain the mass balance. As FY-03 slowly lowers the set point to the new user flow rate (plus bias), FIC-03 recloses the valve. The bias b is used to keep the set point of FIC-03 above its measurement, thus keeping the surge valve closed when conditions are normal.

Figure 4.38 illustrates another method of maintaining mass balance as users are suddenly turned off. The velocity limiter (VY-03) limits the rate at which the surge signal (h) can be lowered. If the surge signal drops off suddenly, the VY-03 output remains high and, through the high signal selector, keeps the surge controller set point up. This opens the surge valve to maintain the mass balance until it is reclosed by the limiter. This method is called an *anticipator algorithm* because it opens the surge valve sooner than the time when the operating point (h) would otherwise drop to the set point, computed by the ratio station (ΔPY).

g. Installation considerations

A check valve in the discharge line as close to the compressor as possible will protect it from surges. On motor-driven compressors, it is helpful to close the suction valve during starting to prevent overload of the motor. After the unit is operating, it should be brought to stable operating

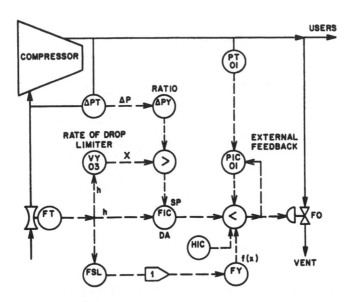

FIGURE 4.38 Velocity limiter is one type of protection that can be provided against sudden load changes.[7] (Sudden drop in h does not immediately lower x, which the high selector picks as set point. This temporarily opens surge valve.)

range as soon as possible to prevent overheating. The recycle valve control should be fast opening and slow closing to come out of surge quickly and then stabilize the flow. The transmitter ranges should be such that the recycle valve is able to open in a reasonable amount of time with reasonable proportional and integral control constants of the antisurge controller. The importance of proper controller tuning cannot be underestimated. Typical settings are PB = 50%, I = 1 to 3 sec/repeat.

h. Multiple compressor systems

Compressors are connected in series to increase their discharge pressure (compression ratio) or they are connected in parallel to increase their flow capacity. Series compressors on the same shaft can sometimes be protected by an overall surge bypass valve common to all surge controllers through the use of a low signal selector.[3] When driven by different shafts, they require separate anti-surge systems. An overall bypass eliminates the interaction that otherwise would occur between surge valves, as the opening of a bypass around one stage not only would increase the flow through the higher stages but would also decrease it through the lower ones. In some installations, this interaction has been found to be of less serious consequence than the time delay caused by the use of a single overall bypass surge valve, which cannot quickly increase flow in the upper stages.[1] In such installations it is best to duplicate the complete surge controls around each compressor in series.

If streams are extracted or injected between the compressors in series and their flows are not equal, a control loop needs to be added to keep all compressors away from surge by automatically distributing the total required compression ratio between them. Such a control system is shown in Figure 4.39. In this system, PIC-01 controls the total discharge pressure by adjusting the speed of compressor #1. The speed of compressor #2 is set by FFIC-02; both compressors are thus kept at equal distances from their respective surge lines. This is accomplished by maintaining the following equality:

$$(K_1(P_1)(\Delta P_1)/(T_1)) + B_1 = (K_2(P_2)(\Delta P_2)/(T_2)) + B_2 \tag{6}$$

In this equation, K is the slope and B is the bias of the surge set points of the respective compressors.

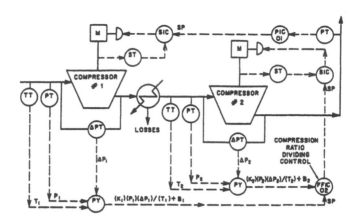

FIGURE 4.39 The compression ratio distribution is controlled between two compressors when they operate in a series.[1]

Parallel Control — Controlling two or more compressors operating in parallel and having identical characteristics would be relatively simple. It is very difficult, if not impossible, to find two compressors having identical performance characteristics. Slight variations in flow can cause one compressor to be fully loaded. The parallel machine then has useless recycle. The control scheme shown in Figure 4.40 alleviates that problem. Typically, the suction valve that receives the lower flow is kept 100% open. This prevents both suction valves from going fully closed to balance the flow.

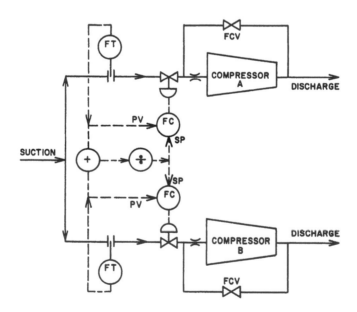

FIGURE 4.40 Multi-inlet compressor control.

Figure 4.41 illustrates how two compressors can be proportionally loaded and unloaded, while keeping their operating points at equal distance from the surge curve. The lead compressor (#1 in Figure 4.41) is selected either as the larger unit or as the one that is closer to the surge curve when the load rises or is further from it when the load drops. In Figure 4.41 it is assumed that the compressors were so chosen that their ratio of bias to slope (b/K) at the surge set point is equal. In that case,

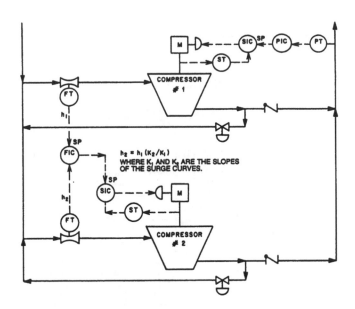

FIGURE 4.41 Two parallel compressors can be proportionally loaded.[1]

$$h_2 = h_1(K_2/K_1) \tag{7}$$

where h is the flowmeter differential and K is the surge set point slope of the respective compressors.

Because of age, wear, or design differences, no two compressors are identical. A change in load will not affect them equally and each should therefore be provided with its own anti-surge system. Another reason for individual surge protection is that check valves are used to prevent backflow into idle compressors. Therefore, the only way to start up an idle unit is to let it build up its discharge head while its surge valve is partially open. If this is not done and the unit is started against the head of the operating compressors, it will surge immediately. The reason why the surge valve is usually not opened fully during start-up is to protect the motor from overloading.

Improper distribution of the load is prevented by measuring the total load and assigning an adjustable percentage of it to each compressor by adjusting the set points of flow ratio controllers.

III. OPTIMIZING CONTROLS

A. ADAPTING THE SURGE CURGE

The surge curve shifts with wear and with operating conditions. Figure 4.42 describes a surge control loop that recognizes such shifts and automatically adapts to the new curve. The adaptation subroutine consists of two segments, the set point adaptation section (blocks ① to ⑥ and the output backup section (blocks⑦to ⑪).

The purpose of the output backup section is to recognize the approach of a surge condition that the feedback controller (FIC) was unable to arrest and to correct such a condition when it occurs. As long as the surge measurement (h) is not below the set point (SP), no corrective action is needed, and therefore blocks ⑦ and ⑨ will set the FIC output modifier signal (A) to zero. Once h is below SP, the operating point is to the left of curve ④ in Figure 4.35, and the output of block ⑦ in Figure 4.42 is switched to "Yes." Next block ⑧ checks if the adjustable time delay D1 — having typical values of 0.3 to 0.8 seconds[1] — has passed. If it has not, signal A remains at its last value. If D1 has passed, A is increased by an increment X. Typical values of the X increment range

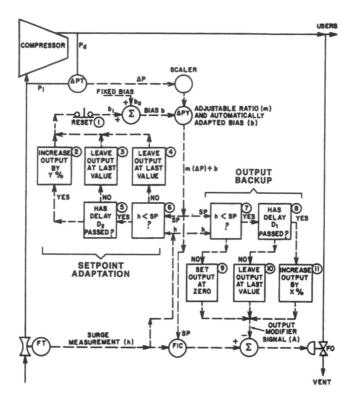

FIGURE 4.42 This control system recognizes changes in the surge curve and adapts the surge controller to the new curve. (Adapted from Reference 1, describing U.S. Patent 856.302, owned by Naum Staroselsky)

from 15 to 30%.[1] As A is subtracted from the FIC output signal, this backup loop will open up the surge valve at a speed of 15 to 30% per 0.3 to 0.8 second. When h is restored to above set point, the signal A is slowly returned to zero, allowing the surge valve to reclose.

The purpose of the set point adaptation loop (blocks ① to ⑥) is to recognize shifts in the surge curve and, as a response, to move the control line ④ in Figure 4.35 to the right by increasing total bias b of the ratio relay ΔPY. The logic of blocks 2 to 6 is similar to that described for blocks ⑦ to ⑪, except that the resulting variable bias signal (b_1) is added to the fixed bias (b_0) to arrive at the adapted new bias (b). The speed of set point adaptation does not need to be as fast as the opening of the surge valve. Therefore, the time delay D2 tends to be longer than D1, and the increment Y is smaller than X. The purpose of the reset button in block ① is to provide a means for the operator to reset the b_1 signal back to zero.

B. Optimized Load-Following

When a compressor is suppling gas to several parallel users, the goal of optimization is to satisfy all users with the minimum investment of energy. The minimum required header pressure is found by the valve position controller (VPC) in Figure 4.43. It compares the highest user valve opening (the needs of the most demanding user) with a set point of 90%, and even if the most open valve is not yet 90% open, it lowers the header pressure set point. This supply-demand matching strategy not only minimizes the use of compressor power but also protects the users from being undersupplied, because it protects any and all supply valves from the need to be open fully. On the other hand, as the VPC causes the user valves to open farther, these valves not only reduce their pressure drops but also become more stable.

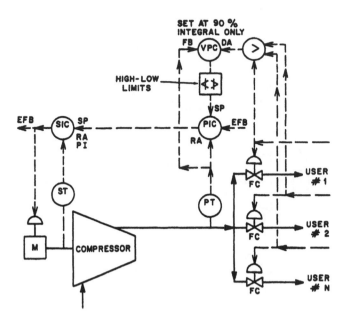

FIGURE 4.43 Optimized load-following controls are designed to satisfy all users with a minimum energy investment.

The VPC is an integral-only controller, with its integral time set for about 10 times that of the PIC. This guarantees that the VPC will act more slowly than all the user controllers, thus giving stable control even if the user valves are unstable. The external feedback (FB) protects the VPC from reset wind-up when its output is limited or when the PIC has been switched to manual.

In addition to following the load, it is also necessary to protect the equipment. In Figure 4.44, one protective override prevents the development of excessively low suction pressures (PIC-02), which could result in drawing oil into the compressor. The other override (KWIC-03) protects from overloading the drive motor and thereby tripping the circuit breaker.

In order to prevent reset wind-up when the controller output is blocked from affecting the SIC set point, external feedback (EFB) is provided for all three controllers. This arrangement is typical for all selective or selective-cascade control systems.

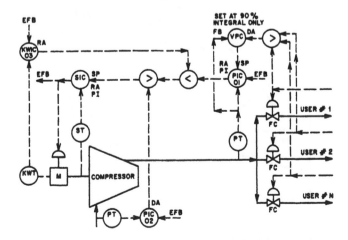

FIGURE 4.44 Protective overrides can be added to optimized load following controls.

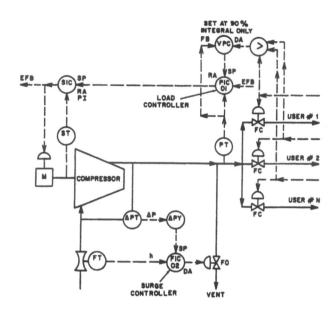

FIGURE 4.45   The load and surge controllers are both affecting compressor throughput.

C. Interaction and Decoupling

The load and surge control loops together, as shown in Figure 4.45, both affect the same variable: compressor throughput. Under normal conditions, there is no problem of interaction; because the surge loop is inactive, its valve is closed. Yet, when point "A" in Figure 4.46 is reached, FIC-02 quickly opens the surge valve, which causes the discharge pressure to drop off as the flow increases. PIC-01 responds by increasing the speed of the machine. The faster the PIC-01 loop instrumentation and the higher its gain (the narrower the proportional band), the larger and faster will be the increase in speed of the compressor. If the load curve is steeper than the surge curve (as in Figure 4.46),

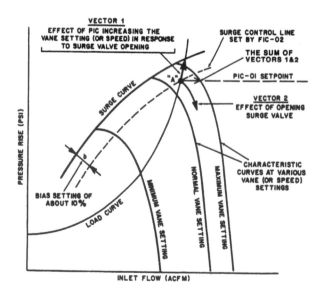

FIGURE 4.46   Load and surge loops conflict if the load curve is steeper than the surge curve and if the controlled variable is pressure.[1]

the action of PIC-01 will bring the operating point closer to the surge line. In this case, a conflict exists between the two loops, because as FIC-02 acts to correct an approaching surge situation, PIC-01 responds by worsening it. The better tuned (narrower proportional band) and faster (electronic or hydraulic speed governors) the PIC-01 loop is, the more dangerous is its effect of worsening the approach of surge.

The throughput controller (PIC-01 in Figure 4.45) moves the operating point on the path of the load curve. The surge controller (FIC-02 in Figure 4.45) moves the operating point on the path of the characteristic curve at constant speed or vane settings. If the characteristic curve is steep, the action of the surge controller will cause a substantial upset in the discharge pressure controlled by PIC-01 (Figure 4.45). If the load curve is flat, the effect of the pressure controller on the surge controller will be the greatest.

If the throughput is under flow control, the opposite effects will be observed. The effect of the surge controller on the flow controller will be the greatest when the characteristic curves of the compressor are flat, and the effect of the flow controller on the surge controller will be the greatest when the load curve is steep.

In determining whether the loops will assist or conflict with each other, both the relative slopes of the load and surge curves and the variable selected for process control need to be considered:

Controlled Variable	Load Curve Steeper than Surge Curve	Nature of Interaction
Pressure	Yes	Conflict
	No	Assist
Flow	Yes	Assist
	No	Conflict

As both loops try to position the operating point on the compressor map, the resulting interaction can cause the type of inverse response that was described in Figure 4.46, or it can cause oscillation and noise. The oscillating interaction is worst if the proportional bands and time constants or periods of oscillation are similar for the two loops. Therefore, if tight control is of no serious importance, one method of reducing interactions is to reduce the response (widen the proportional band) of the load controller. Similarly, the use of slower actuators (such as pneumatic ones) to control compressor throughput will also reduce interaction, but at the cost of less responsive overall load control.

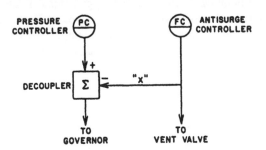

FIGURE 4.47 The opening of the vent valve is automatically converted into a proportional increase in speed to avoid upsetting pressure.[3]

1. Relative Gain

In evaluating the degree of conflict and interaction between the load and surge loops, it is desirable to calculate the relative gain between the two loops. The relative gain is the ratio between the two loops. The relative gain is the ratio between the open-loop gain when the other loop is in manual, divided by the open-loop gain when the other loop is in automatic. The open-loop gain of PIC-01 (Figure 4.45) is the ratio of the change in its output to a change in its input, which caused it. Therefore, if a 1% increase in pressure results in a 0.5% decrease in compressor speed, the open-loop gain is said to be –0.5. Assuming that the open-loop gain of PIC-01 is –0.5 when FIC-02 is in manual and that it drops to –0.25 when FIC-02 is in automatic, the relative gain is 0.5/0.25 = 2. The relative gain (RG) values can be interpreted as follows:

RG Value	Effect of Other Loop
0 to 1.0	Assists the primary loop
Above 1.0	Conflicts with the primary loop
Below 0	Conflict that also reverses the action of primary loop

If the calculated relative gain values are put in a 2-by-2 matrix, the best pairing of controlled and manipulated variables is selected by choosing those that will give the least amount of conflict. These are the RG values between 0.75 and 1.5, preferably close to 1.0.

2. Decoupling

Decoupling is provided to reduce the interaction between the surge and the load control loops. If the goal of decoupling is to maintain good pressure control even during a surge episode, the system shown in Figure 4.47 can be used. In this system, as the antisurge controller opens the vent valve, a feedforward signal (X) proportionately increases the speed of the compressor. The negative sign at the summing device is necessary because the surge valve is a fail open valve. If the two vectors shown in Figure 4.46 are correctly weighed in the summer, the end result of their summation will be a horizontal vector to the right, and PIC-01 will stay on set point.

On the other hand, if the main goal of decoupling is to temporarily reduce the size of vector #1 in Figure 4.46 so that it will not contribute to the worsening of the surge condition, then the half decoupling shown in Figure 4.48 can be considered. Here the decoupling summer also receives a feedforward signal (X) from the surge valve, but it acts to slow down (not speed up) the machine. As the output of FIC-02 drops, the compressor speed is temporarily lowered, because the added value of X is reduced. This brings the operating point further away from the surge line, as if in Figure 4.46 vector #1 were pointing downward. The lagged signal Y later on eliminates this bias, because when its time constant has been reached, the values of X and Y will be equal and will cancel each other. Therefore, this decoupler will serve to temporarily desensitize a fast and tightly tuned PIC-01 loop, which otherwise might worsen the situation by overreacting to a drop in pressure when the vent valve opens.

D. Optimized Load Distribution

The load distribution can be computer-optimized by calculating compressor efficiencies (in units of flow per unit power) and loading the units in the order of their efficiencies. The same goal can be achieved if the operator manually adjusts the ratio settings of the loading controllers.

In the control system of Figure 4.49, the pressure controller (PIC-01) directly sets the set points of SIC-01 and 02 while the balancing controllers (FFIC-01 and 02) slowly bias that setting. This is a more stable and responsive configuration than a pressure-flow cascade, because the time constants of the two loops are similar.

The output of PIC-01 must be corrected as compressors are started or stopped. One method of handling this is illustrated in Figure 4.49. Here a high-speed integrator (item #5) is used on the summed speed signals to assure a correspondence between the PIC output signal and the number

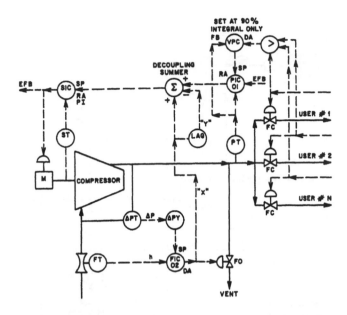

FIGURE 4.48 The interaction between surge and load control loops can also be decoupled by temporarily reducing speed when the surge valve opens.[1]

of compressors (and their loading) used. The integrator responds in a fraction of a second and therefore does not degrade the speed of response of the PIC loop.

Figure 4.49 also illustrates the automatic starting and stopping of individual compressors as the load varies. When the total flow can be handled by one fewer compressor or when any of the surge valves is open, FSL-03 triggers the shutdown logic interlock circuit #1 after a time delay. Operation with an open surge valve would be highly inefficient, because the recirculated gas is redistributed among all the operating units. When a compressor is to be stopped, item #15 (or #16) is switched to the stop position, causing the integrator #6 (or #14) to drive down until it overrides the control signal in low selector #3 (or #11) and reduces the speed until the unit is stopped.

Automatic starting of an individual compressor is also initiated by interlock #1 when PSH-04 signals that one of the compressors has reached full speed. When a compressor is to be started, interlock #1 switches item #15 (or #16) to the start position, causing the integrator #6 (or #14) to drive up (by applying supply voltage to the integrator) until PIC-01 takes over control through the low selector #3 (or #11). The integrator output will continue to rise and then will stay at maximum, so as not to interfere with the operation of the control loop. The ratio flow controllers (FFIC-01 and 02) are protected from reset wind-up by receiving external feedback signals through the low selector #7 (or #12), which selects the lower of the FFIC output and the ramp signal.

Interlock #1 is also provided with "rotating sequencer" logic, which serves to equalize run times between machines and protects the same machine from being started and stopped frequently. A simple approximation of these goals is achieved if the machine that operated the longest is stopped and the one that was idle the longest is started.

If only one of the compressors is variable-speed, the PIC-01 output signal can be used in a split-range manner. For example, if there were five compressors of equal capacity, switches would be set to start an additional constant-speed unit as the output signal rises above 20%, 40%, 60% and 80% of its full range. The speed setting of the one variable-speed compressor is obtained by subtracting from the PIC output the sum of the flows developed by the constant-speed machines and multiplying the remainder by five. The gain of five is the result of the capacity ratio between that of the individual compressor and the total.

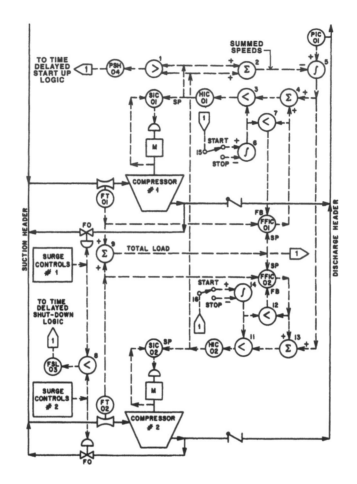

FIGURE 4.49   A flow-balancing bias can be superimposed on direct pressure control.[3]

IV. CONCLUSIONS

As was shown in Figure 2.36, the energy cost of operating a single compressor at 60% average loading can be cut in half by optimized variable-speed design. In the case of multiple compressor systems, savings can be obtained by optimized load-following and supply-demand matching.

The full automation of compressor stations—including automatic start-up and shutdown—not only will reduce operating cost but will also increase operating safety as human errors are eliminated. Figure 4.50 illustrates the use of microprocessor-based compressor unit controllers. Such units can be provided with all the algorithms that were described in this chapter; therefore, they provide flexibility by allowing changes in validity checks, alteration of limit stops, or reconfiguration of control loops.

V. TERMINOLOGY

h=   differential head (ft or m)
H=   polytropic compressor head (ft or m)
$K_{1,2,3}$= flow constant
m=   mass flow (lbs/hr or kg/h) = $c\sqrt{h/\zeta}$
n=   polytropic coefficient

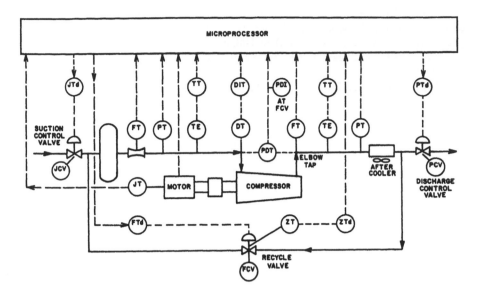

FIGURE 4.50   Microprocess control can be used for compressor optimization.

P_D= discharge pressure (PSIA or Pa)

P_I= inlet pressure (PSIA or Pa)

Q= volume flow rate (ACFH or m³/h) = $c\sqrt{h/\zeta}$

R= gas constant

T_I= inlet temperature

u= rotor tip speed (ft/sec or m/s)

W= weight flow (lbm/hr or kg/h)

Z= gas compressibility factor

τ= motor torque (ft-lbf or J)

ψ= head coefficient

ω= angular velocity (radians/hr)

ζ= density of gas (lbs/ft³ or kg/m³)

REFERENCES

1. McMillan, G.K., *Centrifugal and Axial Compressor Control*, Instrument Society of America, 1983.
2. Lipták, B.G., Integrating compressors into one system, *Chem. Eng.*, January 19, 1987.
3. Shinskey, F.G., *Energy Conservation Through Control*, Academic Press, 1978.
4. McMillan, G.K., *Tuning and Control Loop Performance*, Instrument Society of America, 1996.
5. *Mark's Standard Handbook for Mechanical Engineers*, Baumeister, T., ed., McGraw-Hill, New York, 1978.
6. Waggoner, R.C., Process Control for Compressors, paper given at Houston Conference of ISA, October 11–14, 1976.
7. Baker, D.F., Surge control for multistage centrifugal compressors, *Chem. Eng.*, May 31, 117—122, 1982.

BIBLIOGRAPHY

Chopey, N.P., *Handbook of Chemical Engineering Calculations*, 2nd ed., McGraw-Hill, New York, 1994.
Higgins, L.R., *Maintenance Engineering Handbook*, 5th ed., McGraw-Hill, New York, 1995.

Katz, D.L. and Lee, R.L., *Natural Gas Engineering*, McGraw-Hill, New York, 1990.

McMillan, G.K., *Tuning and Control Loop Performance*, 3rd ed., Instrument Society of America, 1994.

O'Callaghan, P.W., *Energy Management*, McGraw-Hill, New York, 1992.

Rammler, R., Advanced Centrifugal Compressor Control, ISA/93, Technical Conference in Chicago, September 19-24, 1993.

Shinskey, F.G., *Process Control Systems*, 4th ed., McGraw-Hill, New York, 1996.

5 Cooling Towers

I. THE PROCESS

Cooling towers are water-to-air heat exchangers that are used to discharge waste heat into the atmosphere. Their cost of operation is a function of the water and air transportation costs. The goal of cooling tower optimization is to maximize the amount of heat discharged into the atmosphere per unit of operating cost invested.

In this chapter the types of cooling tower design are described and the various techniques of load-following capacity controls are discussed. The optimization strategies discussed here are not limited to standard cooling towers but also include the evaporative types and the free cooling configurations. Cooling tower winterizing and blowdown controls are also covered.

A. DEFINITIONS

Before the control aspects of cooling towers can be discussed, some related terms must be defined:[1]

Approach   The difference between the wet-bulb temperature of the ambient air and the water temperature leaving the tower. The approach is a function of cooling tower capability; a larger cooling tower will produce a closer (smaller) approach (colder leaving water) for a given heat load, flow rate, and entering air condition (Figure 5.1). (Units: °F or °C.)

Blowdown   Water discharged to control concentration of impurities in circulated water. (Units: percentage of circulation rate.)

Cold Water Basin   A device underlying the tower to receive the cold water and direct its flow to the suction line or sump.

Counterflow Water Cooling Tower   A cooling tower in which airflow is in opposite direction from the fall of water through the water cooling tower.

Cross-Flow Water Cooling Tower   A cooling tower in which airflow through the fill is perpendicular to the plane of falling water.

Dew-Point Temperature (TDP)   The temperature at which condensation begins if air is cooled under constant pressure (see Figure 5.2).

Double-Flow Water Cooling Tower   A cross-flow tower with two fill sections and one plenum chamber that is common to both fill sections.

Drift   Water loss due to liquid droplets entrained in exhaust air. Usually under 0.2% of circulated flow rate.

Drift Eliminator   An assembly constructed of wood or honeycomb materials which serves to remove entrained moisture from the discharged air.

Dry-Bulb Temperature (TDB)   The temperature of air measured by a normal thermometer (see Figure 5.2).

Evaporation Loss   Water evaporated from the circulating water into the atmosphere in the cooling process. (Unit: percentage of total GMP.)

Fan-Drive Output   Actual power output (BHP) of drive to shaft.

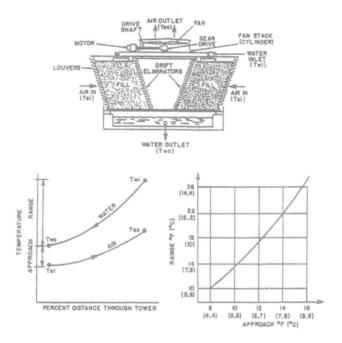

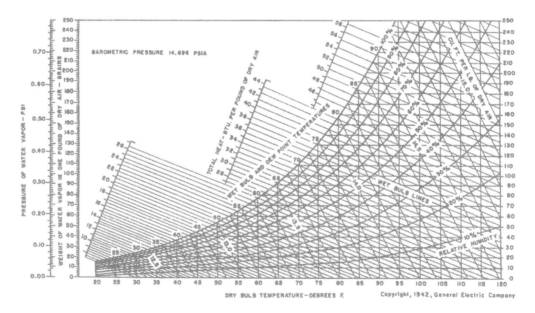

FIGURE 5.1   The mechanical cooling tower is a water-to-air heat exchanger. A direct relationship exists between water and air temperature in a counterflow cooling tower. An increase in range will cause an increase in approach, if all other conditions remain unaltered.

FIGURE 5.2   The psychrometric chart describes the properties of air.

$$\text{BHP} = \frac{(\text{motor efficiency})(\text{amps})(\text{volts})(\text{power factor})\ 1.73}{746} \tag{1}$$

Fan Pitch   The angle a fan blade makes with the plane of rotation. (Unit: degrees from horizontal.)

Fan Stack (Cylinder)   Cylindrical or modified cylindrical structure in which the fan operates. Fan cylinders are used on both induced draft and forced draft axial-flow propeller type fans.

Filling   That part of an evaporative tower consisting of splash bars, vertical sheets of various configurations, or honeycomb assemblies that are placed within the tower to effect heat and mass transfer between the circulating water and the air flowing through the tower.

Forced Draft Cooling Tower   A type of mechanical draft water cooling tower in which one or more fans are located at the air inlet to force air into the tower.

Heat Load   Heat removed from the circulating water within the tower. Heat load may be calculated from the range and the circulating water flow. (Unit: BTU per hour = GPM × 500 × [Tctwr-Tctws].)

Liquid Gas Ratio (L/G)   Mass ratio of water to airflow rates through the tower. (Units: lbs/lbs or kg/kg.)

Louvers   Assemblies installed on the air inlet faces of a tower to eliminate water splash-out.

Makeup   Water added to replace loss by evaporation, drift, blowdown, and leakage. (Unit: percentage of circulation rate.)

Mechanical Draft Water Cooling Tower   A tower through which air movement is effected by one or more fans. There are two general types of such towers: those that use forced draft with fans located at the air inlet and those that use induced draft with fans located at the air exhaust.

Natural Draft Water Cooling Tower (Air Movements)   A cooling tower in which air movement is essentially dependent upon the difference in density between the entering air and the internal air. As the heat of the water is transferred to the air passing through the tower, the warmed air tends to rise and draw in fresh air at the base of the tower.

Nominal Tonnage   One nominal ton corresponds to the transfer of 15,000 BTU/hr (4.4 kW) when water is cooled from 95 to 85°F (35 to 29.4°C) by ambient air having a wet-bulb temperature of 78°F (25.6°C) and when the water circulation rate is 3 GPM (11.3 lpm) per ton.

Performance   The measure of a cooling tower's ability to cool water. Usually expressed in gallons per minute cooled from a specified hot water temperature to a specified cold water temperature with a specific wet-bulb temperature. Typical performance curves for a cooling tower are shown in Figure 5.3. This is a "7500 GPM at 105-85-78" tower, meaning that it will cool 7500 GPM of water from 105 to 85°F when the ambient wet-bulb temperature is 78°F.

Plenum   The enclosed space between the eliminators and the fan stack in induced draft towers, or the enclosed space between the fan and the filling in forced draft towers.

Power Factor   The ratio of true power (watts) to the apparent power (amps × volts).

Psychrometer   An instrument used primarily to measure the wet-bulb temperature.

Range (Cooling Range)   The difference between the temperatures of water inlet and outlet, as shown in Figure 5.1. For a system operating in a steady state, the range is the same as the water temperature rise through the load heat exchanger. Accordingly, the range is determined by the heat load and water flow rate, not by the size or capability of the cooling tower. On the other hand, the range does affect the approach, and as the range increases, a corresponding increase in the approach will occur if all other factors remain unaltered (see Figure 5.1).

Riser   Piping that connects the circulating water supply line from the level of the base of the tower or the supply header to the tower inlet connection.

Standard Air   Dry air having a density of 0.075 lbs/cu. ft. at 70°F and 29.92 in. Hg.

Sump (Basin)   Lowest portion of the basin to which cold circulating water flows: usually the point of suction connection.

Tonnage   *See* Nominal Tonnage.

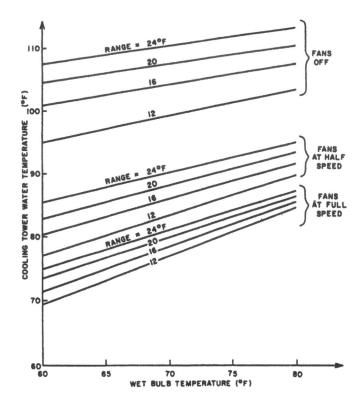

FIGURE 5.3 Performance curves describe a cooling tower's ability to cool water.

Total Pumping Head   The total head of water, measured above the basin curb, required to deliver the circulating water through the distribution system.

Tower Pumping Head   Same as the total pumping head minus the friction loss in the riser. It can be expressed as the total pressure at the centerline of the inlet pipe plus the vertical distance between the inlet centerline and the basin curb.

Water Loading   Water flow divided by effective horizontal wetted area of the tower. (Unit: GPM/ft² or m³/hr · m².)

Wet-Bulb Temperature (TWB)   If a thermometer bulb is covered by a wet, water-absorbing substance and is exposed to air, evaporation will cool the bulb to the wet-bulb temperature of the surrounding air. This is the temperature read by a psychrometer. If the air is saturated with water, the wet-bulb, dry-bulb, and dew-point temperatures will all be the same. Otherwise, the wet-bulb temperature is higher than the dew-point temperature but lower than the dry-bulb temperature (see Figure 5.2).

B. OPERATION

In a cooling tower, heat and mass transfer processes combine to cool the water. The mass transfer due to evaporation does consume water, but the amount of loss is only about 5% of the water requirements for equivalent once-through cooling by river water.

Cooling towers are capable of cooling to within 5 to 10°F (2.8 to 5.6°C) of the ambient wet-bulb temperature. The larger the cooling tower for a given set of water and airflow rates, the smaller this approach will be. The mass transfer contribution to the total cooling is illustrated in the psychrometric chart in Figure 5.4. Ambient air might enter in various conditions, such as illustrated by points A or D. After it transfers heat with and absorbs mass from the evaporating water, the air

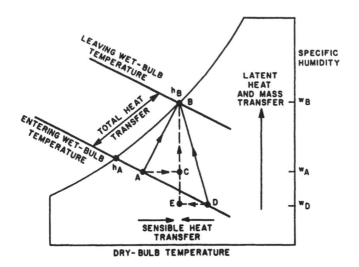

FIGURE 5.4   Heat and mass transfer processes combine to cool water.[2]

leaves in a saturated condition at point B. The total heat transferred in terms of the difference between entering and leaving enthalpies ($h_B - h_A$) is the same for both the AB and the DB process.

The AC component of the AB vector represents the heat transfer component (sensible air heating), and the CB vector represents the evaporative mass transfer component (latent air heating). If the air enters at condition D, the two component vectors do not complement each other, because only one of them heats the air (EB). Therefore, the total vector DB results from the latent heating (EB) and sensible cooling (DE) of the air. In terms of the water side of the DB process, the water is being sensibly heated by the air and cooled by evaporation. Therefore, as long as the entering air is not saturated, it is possible to cool the water with air that is warmer than the water.

C. COOLING PROCESS

The basic cooling tower designs are described in Figure 5.5. For operational purposes, tower characteristic curves for various wet-bulb temperatures, cooling ranges, and approaches can be plotted against the water flow to air-flow ratio as shown in the following equation:

$$\frac{KaV}{L} \sim \frac{(N)^n}{(G)} \qquad (2)$$

where K = overall unit of conductance, mass transfer between saturated air at mass water temperature, and main air stream: lb per hour (ft^2) (lb/lb), a = area of water interface per unit volume of tower: ft^2 per ft^3, V = active tower volume per unit area: ft^3 per ft^2, L = mass water flow rate: lb per hour, G = airflow rate: lb dry air per hour, and n = experimental coefficient: varies from –0.35 to –1.1 and has an average value of –0.55 to –0.65.

From these curves, it is possible to determine tower capability, effect of wet-bulb temperature, cooling range, water circulation, and air delivery, as shown in Figure 5.6. The ratio KaV/L is determined from test data on hot water and cold water, wet-bulb temperature, and the ratio of the water flow to the airflow.

The thermal capability of cooling towers for air conditioning applications is usually stated in terms of nominal tonnage based on heat dissipation of 4.4 kW (15,000 BTU/h) and a water circulation rate of 3.24 l/s per kW (3 GPM per ton) cooled from 95 to 85°F (35 to 29.4°C) at 78°F (25.6°C) wet-bulb temperature. It may be noted that the subject tower would be capable of handling

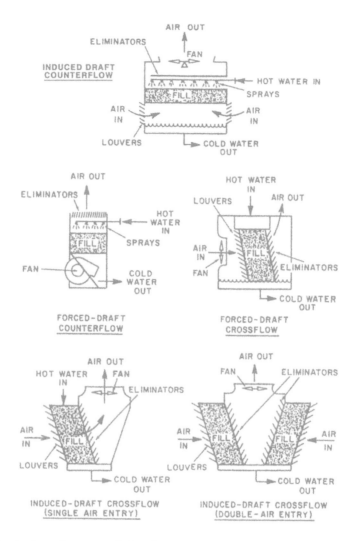

FIGURE 5.5   Five basic cooling tower designs.[2]

a greater heat load (flow rate) when operating in a lower ambient wet-bulb region. For operation at other flow rates, tower manufacturers usually provide performance curves covering a range of at least 2 to 5 GPM per nominal ton (2.16 to 5.4 l/s per kW).

The initial investment required for a cooling tower is essentially a function of the required water flow rate, but this cost is also influenced by the design criteria for approach, range, and wet bulb. As shown in Figure 5.7, cost tends to increase with range and to decrease with a rise in approach or wet bulb. The initial investment is about $35 per GPM of tower capacity, whereas the energy cost of operation is approximately 0.01 BHP/GPM, or about $7 per year per GPM. This means that the cost of operation reaches the cost of initial investment in five to ten years, depending on energy costs.

II. BASIC CONTROLS

The cooling tower is an air-to-water heat exchanger. The controlled variables are the supply and return water temperatures, and the manipulated variables are the air and water flow rates.

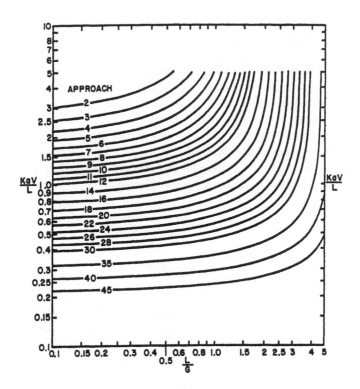

FIGURE 5.6   Characteristic curves of cooling towers (based on 64°F [17.8°C] air wet-bulb temperature and 18°F [10°C] cooling range).

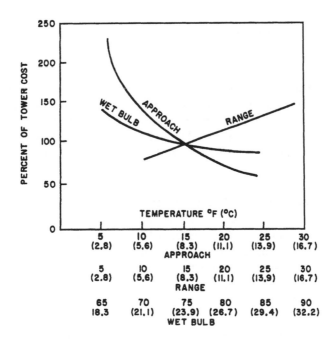

FIGURE 5.7   Comparative cooling-tower costs (Courtesy of Foster Wheeler Corp.)

Manipulation of these flow rates can be continuous, through the use of variable speed fans and pumps, or incremental, through the cycling of single- or multiple-speed units.

If the temperature of the water supplied by the cooling tower is controlled by modulating the fans, a change in the cooling load will result in a change of the operating level of the fans. If the fans are single-speed devices, the only control available is to cycle the fan units on or off as a function of load, as illustrated in Figure 5.8. This can be done by simple sequencers or by more sophisticated digital systems. In most locations, the need to operate all fans at full speed applies for only a few thousand hours every year. Fans running at half speed consume approximately one-seventh of design air horsepower but produce over 50 percent of the design air rate (cooling effect). Because 98 percent of the time the entering air temperature is less than the design, two-speed motors are a wise investment for minimizing operating costs.

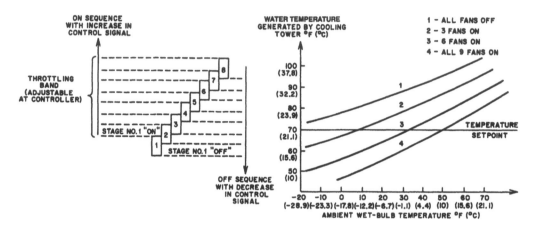

FIGURE 5.8 Load following cooling tower control by fan cycling. Left: The output of the approach controller (TDIC-1 in Fig. 2.21) can be sent to a sequencer, which turns fan stages on and off. Right: Example showing the effect of fan cycling on a nine-cell tower designed for 160,000 GPM (608,000 l/m) with an 84°F (29°C) inlet and 70°F (21°C) outlet temperature, operating at 50% load.

A. Interlocks

Figure 5.9 describes a simple fan starter. It is controlled by the three-position "hand/off/automatic" switch. When the switch is placed in "automatic" (contacts 3 and 4 connected), the status of the interlock contact I determines whether or not the circuit is energized and the fan is on. The sequencing of this contact I was described in Figure 5.8. The purpose of the auxiliary motor contact M is to energize the running light R whenever the fan motor is on. The parallel hot lead to contact 6 allows the operator to check quickly to see whether the light has burned out.

The amount of interlocking is usually greater than that shown in Figure 5.9. Some of the additional interlock options are described in the paragraphs below.

Most fan controls include some safety overrides in addition to the overload (OL) contacts shown in Figure 5.9. These overrides might stop the fan if fire, smoke, or excessive vibration is detected. They usually also provide a contact for remote alarming.

Most fan controls include a reset button that must be pressed after a safety shutdown condition is cleared before the fan can be restarted.

When a group of starters is supplied from a common feeder, it may be necessary to make sure that the starters are not overloaded by the inrush currents of simultaneously started units. If this feature is desired, a 25-second time delay is usually provided to prevent other fans from starting until that time has passed.

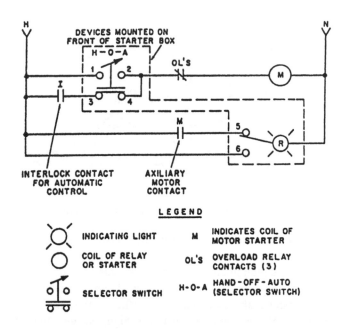

FIGURE 5.9   This simple fan starter circuit is controlled by a three-position switch.

Larger fans should be protected from overheating caused by too frequent starting and stopping. This protection can be provided by a 0 to 30 minute time delay guaranteeing that the fan will not be cycled at a frequency faster than the delay time setting.

When two-speed fans are used, added interlocks are frequently provided. One interlock might guarantee that even if the operator starts the fan in high, it will operate for 0 to 30 seconds in low before advancing automatically to high. This makes the transition from off to high speed more gradual. Another interlock might guarantee that when the fan is switched from high to low, it will be off the high speed for 0 to 30 seconds before the low-speed drive is engaged. This will give time for the fan to slow before the low gears are engaged.

When the fan can also be operated in reverse, further interlocks are needed. One interlock should guarantee that the fan is off for 0 to 2 minutes before a change in the direction of rotation can take place. This gives time for the fan to come to rest before making a change in direction. Another interlock will guarantee that the fan can operate only at slow speed in the reverse mode and that reverse operation cannot last more than 0 to 30 minutes. This limitation is desirable because in reverse ice tends to build up on the blades in winter and the gears wear excessively.

If the same fans can be controlled from several locations, interlocks are needed to resolve the potential problems with conflicting requests. One method is to provide an interlock that determines the location that is in control. This can be a simple local/remote switch or a more complicated system. When many locations are involved, conflicts are resolved by interlocks that either establish priorities between control locations or select the lowest, highest, safest, or other specified status from the control requests. In such installations with multiple control centers, it is essential that feedback be provided so that the operator is always aware not only of the actual status of all fans but also of any conflicting requests coming from other operators or computers.

B. EVAPORATIVE CONDENSERS

A special cooling tower type is the evaporative condenser illustrated in Figure 5.10. Induced air and sprayed water flow concurrently downward, and the process vapors travel upward in the multipass condenser tubes. The PIC modulates the amount of cooling to match the load. The

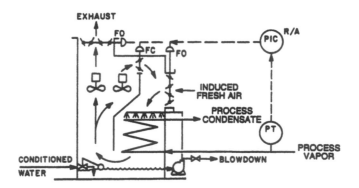

FIGURE 5.10   Evaporative condensers can be controlled by the recirculation of humid air.

continuous water spray on the tube surfaces improves heat transfer and maintains tube temperatures. The approach of these cooling towers is around 20°F (11°C), and the wet-bulb temperature of the cooling air can be varied by the recirculation of humid (or saturated) air. The resulting condensing temperatures are lower than those in dry-air—cooled condensers. The higher heat capacity and lower resistance to heat transfer make these units more efficient than the dry-air coolers.

The turndown capability of these units is also very high, because the fresh-air flow can be reduced all the way to zero. Both the exhaust and the fresh air dampers close while the recirculation damper simultaneously opens to maintain a constant internal air flow rate. As more humid air is recirculated, the internal wet-bulb temperature rises and the heat flow drops off. In Figure 5.10 the heat flow is effectively linear with fresh-air flow.

III. OPTIMIZING CONTROLS

The optimization of cooling towers as parts of chiller systems is discussed in Chapter 2. The discussion here will concentrate on cooling towers that directly cool the process. The goal of optimization is to minimize the fan and pump operating costs and thereby minimize the unit cost of cooling.

The cost of fan operation can be reduced by allowing the cooling tower temperature (Tctws in Figure 5.11) to rise, thereby increasing the approach (Tctws – Twb). As was shown in Figure 5.3, the approach can be increased to a point at which the fan cost is zero (the fans are off). Under most load conditions, however, this would not produce a low enough temperature for the process. The cooling tower water temperature must always be lower than the desired process temperature (Tp) on the other side of the process heat exchangers.

As shown in Figure 5.11, as the approach, and therefore Tctws, rises, the exchanger ΔT (Tp – Tctws) is reduced; thus, the process temperature controller (TIC-4) must open its coolant valve, CV-4. As the exchanger ΔT is reduced, more and more water must be pumped. Consequently, as the approach increases, the pumping costs rise.

If the actual total operating cost is plotted against approach (as in the right side of Figure 5.11), the fan costs tend to drop and the pumping costs tend to rise with an increase in approach. If the operating cost of some of the users is also affected by cooling water temperature (this was the case with chiller condensers in Figure 2.22), then the total cost model should also consider that effect. Once a total operating cost curve is found, the optimum approach is that which corresponds to the minimum point on that curve.

The total operating cost curve of Figure 5.11 tends to shift downward when the wet-bulb temperature of the air or the cooling load generated by the process drops and tends to shift up when they rise. These changes do not necessarily affect the optimum approach (the location of the minimum point on the total cost curve), yet it is likely that a drop in cooling load (Figure 5.12) or

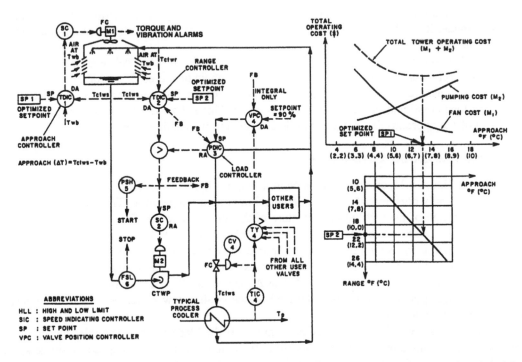

FIGURE 5.11   Cooling tower optimization can reduce the fan and pump operating costs, thereby minimizing the unit cost of cooling.

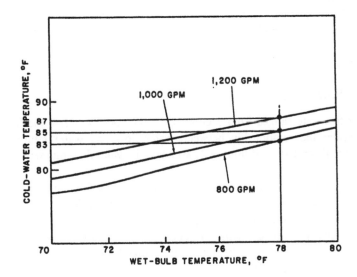

FIGURE 5.12   Performance of a cooling tower is described under the following design conditions: 85°F cold water, 95°F hot water, 78°F wet bulb, 10°F range, and 1000 gal/min. cell.

in air temperature will also lower the approach. This is because the cooling tower heat transfer surface area is constant and therefore the heat transfer efficiency rises if the load drops or if the ΔT (Tctws − Twb) across the heat transfer surface rises (Twb drops).

The data for the cost curves are empirically collected and continually updated through the actual measurement of fan and pump operating costs. Consequently, for any combination of load and ambient conditions there is a reliable prediction of optimum approach setting. Once the initial setting is made, it can be refined by adjusting it in 0.5°F increments in the direction that lowers the total operating cost.

As is also shown in the right side of Figure 5.11, there is an empirical relationship between the values of approach and range, if all other conditions remain unaltered. Therefore, if the optimum value of approach (SP1) has been determined, the corresponding range value (SP2) under the prevailing load conditions can be read from this curve. This is how the optimized set points (SP1 and SP2) of the approach and range controllers are determined and updated.

A. SUPPLY TEMPERATURE OPTIMIZATION

As shown in Figure 5.11, an optimization control loop is required in order to maintain the cooling tower water supply continuously at an economical minimum temperature. This minimum temperature is a function of the wet-bulb temperature of the atmospheric air. The cooling tower cannot generate a water temperature that is as low as the ambient wet bulb, but it can approach it.

Figure 5.11 illustrates the fact that as the approach increases, the cost of operating the cooling tower fans drops and the cost of pumping increases. The optimum approach is the one that will allow operation at an overall minimum cost. This ΔT automatically becomes the set point of TDIC-1. The optimum approach increases if the load on the cooling tower increases or if the ambient wet-bulb temperature rises.

If the cooling tower fans are centrifugal units or if the blade pitch is variable, the optimum approach is maintained by continuous throttling. If the tower fans are two-speed or single-speed units, the output of TDIC-1 will incrementally start and stop the fan units in order to maintain the optimum approach as was described in Figure 5.8.

In cases in which a large number of cooling tower cells constitute the total system, it is also desirable to balance the water flows to the various cells automatically as a function of the operation of the associated fans. In other words, the water flows to all cells whose fans are at high speed should be controlled at equal high rates; cells with fans operating at low speeds should receive water at equal flow rates; and cells with their fans off should be supplied with water at equal minimum flow rates.

B. WATER FLOW BALANCING

The normal water flow rate ranges from 2 GPM to 5 GPM per ton when the fan is at full speed. Figure 5.12 illustrates the relationship between water flow and water temperature (Tctws) or approach. Under the illustrated conditions, a 20% change in water flow rate will affect the approach by about 2°F.

Water balancing is usually done manually, but it can also be done automatically, as shown in Figure 5.13. Here the total flow is used as the set point of the ratio flow controllers (FFICs). If the ratio settings are the same, the total flow is equally distributed. The ratio settings can be changed manually or automatically to reflect changes in fan speeds. Naturally, the total of the ratio settings must always be 1; therefore, if one ratio setting is changed, all others should also be modified. This, too, can be done automatically.

The purpose of the control system in Figure 5.13 is not only to distribute the returning water correctly between cells but also to make sure that this is done at minimum cost. The cost in this case is pumping cost, and it will be minimum when the pressure drop through the control valves is minimum. This is the function of the valve position controller (VPC) in Figure 5.13. As long as even the most open valve is not nearly fully open, the VPC adds a positive bias to all the set point signals of all the FFICs. As a result, all valves open and keep opening until the most open valve reaches the desired 95% opening. This technique enables correct flow distribution to be achieved at a minimum cost in pumping energy.

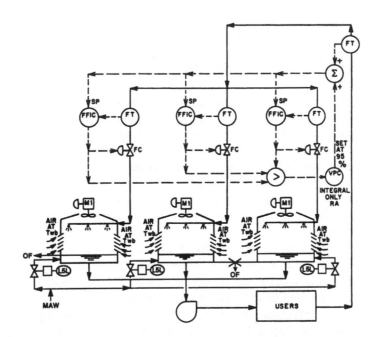

FIGURE 5.13 Water distribution can be controlled automatically. (MAW = Makeup Water, OF = Overflow)

When cells are manually balanced, it is not unusual to find all balancing valves throttled and to see the same water flow when the fan is on or off. Both of these conditions will increase operating costs, and the savings from automatic balancing can more than justify the cost of instruments required.

C. RETURN TEMPERATURE OPTIMIZATION

Figure 5.11 also shows the controls for return water temperature optimization. TDIC-2 is the range controller having the optimized set point of SP2, which is the range value corresponding to the optimum approach. TDIC-2 throttles the water circulation rate in order to maintain the range at its optimum value. The output signals of TDIC-2 and PDIC-3 are sent through a high signal selector, which guarantees that the needs of the users take priority over range control. The user demand is established by PDIC-3, and if its output is higher than that of the range controller, it will be selected for setting the pump speed. The load-following optimization loop guarantees that all cooling water users in the plant will always be satisfied while the range is being optimized. This is done by selecting (with TY-4) the most-open cooling water valve in the plant and comparing that signal with the 90% set point of the valve position controller, VPC-4. If even the most open valve is less than 90% open, the set point of PDIC-3 is decreased; however, if the valve opening exceeds 90%, the PDIC-3 set point is increased. In this way, a condition is maintained that allows all users to obtain more cooling (by further opening their supply valves) if needed while the differential pressure of the water is continuously optimized. The VPC-4 set point of 90% is adjustable. Lowering it gives a wider safety margin, which might be required if some of the cooling processes served are very critical. Increasing the set point maximizes the energy conservation at the expense of the safety margin.

An additional benefit of this load-following optimization strategy is that because all cooling water valves in the plant are opened up as the water ΔP across the users is minimized, valve cycling is reduced, and pumping costs are lowered. The reduction in pumping costs is a direct result of the opening of all cooling water valves, which require less pressure drop in a more open condition. Valve cycling is eliminated when the valve opening is moved away from the unstable region near the closed position.[3]

In order for the control system in Figure 5.11 to be stable, it is necessary to use an integral-only controller for VPC-4, with an integral time that is tenfold that of the integral setting of PDIC-3. This mode selection is needed to allow the optimization loop to be stable when the most open valve selected by TY-4 is either cycling or noisy. A high limit setting on the output of VPC-4 guarantees that it will not drive the cooling water pressure to unsafe or undesirable levels. Because this limit can block the VPC-4 output from effecting the pump speed, it is necessary to protect against reset wind-up in VPC-4. This is done through the external feedback signal (FB) shown in Figure 5.11, which protects TDIC-2, PDIC-3, and VPC-4 from reset wind-up.

A side benefit of this optimization system is that it brings attention to design errors. For example, if PDIC-3 is in control most of the time, it shows that the control valves and water pipes are undersized. If TY-4 consistently selects the same user, it shows that the water supply to that user is undersized. When such design errors are corrected, either by adding local booster pumps or by replacing undersized valves and pipes, control will automatically return to TDIC-2. Therefore, in a well-designed water distribution network, the range optimizer TDIC-2 will be in control most of the time; PDIC-3 will act only as a safety override that becomes active under emergency conditions, guaranteeing that no user will ever run out of cooling water.

When the cooling tower water pump station consists of several pumps, only one of which is variable-speed, additional pump increments are started when PSH-5 signals that the pump speed controller set point is at its maximum. When the load is dropping, the excess pump increments can be stopped on the basis of a drop in flow, detected by FSL-6. In order to eliminate cycling, the excess pump increment is turned off only when the actual total flow corresponds to less than 90% of the capacity of the remaining pumps.

This load-following optimization loop will float the total cooling tower water flow to achieve maximum overall economy.

IV. SAFETY AND MAINTENANCE

A. WINTER OPERATION

To protect the tower from power failure during subfreezing weather, electrically heated tower basins should be provided with an emergency draining system. Additional freeze protection can be provided by the bypass circulation method illustrated in Figure 5.14. In this system a thermostat (TSL) detects the outlet temperature of the cooling tower water, and when it drops below 40°F (4.4°C), the bypass circulation (located in a protected indoor area) is started. The circulation is terminated when the water temperature rises to approximately 45°F (7.2°C).

The thermostat is wired to open the solenoids S1, S2, and S3 first and then to start the circulating pump (P2), which usually is a small, 1/4- or 1/3-HP (186.5 or 248.7 W) unit. The bypass line containing solenoid S1 is sized to handle the low capacity of P2, with a pressure drop (h) that is less than the height of the riser pipe (h). This makes sure that when P1 is off, no water will reach the top of the column.

In the northern regions, in addition to the S1 bypass shown in Figure 5.14, there is a full-sized bypass 54 also shown in Figure 5.14. When the main (P1) pump is started in the winter, this full bypass is opened; it does not allow the water to be sent to the top of the tower until its temperature is approximately 70°F (21°C). Once the water reaches that temperature, the bypass is closed because the water is warm enough to be sent to the top of the tower without danger of freezing. As the water temperature rises further, fans are turned on to provided added cooling.

When open cooling towers are operated at freezing temperatures, the induced draft fans must be reversed periodically in order to de-ice the air intakes. The need for de-icing can be determined through visual observation, through remote closed-circuit TV inspection, or by comparing the load and ambient conditions to past operating history. The safety requirements of reverse operation have already been discussed under interlocks.

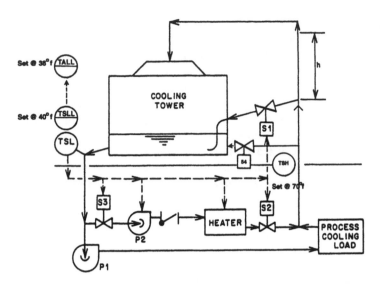

FIGURE 5.14   The bypass circulation method can be used to protect against freezing in cooling towers.

All exposed pipes that will contain water when the tower is down should be protected by electric tape or cable tracing and by insulation. The sump either should be drainable to an indoor auxiliary sump or should be provided with auxiliary steam or electric heating.

Closed cooling towers should either operate on antifreeze solutions or be provided with supplemental heat and tracing. If the second method is used, the system must be drained if power failure occurs.

B. Blowdown Controls

The average water loss by blowdown is 0.5 to 3.0% of the circulating water rate. The loss is a function of the initial quality of water and the amount and concentration of the dissolved natural solids and chemicals added for protection against corrosion or buildup of scale on the heat transfer surfaces. Because the cooling tower is a highly effective air scrubber, it continuously accumulates the solid content of the ambient air on its wet surfaces, which are then washed off by the circulated water.

The "normal" condition of the circulated water is generally as follows:

pH	6 to 8
Chloride as NaCl	under 750 ppm
Total dissolved solids	under 1500 ppm

The blowdown requirement can be determined as:

$$B = \frac{([E+D]/N) - D}{1 - (1/N)} \tag{3}$$

where B = blowdown in percentages (typically, for N = 2, blowdown is 0.9%; for N = 3, 0.4%; for N = 4, 0.24%, etc.); E = evaporation in percentages (typically 1% for each 12.5°F (7°C) of cooling range); D = drift in percentages (typically 0.1%); and N = number of concentrations relative to initial water quality.

In cooling towers, the major causes for concern are delignification (loss of the binding agent for the cellulose) caused by the use of oxidizing biocides, such as chlorine; excessive bicarbonate alkalinity; biological growth, which can clog the nozzles and foul the heat exchange equipment; corrosion of the metal components (this corrosion should be less than 3 mils per year without pitting); general fouling by a combination of silt, clay, oil, metal oxides, calcium and magnesium salts, organic compounds, and other chemical products that can cause reduced heat transfer and enhanced corrosion; and scaling by crystallization and precipitation of salts or oxides (mainly calcium carbonate and magnesium silicate) on surfaces. The treatment techniques to prevent these conditions from occurring are listed in Table 5.1. The blowdown must meet the water quality standards for the accepting stream and must not be unreasonably expensive. It is also possible to make the water system a closed-cycle system (no blowdown) by ion-exchange treatment.

TABLE 5.1
Cooling Tower Problems and Solutions

Problem	Factors	Causative Agents	Corrective Treatments
Wood deterioration	Microbiological	Cellulolytic fungi	Fungicides
	Chemical	Chlorine	Acid
Biological growths	Temperature	Bacteria	Chlorine
	Nutrients	Fungi	Chlorine donors
	pH	Algae	Organic sulfurs
	Inocula		Quaternary ammonia
General fouling	Suspended solids	Silt	Polyelectrolytes
	Water	Oil	Polyacrylates
	Velocity		Lignosulfonates
	Temperature		Polyphosphates
	Contaminants		
	Metal oxides		
Corrosion	Aeration	Oxygen	Chromate
	pH	Carbon dioxide	Zinc
	Temperature	Chloride	Polyphosphate
	Dissolved solids		Tannins
	Galvanic couples		Lignins
			Synthetic organic compounds
Scaling	Calcium	Calcium carbonate	Phosphonates
	Alkalinity	Calcium sulfate	Polyphosphates
	Temperature	Magnesium silicate	Acid
	pH	Ferric hydroxide	Polyelectrolytes

C. MISCELLANEOUS CONTROLS

The minimum basin level is maintained by the use of float-type makeup level control valves (Figure 5.10). The maximum basin level is guaranteed by overflow nozzles sized to handle the total system flow (Figure 5.13). In critical systems, additional safety is provided by the use of high/low level alarm switches.

The operational safety of the fan and the fan drive can be safeguarded by torque and vibration detectors (Figure 5.11). Low-temperature alarm switches can also be furnished as warning devices signaling the failure of freeze protection controls (TALL in Figure 5.14).

The need for flow balancing among multiple cells is another reason for using added valving and controls. The warm water is usually returned to the multicell system through a single riser, and a manifold is provided at the top of the tower for water distribution to the individual cells.

Balancing valves should be installed to guarantee equal flow to each operating cell. If two-speed fans are used, further economy can be gained by lowering the water flow when the fan is switched to low speed (Figure 5.13). It is also advisable to install equalizer lines between tower sumps to eliminate imbalances caused by variations in flow rates or pipe layouts.

Another reason for considering the use of two-speed fans is to have the ability to lower the associated noise level at night or during periods of low load. Switching to low speed will usually lower the noise level by approximately 15 dB.

V. CONCLUSIONS

One of the least understood and most neglected unit operations in the processing industries is the cooling tower. Its pumps are frequently constant speed with three-way or bypass valves used to circulate the excess water, which should not have been pumped in the first place because the process does not require it. Meeting a variable load with a constant supply is also frequently practiced in fan operation. In some installations, fan speeds cannot be changed at all; in others, they are changed only manually, which can mean seasonal adjustments in some extreme cases.

As the rating of the tower fans and pumps usually adds up to several hundred horsepowers, their yearly operating cost is high. As optimization can cut this in half, the added cost for controls is justified.

REFERENCES

1. Dickey, J.B., Evaporative Cooling Towers, Marley Cooling Tower Co.
2. American Society of Heating, Refrigerating, and Air-Conditioning Engineers (ASHRAE), Cooling Towers, in Equipment Volume, latest edition of ASHRAE Handbook, Atlanta, Georgia.
3. Lipták, B.G., Optimization controls for chillers and heat pumps, *Chem. Eng.*, October 17, 1983.

BIBLIOGRAPHY

American Society of Heating, Refrigerating, and Air-Conditioning Engineers (ASHRAE), Cooling Towers, in Equipment Volume, latest edition of ASHRAE Handbook, Atlanta, Georgia.
Anderson, E.P., *Air Conditioning: Home and Commercial*, 5th ed., Macmillan, New York.
Baechler, R.H., Cause and Prevention of Decay of Wood in Cooling Towers, ASME Petroleum Division Conference, New York, 1961.
Baker, D.R., Durability of Wood in Cooling Towers, The Marley Co., Kansas City, Missouri.
Brumbaugh, J.E., *Heating, Ventilating and Air Conditioning Library*, rev. ed., Macmillan, New York.
Chopey, N.P., *Handbook of Chemical Engineering Calculations*, 2nd ed., McGraw-Hill, New York, 1994.
Donohue, J.M., Cooling towers–chemical treatment, *Ind. Water Eng.*, May 1970.
Lipták, B.G., Optimizing controls for chillers and heat pumps, *Chem. Eng.*, October 17, 1983.
McMillan, G.K., *Tuning and Control Loop Performance*, 3rd ed., Instrument Society of America, 1994.
Nussbaum, O.J., Dry type cooling towers for packaged refrigeration and air conditioning systems, *Refrig. Sci. Technol.*, 1972.
O'Callaghan, P.W., *Energy Management*, McGraw-Hill, New York, 1992.
Shinskey, F.G., *Process Control Systems*, 4th ed., McGraw-Hill, New York, 1996.
Wang, S.K., *Handbook of Air Conditioning and Refrigeration*, McGraw-Hill, New York, 1994.
Whitman, W.C., *Refrigeration and Air Conditioning Technology*, Delmar, 1991

6 CSTR and BSTR Chemical Reactors

Chemical reactor designs include the continuous stirred tank reactors (CSTR), the batch stirred tank reactors (BSTR), the tubular reactors, and the packed bed reactors. This chapter describes the control and optimization of the basic operating parameters of chemical reactors: pressure, temperature, endpoint, safety, and batch recipe charging.

The optimization of batch and continuous chemical reactors has many potential benefits, including increase in productivity and improvement in safety, product quality, and batch-to-batch uniformity. The combined impact of these factors on plant productivity can approach a 25% improvement. Such overall results are the consequences of many control strategies. These loops will program temperature and pressure and maintain concentration and safety, while providing sequencing and record-keeping functions. All elements of the overall chemical reactor control system are discussed in this chapter.

I. THE PROCESS

Two types of reactors are used in chemical plants: continuous reactors and batch reactors. Continuous reactors are designed to operate with constant feed rate, withdrawal of product, and removal or supply of heat. If properly controlled, the composition and temperature can be constant with respect to time and space. In batch reactors, measured quantities of reactants are charged in discrete quantities and allowed to react for a given time, under predetermined controlled conditions. In this case composition is the function of time.

Most reactions are reversible—that is, there is a ratio in product to reactant concentration that brings about equilibrium. Under equilibrium conditions the production rate is zero, because for each molecule of product formed there is one that converts back to its reactant molecules. The equilibrium constant (K) is the ratio of product/feed concentration at equilibrium or forward to reverse reaction rate coefficients ($K = K_f/K_r$). The value of K is also a function of the reaction temperature and the type of catalyst used. K naturally places a limit on the conversion that can be achieved within a particular reactor, but conversion can usually be increased, if at least one of the following conditions can be achieved:

- Reactant concentration can be increased
- Product concentration can be decreased through separation or withdrawal
- Temperature can be lowered by increased heat removal in reversible exothermic reactions
- A change in operating pressure can be effected (this increase conversion only in certain reactions)

The catalyst does not take part in the reaction, but it does affect the reaction rate (k). Some catalysts are solids and are packed in a bed; others are fluidized, dissolved, or suspended. Metal catalysts are frequently formed as flow-through screens. Whatever their shape, the effectiveness of the catalyst is a function of its active surface, because all reactions take place on that surface. When it is fouled, its lost effectiveness can first be compensated by increasing reaction temperature, but eventually the catalyst must be reactivated or replaced.

The reaction rate coefficient exponentially increases with temperature. In some processes reaction rate can double with every 18°F (10°C) increase in reaction temperature.[2] The activation energy (E) determines its degree of temperature dependence according to the Arrhenius equation:

$$k = \alpha e^{-(E/RT)} \tag{1}$$

where k = specific reaction rate coefficient (min^{-1}), α = preexponential factor (min^{-1}), E = activation energy of reaction (BTU/mole), R = perfect gas constant (1.99 BTU/mole -°R), and T = absolute temperature (°R).

A. REACTOR TYPES

The reactions can take place in gas, liquid or multiple phases, can be batch, semibatch, or continuous, can be endothermic or exothermic, and can occur in stirred tank, pipe or fluid bed reactors. This chapter will concentrate on the control of exothermic, stirred, batch and continuous (BSTR and CSTR) reactors.

In a batch reactor, after the initial charge there is no inflow or outflow. Therefore, an isothermal batch reactor is similar in its conversion characteristics to a plug-flow tubular reactor. If the residence times are similar, both reactors will provide the same conversions. Batch reactors are usually selected when the reaction rates are low, when there are many steps in the process, when isolation is required for reasons of sterility or safety, when the materials involved are hard to handle, and when production rates are not high. As shown in Figure 6.1, the batch (or tubular) reactor is kinetically superior to the continuous stirred tank reactor.[1] The batch reactor has a smaller reaction time and can produce the same amount of product faster than the back-mixed one.

Because the continuous plug flow type reactor is dominated by dead time, its temperature control is difficult. On the other hand (as shown in Figure 6.2), the plug flow reactor gives higher conversion of reactants into products than a back-mixed reactor operating under the same conditions.

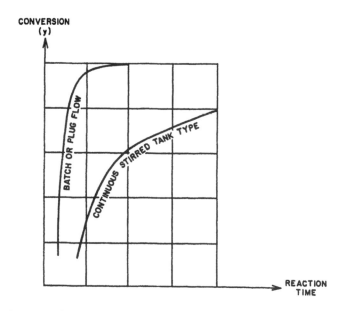

FIGURE 6.1   Batch reactors have better conversion efficiencies than back-mixed reactors.[1]

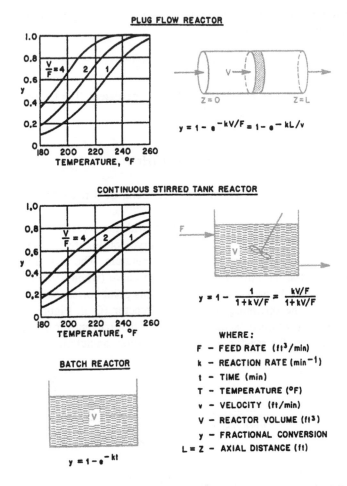

FIGURE 6.2 Conversion equations and conversion vs. temperature characteristics differ depending on the type of reactor.[1,2]

If the reaction rate is low, a long tubular reactor or a large back-mixed reactor is required to achieve reasonable conversions.

The amount of heat generated by an exothermic reactor increases as the reaction temperature rises. If the reactor is operated without a temperature controller (in an open loop), an increase in the reaction temperature will also increase heat removal, because of the increase in ΔT between process and coolant temperatures. If an increase in reaction temperature results in a greater increase in heat generation than heat removal, the process is said to display positive feedback; as such, it is considered to be "unstable in the open loop."

The positive feedback of the open-loop process can be compensated for by the negative feedback of a reactor temperature controller, which will increase the heat removal rate as the temperature rises. The addition of such a feedback controller can stabilize an open-loop unstable process only if the control loop is fast and does not contain too much dead time. Cascade control can increase speed, and maximized coolant flow can reduce dead time.

B. Reactor Stability

Endothermic processes are inherently stable. On the other hand, the exothermic process of converting reactants into products can be stable or unstable in the steady state. The reaction is stable if the rate of heat removal (cooling – Q_c) exceeds the rate of heat generation (reaction heat – Q_r) in the reactor. The rate of cooling is

$$Q_c = AU(T - T_c) \tag{2}$$

where A = heat transfer area, U = overall heat transfer coefficient, T = temperature of the reacting chemicals, and T_c = temperature of the coolant.

The rate of heat generation can be calculated as

$$Q_r = -WkCrHr \tag{3}$$

where W = the reacting mass, k = reaction rate coefficient, Cr = reactant concentration, and Hr = heat of reaction.

Because the reaction rate increases exponentially with temperature (equation 1), Q_r is also a function of temperature. The sigmoid (S-shaped) curve[3] of Figure 6.3 shows the relationship between the rate of heat evolution and reaction temperature. The heat evolution sensitivity to temperature is the highest at the inflection point of the curve (when $kV/F = 1$)[2], where the slope of its tangent is the greatest. In back-mixed reactors, this point corresponds to 50% conversion. If that point is selected as the operating point of the reactor (O in Figure 6.3), the temperature sensitivity of cooling rate (Q_c) must exceed the slope of that tangent for the reactor to be stable in the steady state. In that case the reactor could be left without an automatic temperature controller.

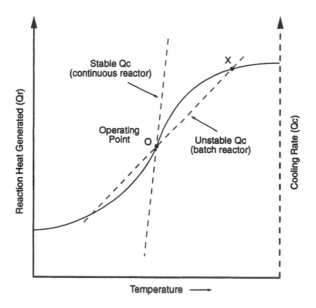

FIGURE 6.3 For stability, the heat removal rate must exceed the heat evolution rate at the operating point.

If the cooling rate is less, and if the slope is below this tangent, the reactor is *unstable in the steady state* and must be provided with automatic temperature controls. Otherwise, as the temperature rises above the operating point (O), heat generation will exceed cooling ($Q_r > Q_c$) and the reaction temperature will rise until the next equilibrium point at X.

Most *continuous* reactors are provided with enough heat-transfer area (relative to their volume) for cooling to be stable in the steady state, and therefore these reactors are self-regulating. On the other hand, the majority of exothermic *batch* reactors are neither self-starting, nor self-regulating. They are unstable in the steady state (their heat transfer capability is below the maximum potential heat generation). Therefore, they must be provided with automatic temperature control. Automatic control of batch reactors is usually successful because of the large back-mixed volume of the reactor, which makes them slower to increase the rate of heat release than the time it takes for the control system to increase cooling.

In other words, for an unstable reactor to be controllable, its time constant (T_q) must exceed its dead time (T_{dq} in Figure 10.2). If this is not the case, if the reactor is "dead-time dominant," it will be uncontrollable. In a continuous reactor, one can increase the time constant (T_q) by reducing feed rate or reactant concentration in the feed, but that also reduces production rate. Another option is to increase the overall heat transfer coefficient (U), which can be done by increasing the ΔT or replacing the circulating coolant with a boiling fluid, which increases U about threefold.[2]

C. Batch Process Characteristics

Most batch cycles are started by purging and pressure testing followed by charging reactants into the reactor and then mixing and heating them until the reaction temperature is reached. The reaction itself is frequently started by the addition of a catalyst or other initiator. *Exothermic reactions* produce heat, and *endothermic reactions* consume it. The reactor itself can be *isothermal*, meaning that it is operated at constant temperature, or *adiabatic*, meaning that heat is neither added nor removed within the reactor; the reaction is controlled by other means, such as the manipulation of pressure, catalyst, reactants, etc. The chemical reactions can be irreversible, reversible, consecutive, or simultaneous.

Batch-to-batch uniformity is a function of many factors, from the purity of reactants, catalysts, and additives to the repeatability of controllers serving to maintain heat and material balance. In a batch cycle, there is no steady state and therefore no "normal" condition at which controllers could be tuned (Figure 6.4). The batch process is always in a transient state; thus, the process variables, the process gains, and time constants also very during the batch cycle. In addition, there are the problems of runaway reactions and batch-to-batch product uniformity. Runaway reactions occur in exothermic reactions, in which an increase in temperature speeds up the reaction, which in turn releases more heat and raises the temperature further. In order to counter this positive feedback cycle, highly self-regulating cooling systems are required. One of the most self-regulating cooling systems is a bath of boiling water, because it needs no rise in temperature to increase its rate of heat transfer.

Figure 6.4 describes the concentration, temperature, and heat-release profiles of an exothermic batch reaction cycle. After the heat-up cycle, when the reaction temperature has been reached, the reaction is initiated, usually by the addition of a catalyst. As ingredients A and B are converted into product P, heat is released. As the concentrations of A and B drop, the heat release also drops and a side reaction is started, which converts some of the product into waste (W). Therefore, the reaction is stopped before the product concentration would start to diminish. The jacket temperature (T_2) exceeds the batch temperature (T_1) during heat-up and drops below it after the reaction has started generating heat. As the heat release rate drops, so does the temperature difference ($T_1 - T_2$).

1. Gains, Time Constants and Tuning

The unsteady-state heat balance of a continuous reactor can be obtained by

Heat accumulation = heat evolution – (heat absorbed by feed and coolant)

The thermal steady-state gain is the change in process temperature (T_1 in Figure 6.4) resulting from a unit change in coolant temperature (T_2) or coolant flow. The open loop gain of a reactor temperature loop ranges from 0.5 to 5.

As shown in Figure 10.3, the process gain of all heat-transfer processes drops as the load rises. This is because the fixed heat-transfer area becomes less efficient as it needs to transfer more heat.

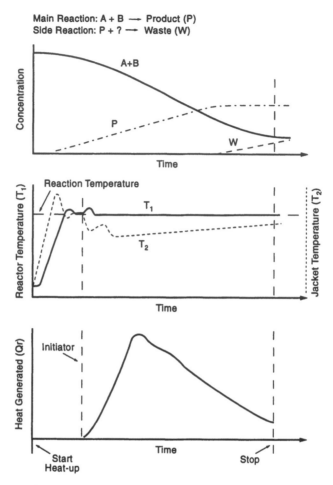

Main Reaction: A + B ⟶ Product (P)
Side Reaction: P + ? ⟶ Waste (W)

FIGURE 6.4   Concentration, temperature, and heat release profiles of a batch reactor.

As suggested by Figure 6.5, this drop in gain can be partially compensated by the use of an equal percentage valve, whose gain increases with rising load and therefore keeps the total loop gain more or less constant. Experience has shown that this is only an approximate correction, and, if the coolant valve ΔP is constant, a modified percentage valve is a better fit.[3]

D. Reactor Time Constants

The cooling of an exothermic reactor involves the heat capacity of the reactor contents of the jacket wall, and of the coolant, each producing a lag (Figure 6.6). In addition, the temperature sensor also behaves as a first-order lag. The dominant time constant usually is the thermal time constant:

$$T_1 = K_T(W_1C_1)/(k_1A) \tag{4}$$

This shows that anything that affects the steady-state gain (K_T), such as feed rate or composition, will also change T_1 in direct proportion. Also, according to Shinskey,[2] when K_T (the thermal steady-state gain) is negative, the reactor is steady-state unstable; when it is positive, the reactor is self-regulating, and if it is very high, the reactor is no longer self-regulating. When $K_T = 1$ the thermal time constant is determined by Equation 5.

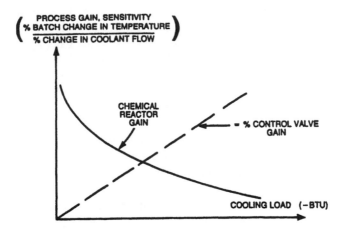

FIGURE 6.5 Equal-percentage coolant control valve partially compensates for the nonlinearity of a jacketed reactor.

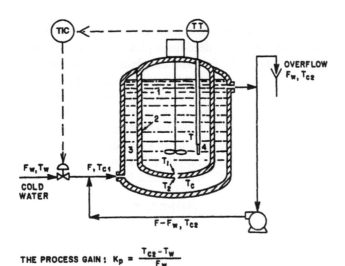

FIGURE 6.6 There are four interacting time lags in a chemical reactor process. The smaller the process gain, the narrower the proportional band and the more sensitive the control loop.[2]

$$\text{Thermal time constant}: \tau_1 = \frac{W_1 C_1}{k_1 A} = \frac{W_1 C_1}{Q}\left(T - T_1\right) \tag{5}$$

$$\text{Reactor wall time constant}: \tau_2 = \frac{W_2 C_2 \ell}{k_2 A} = \frac{W_2 C_2}{Q}\left(T_1 - T_2\right) \tag{6}$$

$$\text{Coolant time constant}: \tau_3 = \frac{W_3 C}{k_3 A} = \frac{W_3 C}{Q}\left(T_2 - T_c\right) \tag{7}$$

$$\text{Thermal bulb time constant}: \tau_4 = \frac{W_4 C_4}{k_1 A_4} \tag{8}$$

where A = heat-transfer area, ft^2, A_4 = surface area of bulb, ft^2, C = specific heat of coolant, BTU/(lb)(°F), C_1 = specific heat of reactants, BTU/(lb)(°F), C_2 = specific heat of wall, BTU/(lb)(°F), C_4 = specific heat of bulb, BTU/(lb)(°F), K_T = thermal steady-state gain, k_1 = heat-transfer coefficient, BTU/(h)(ft^2)(°F), k_2 = thermal conductivity, BTU/(h)(ft^2)(°F/in), k_3 = heat-transfer coefficient BTU/(h)(ft^2)(°F), ℓ = wall thickness, in., Q = rate of heat evolution, BTU/h, T = reactor temperature, °F, T_1 = wall temperature, °F, T_2 = outside wall temperature, °F, T_c = average coolant temperature, °F, W_1 = weight of reactants, lb, W_2 = weight of wall, lb, W_3 = weight of jacket contents, lb, and W_4 = weight of bulb, lb.

Typical values of these time constants are τ_1 = 30 to 60 min, τ_2 = 0.5 to 1.0 min, τ_3 = 2 to 5.0 min, and τ_4 = 0.1 to 0.5 min (can be minimized by the use of bare bulbs). The overall time constant of the reactor process ranges from 20 to 100 minutes.

These time constants do vary with reactor design. For example, if the jacket is not heated directly, or steam is not injected into the circulating water but an external steam-heated exchanger is used, this will add another thermal lag during the switch from heating to cooling if no bypass is provided around the exchanger. Similarly, if oil is the heat-transfer media, the time constant of the dedicated oil system must also be considered in the total time constant of the reactor system.

E. REACTOR DEAD TIMES

The total dead time in a reactor temperature control loop is the sum of two major dead times: the jacket transport and the imperfect mixing dead times plus: a number of smaller lads. These include the valve stroke dead time (dead-band), feed and sample transportation delays, configuration scan times and others. The dead time due to jacket replacement can be reduced by increasing the pumping rate. This should be kept under 2 minutes in a well-designed reactor. The dead time caused by imperfect mixing can be reduced by increasing the agitator pumping capacity. In a well-designed reactor it should be held to less than 10% of the thermal time constant τ_1.

To minimize the deadtime due to backmixing in a reactor, the agitator should be the axial type and the tank diameter should equal the liquid depth. If in a continuous reactor, the agitator pumping rate is F_a, the production rate is F, and the tank volume is V, the dead time is

$$T_d = V/2(F_a + F) \tag{9}$$

and the time constant is

$$T_1 = V/F - T_d \tag{10}$$

To minimize the dead time of jacket displacement, the time it takes for transporting the coolant from the heat-transfer surface to the temperature sensor at the jacket outlet must be minimized.

Figure 6.7 illustrates the temperature response of an uncontrolled chemical reactor to a step change in load, assuming that the coolant is applied in a once-through manner (Figure 6.8), without recirculation. The solid line depicts the temperature response at low loads, and the dotted line depicts temperature response at high load. Both the dead time and the process gain increase as the load drops. In other words, at low loads it takes longer for the process to start responding, but once it has, the full response develops quickly. At high loads the opposite is the case.

The dead time of a well-designed chemical reactor ranges from 1 to 5 minutes. The actual dead time of the temperature control loop can be approximated as one quarter of its period of oscillation. The "once-through" cooling system shown in Figure 6.8 is undesirable because only constant dead-time processes can be properly tuned (by setting a noninteracting PID controller so that its integral and derivative will approximately equal the dead time) and here the dead time varies with load.

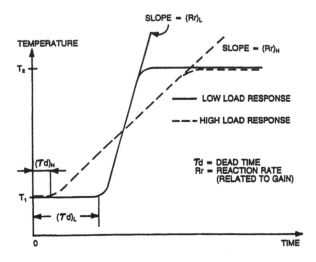

FIGURE 6.7 In a chemical reactor without a water circulating pump, both the dead time and the gain of the process rises as the process load drops.

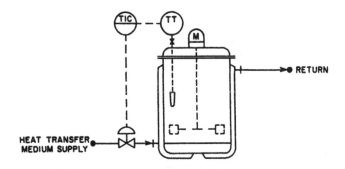

FIGURE 6.8 In once-through cooling of chemical reactors, the coolant temperature is not uniform.

As the water valve closes, the time required to displace the jacket, and with it the dead time, rises. In addition to the variable residence time, this "once-through" method of cooling is also undesirable because the coolant temperature is not uniform. This can cause cold spots near the inlet and hot spots near the outlet. In addition, when the water flow is low, the Reynolds number will drop off, and with it, the heat transfer efficiency will also diminish. Lower water velocity can also result in fouling of the heat transfer surfaces.

For all the above reasons, the recirculated cooling water configuration shown in Figure 6.9 is more desirable because it guarantees a constant and high rate of water circulation. This keeps the jacket dead time constant, the heat transfer coefficient high, and the jacket temperature uniform, thereby eliminating cold and hot spots.

The fluid velocity in the reactor jacket is maintained high enough to produce satisfactory film coefficients for heat transfer. The fluid velocity can be further increased by additional jets. In addition, the liquid is circulated at a high enough rate to keep the temperature rise of the heat transfer medium, as it passes through the jacket, at a low enough level to maintain the jacket wall temperatures uniform throughout the reactor. This keeps the jacket dead time constant and eliminates fouling of the heat transfer surfaces. The circulating pump in Figure 6.9 is sized to displace the cooling loop at least once every two minutes.

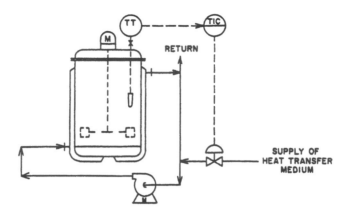

FIGURE 6.9 Recirculated cooling of chemical reactors guarantees a constant and high rate of water circulation.

II. BASIC CONTROLS

A. CONTROLLABILITY AND TUNING

The control requirements for the periods of heat-up, reaction, and stripping or refluxing are quite different. Here, the controls for the reaction phase are discussed; the others will be discussed later. The control variable for the reaction phase is usually temperature (often held within 0.5°C), although, control by reaction rate, side reactions, molecular weight, or molecular weight distribution is also possible.

The period of oscillation of a control loop (T_o), is a function of its dead time (T_o is about $3T_d$). If the period of oscillation during the reaction phase of a "typical" chemical reactor is 15 minutes, its dead time is about 5 minutes.

As was already noted in this chapter, it is the dynamic gain of the process (approximated as the T_d/T_1 ratio)[2] that determines the controllability of a chemical reactor. The smaller this ratio, the more "lag dominant" the process and therefore the easier it is to control. Once the dead time reaches 33% of the thermal time constant, control becomes more difficult, and, if it reaches or exceeds 100%, it becomes very difficult to control without some form of dead-time compensation.

In some cases, the tuning requirements differ when the loop is to respond to a load or to a set-point change. The dilemma of which response to tune for can be resolved through the use of dead time-compensated PID algorithms (Figure 6.10).

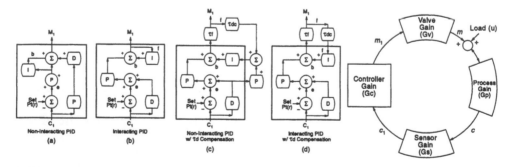

FIGURE 6.10 Control algorithm choices for use in reactor temperature control. P, I, and D refer to the proportional, integral, and derivative modes, T_f is a filter, with a first-order lag set to 10% of the process dead time (T_{dp}), T_{dc} is the dead-time compensation in the controller, which is set to equal T_{dp}, and Gc, Gv, Gp, and Gs stand for the steady-state and dynamic gain products of the controller, process, valve, and sensor.

In tuning the PID controller, its gain setting (K_c = 100/PB) must fall between two limits. The lower limit on the controller gain must exceed the thermal steady-state gain of the process (K_c > K_T). The maximum limit for controller gain setting (K_c = 100/PB) comes from the need to maintain stability, which necessitates that the loop gain (Figure 10.4) must be less than one, usually set around 0.5 for quarter amplitude damping. Therefore, the proportional band setting of the PID controller should fall between the limits given in Equation 11:[2]

$$(100K_T T_d)/T_1 < PB < -100K_T \qquad (11)$$

where T_1 is the thermal time constant, as it was defined in Equation 4 and K_T is the thermal steady-state gain of the process, as illustrated in Figure 10.2. From this, we can conclude that, for a steady-state unstable reactor to be controllable, it must be lag dominant (T_1 must exceed T_d). T_1 can be increased through increasing the thermal steady-state gain of the process (K_T) by either increasing the thermal heat transfer coefficient (k_1) or by reducing either the reactant feed rate (F) or the reactant feed concentration (x). Figure 6.10 describes some of the control algorithms that can be considered for chemical reactor control.

B. CONTROL ALGORITHMS

The noninteracting and interacting PID configurations are shown in parts A and B of Figure 6.10. In both, the derivative action (D) is applied only to the measurement, so that it does not respond to the speed at which the set point is changed. In A, there is no interaction because I and D are in parallel. In B, the interaction is caused by I and D being in series. The performance of the noninteracting PID (the area under the error curve after a load upset) is about 10% better than that of the interacting PID configuration shown in B. The relative advantage of type B is that its integral time setting is shorter (about I = 3D), while type A requires a longer (less effective) integral setting of I = 4D. Another advantage of the interacting (B) configuration is that it is readily suited for using *external* feedback as a protection from integral wind-up.

For a nonself-regulating reactor controlled by a type A controller, the recommended settings[2] are PB = $78 T_d/T_1$, I = $1.9 T_d$, and D = $0.48 T_d$. The settings for the same process, if controlled by a type B controller, are[2] PB = $108 T_d/T_1$, I = $1.57 T_d$, and D = $0.56 T_d$.

The various model-based control algorithms, such as IMC (internal model control), MPC (model predictive control), and DMC (dynamic matrix control), the Smith and Dahlin predictors (a first order MPCs) are some of the attempts to improve upon the basic PIDs. Their common goals are to improve control quality on dead-time-dominant and noisy processes.

Dead-time-compensated PID controllers are effective on nonself-regulated processes and are easier to tune than model-based controllers. They are shown in parts C and D of Figure 6.10. They place a compensating dead time (T_{dc}) into the integral feedback line and set it to match the process dead time T_d. Because it is not possible to match the two dead times perfectly, one will run out before the other, and will cause a pulse. To overcome this upset, a filter (first order lag) is located in such a way that it affects both the process and the controller feedback and is set for about 10% of the dead time. Relative to conventional PID, type C and D controllers give superior performance on exothermic reactors. For type D, Shinskey[2] recommends the settings of PB = $108 T_d/T_1$, I = $0.15 T_d$, and D = $1.5 T_d$.

C. TEMPERATURE CONTROLS

1. Cascade Control

Cascade loops consist of two or more controllers in series, but have only a single, independently adjustable set point, that of the primary (master) controller. The main value of having secondary (slave) controllers is that they act as the first line of defense against disturbances, preventing these upsets from entering the primary process. For example, in Figure 6.11, if there were no slave controller, an upset due to a change in steam pressure or water temperature would not be detected until it had upset the master measurement. A cascade slave detects the occurrence of such upsets and immediately counteracts them, so that the master measurement is not upset and the primary loop is not even aware that an upset occurred in the condition of the utilities.

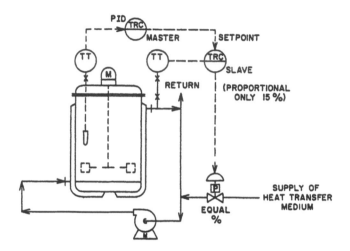

FIGURE 6.11   Cascade control of a reactor with recirculation reduces the period of oscillation of the master loop.

In order for the cascade loop to be effective, it should be more responsive (faster) than the master. Some rules of thumb suggest that the slave's time constant should be under $1/4$ to $1/10$th that of the master loop and the slave's period of oscillation should be under $1/2$ to $1/3$ that of the master loop.

There are two distinct advantages gained with cascade control:

1. Disturbances affecting the secondary variable can be corrected by the secondary controller before a pronounced influence is felt by the primary variable.
2. Closing the control loop around the secondary part of the process reduces the phase lag seen by the primary controller, resulting in increased speed of response.

For optimum performance, the dynamic elements in the process should be distributed as equitably as possible between the two controllers.

A superior method of reactor temperature control, a cascade loop, is depicted in Figure 6.11. Here the controlled process variable (reactor batch temperature, hereafter called reactor temperature), whose response is slow to changes in the heat transfer medium flow (manipulated variable), is allowed to adjust the set point of a secondary loop, whose response to coolant flow changes is rapid. In this case, the reactor temperature controller varies the set point of the jacket temperature control loop.

The purpose of the slave loop is to correct for outside disturbances, without allowing them to affect the reaction temperature. For example, if the control valve is sticking or if the temperature or pressure of the heat transfer media changes, this would upset the reaction temperature in Figures 6.8 or 6.9 but not in Figure 6.11, because in Figure 6.11 the slave would notice the resulting upset at the jacket outlet and would correct for it before it had a chance to upset the master.

It is preferred that the slave controller maintain the jacket outlet (and not inlet) temperature, because this way the jacket and its dynamics are included in the slave loop. Another advantage of this configuration is that it removes the principal nonlinearity of the system from the master loop, since reaction temperature is linear with jacket-outlet temperature. The nonlinear relationship between jacket-outlet temperature and heat-transfer-medium flow is now within the slave loop, where it can be compensated for by an equal-percentage valve, whose gain increases as the process gain drops off (Figure 6.5). In most instances the slave will operate properly with proportional-plus-derivative or proportional control only, set for a proportional band of 10 to 20%.

The period of oscillation of the master loop is usually cut in half as direct control is replaced by cascade. This might mean a reduction from 40 to 20 minutes in the period and a corresponding reduction of perhaps 30 to 15% in the proportional band. The derivative and integral time settings would also be cut in about half. This represents a fourfold overall loop performance improvement.

Using jacket-inlet temperature as an override or measurement for the slave may be useful in cases in which the jacket temperature must be limited, e.g., when the reactor is used as a crystallization unit, or in safety systems which serve to protect glass-lined reactors from thermal shock. Such protection might be needed when hot water is generated by direct injection of steam. If heat needs to be added in some phases of the reaction while in other phases it must be removed, the controls must be configured in a two-directional manner. Figure 6.12 depicts a cascade temperature control system with provisions for batch heat-up. The heating- and cooling-medium control valves are split-range controlled, such that the heating-medium control valve operates between 12 to 20 mA (50% to 100%) and the cooling-medium control valve operates between 4 to 12 mA (0% to 50%) output signal. It is important to match the characteristics of the valves and to avoid nonlinearity at the transition, which can result in cycling. The control system shown in Figure 6.12 is a fail-safe arrangement, because in the case of air failure the heating valve is closed and the coolant valve is opened.

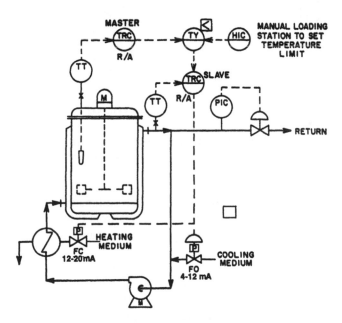

FIGURE 6.12 Two-directional cascade loop with a maximum limit on jacket temperature allows heat to be added in some phases of the operation and removed in others.

Figure 6.12 also shows an arrangement whereby an upper temperature limit is set on the recirculating heat-transfer-medium stream. This is an important consideration if the product is temperature-sensitive or if the reaction is adversely affected by high reactor wall temperature. In this particular case, the set point of the slave controller is prevented from exceeding a preset high-temperature limit. Another feature shown is a backpressure control loop in the heat-transfer-medium return line. This may be needed to impose an artificial backpressure, so that during the heat-up cycle no water will leave the recirculation loop and therefore the pump will not experience cavitation problems.

a. Limitations of cascade control

Figure 6.13 illustrates a cascade loop consisting of three controllers in series. The master is the reaction temperature controller, the primary slave is the jacket temperature controller, and the secondary slave is the valve position controller. A positioner is a position controller that detects the opening of the valve and corrects for any deviations between measurement (the mechanically detected position) and set point (the pneumatic signal received). In such a hierarchical arrangement, *only the master* can control its variable independently; the slave controller set points must be freely adjustable to satisfy the requirements of the master.

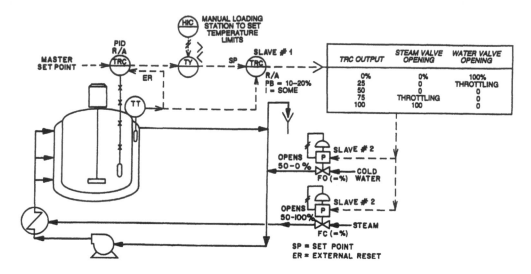

FIGURE 6.13 If the valves have positioners, a cascade loop will consist of three controllers in series. All cascade masters should be provided with external reset from the measurement of the slave.

The accuracy of positioners is between 0.05 and 0.1 millimeter; their dead band (minimum input change required for a detectable change in output) is from 0.1% to 0.5% of span. The open loop gain (the amount of change in output pressure per unit change in input when stem is locked in position) is from 10:1 to 200:1.

Another limitation is that the cascade loop will be stable only if each slave is faster than its master. Otherwise, the slave cannot respond in time to the variations in the master output signal, and a cascade configuration will in fact degrade the overall quality of control. A rule of thumb is that the period of oscillation of the slave should not exceed 30% of the period of oscillation of the master loop. This requirement is not always satisfied. For example, in Figure 6.13 it is important to select valve positioners that are faster than the slave temperature controller on the jacket. Similarly, the jacket temperature control loop should contain less dead time than its master, which might not be possible if a once-through piping configuration (Figure 6.8) is used. One possible method of reducing the dead time of the cascade slave loop is to move the measurement from the jacket outlet to the jacket inlet. This usually is not recommended, because when this is done, the slave will do much less work because the nonlinear dynamics of the jacket (Figure 6.5) have been transferred into the master loop.

b. Reset wind-up

Whenever the master is prevented from modifying the set point of the slave—because of a limiter, such as in Figure 6.13, or because the slave has been manually switched from remote to local set point—reset wind-up can occur. Reset wind-up is the integration of an error that the controller is prevented from eliminating. Consequently, the controller output is saturated at an extreme value. Once saturated, the controller is ineffective when control is returned until an equal and opposite area of error unsaturates it. This problem is eliminated by the external reset (ER) shown in Figure 6.13. The external reset signal converts the contribution of the integral mode to just a bias and thereby stops the integral action whenever the slave is not on set point

The external reset of the cascade master by the slave measurement guarantees bumpless transfer when the operator switches from automatic to full cascade control. The internal logic of the master controller algorithm is such that as long as its output signal (m) does not equal its external reset (ER), the value of m is set to be the sum of the external reset (ER) and the proportional correction (K_c(e)) only.

$$m = ER + K_c e \qquad (12)$$

When m = ER, the integral mode is activated. This feature eliminates the need for switching the master to manual and thereby also eliminates the need for the auto/manual station. In addition, it eliminates reset wind-up upsets due to start-ups, shutdowns, or emergency overrides. Whenever external reset is used the slave must have some integral to eliminate the offset; otherwise, the slave offset would cause an offset in the master.

2. Multiple Heat-Transfer Media

The use of a single coolant and single heating media (shown in Figure 6.13) is often insufficient or uneconomical. If one type of coolant (or heating media) is less expensive than another—for example, the cold water used in the system in Figure 6.14 might be less expensive than the chilled water—it is desirable to fully utilize the first before starting to use the second.

For best performance, the fact that there are three valves should be transparent to the temperature controllers. Their gain should be about the same, and their combined range should appear as the straight line in Figure 6.5. This is not easy to achieve, particularly in the regions where all the valves are nearly closed. This happens to be the case when the controller output is 14.6 mA (11 PSIG) in Figure 6.14. Therefore, some users prefer to provide some overlap so that the water valve might start opening at 14.6 mA (11 PSIG) while the steam valve does not close fully until the signal drops to 13.8 mA (10.5 PSIG). Overlapping at the transition points improves control quality but at the price of energy efficiency.

Figure 6.14 also shows that the destination of the returning water should not be selected on the basis of the origin of that water but rather should be based on the temperature of that water. This will reduce the upset caused in the plant utilities when a reactor switches from heating to cooling (Figure 2.3).

Figure 6.15 describes a reactor with a separate chilled water coil. This coil is inoperative until the cold water valve approaches full opening (the TC output drops to 5%). When the valve position controller (VPC) detects that condition, it starts opening the chilled water valve. The resulting increased heat removal will cause the temperature cascade loop to close the cold water valve until its opening is below the setting of the VPC. The VPC is tuned similarly to the master of the cascade loop.

In certain industries, such as pharmaceuticals, the heat-transfer fluids might include several waters, steam, and methanol. Figure 6.16 illustrates a configuration in which five modes of operation can be configured through the use of two throttling and six on-off valves. The on-off valves are of

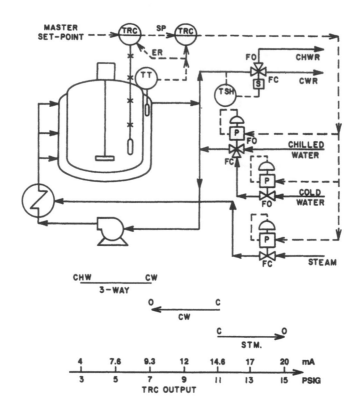

FIGURE 6.14 Split-range sequencing with multiple coolants can be used to minimize cost.

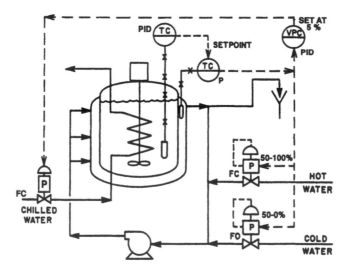

FIGURE 6.15 Higher-cost chilled water is used only when the lower-cost cold water is insufficient to meet the demand for cooling.

the tight shut-off design, to make sure that methanol and water will not intermix. Direct steam heating can lower the heat-up period in applications where high jacket temperatures are acceptable. Interlocks are provided to match the supply and return paths of the utilities, and a back-pressure controller (PIC) is installed to protect against jacket draining.

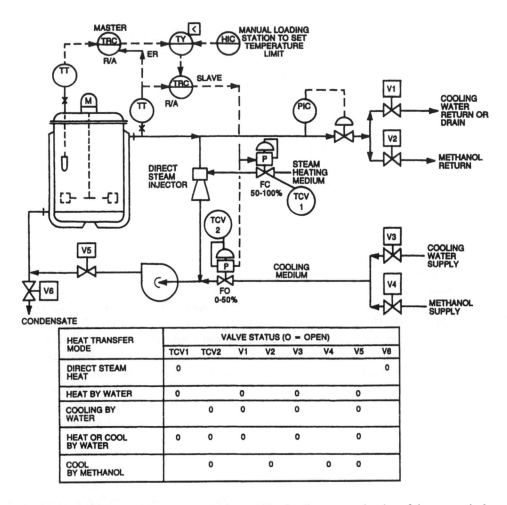

FIGURE 6.16 Multimedia heating system with capability for direct steam heating of the reactor jacket and for using both water and methanol as the cooling media.

3. Cooling by Boiling Jacket Water

Figure 6.17 describes the most effective method of cooling, which is by boiling water in the reactor jacket. In this control system, an increase in reaction temperature results in a lowering of the boiling pressure in the jacket. This causes a fast and linear increase in temperature difference across the heat-transfer area as a response to an increase in reaction temperature. In this control system, the loss in jacket water is made up on level control and the make-up rate is an indication of the rate of reaction heat release. This method of cooling provides a very stable, yet highly responsive heat removal system.

4. Improving the Speed of Response

In a process where the reactor pressure is a function of temperature (e.g., the reactor pressure is essentially the vapor pressure of one of the major components in the reaction), this pressure may be sensed and used to speed up temperature control. In large polymerization reactors having low heat-transfer coefficients and large changes in heat evolution, the conventional temperature cascade loop is not fast enough. On the other hand, pressure management gives an almost instantaneous indication of changes in temperature.

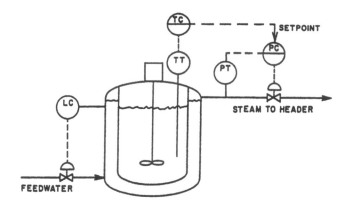

FIGURE 6.17   An internal heat sink is provided by the boiling coolant in the jacket. This stabilizes exothermic reactors.

Figure 6.18 illustrates the application of a pressure-compensated temperature control system to a reactor with both jacket and overhead condenser cooling (U.S. Patent No. 3,708658, assigned to Byrd Hopkins of the Monsanto Company). This same approach can also be applied to reactors with jacket cooling only. Under steady-state conditions the reactor temperature (T_m) is on set point, set by the cam programmer, and therefore the calculated temperature $T_c = T_m$. When an upset occurs, the pressure transmitter (PT) will detect it first, causing the calculated temperature (T_c) to change as the AP part of the expression is changed. This will make the measurement of TC-1 much faster than it otherwise would have been; it also allows the overhead condenser to start removing the excess heat even before the temperature transmitter (TT-1) is able to detect it. After each dynamic upset, the PI controller slowly returns the calculated temperature (T_c) to equal the measured temperature (T_m). This then automatically reestablishes the correct pressure-temperature relationship as the composition in the reactor changes.

The net result is the ability to operate the reactor at a much higher reaction rate, thereby obtaining higher productivity than was possible with temperature control alone.

TC-1 in Figure 6.18 would normally be provided with proportional and derivative control modes only. When TC-1 is on set point, the output signal returns to a value set by an adjustable internal bias. If the preferred means of cooling is through the jacket, then this bias will be set to a low value, but not to zero.

If there is no overhead condenser, the control system could be configured as in Figure 6.19. The control modes are so distributed between the measured variables that integral action acts on the slowly responding temperature while derivative action is applied to the more sensitive pressure. The derivative time setting will be much shorter than the integral, because pressure responds faster than temperature. Integral action cannot be applied to the pressure measurement, because the pressure can vary even under constant temperature conditions (as a result of variation in feed composition or catalyst activity), and the intent here is to use the pressure loop *only in the unsteady state*. Naturally, integral action is applied to the temperature measurement signal, because it is steady-state temperature that determines product quality, and integral action will assure its return to set point.

D. Heat-Up Controls

In most chemical processes, before the reactions can be initiated the ingredients first have to be heated up to the reaction temperature (T_r). If during this heat-up period the temperature is controlled in a way such as shown in Figure 6.11, and, if the cascade master is provided with a PID algorithm, such as A or B in Figure 6.10, the response will be as shown in Figure 6.20. This is because the integral mode (I) will note a negative error as the measurement(c) is rising from ambient to the

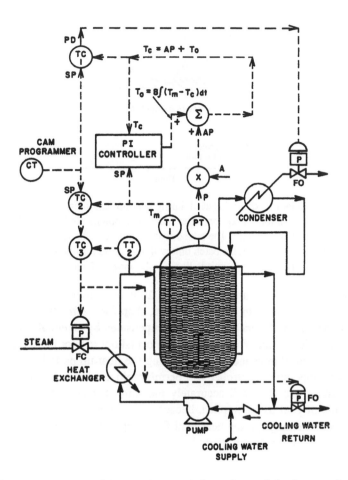

FIGURE 6.18 Pressure-compensated temperature control can be used for improved speed of response. (Source: Adapted from Byrd Hopkins, U.S. Patent No. 3,708658.)

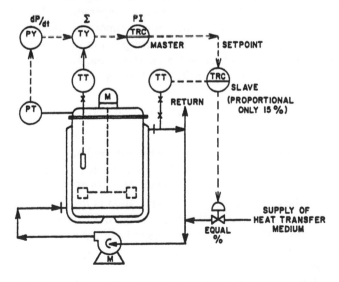

FIGURE 6.19 The derivative mode can be applied to the quickly responding variable, and the integral mode can be applied to the slowly responding variable.

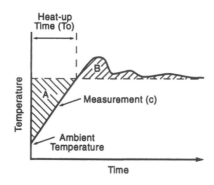

FIGURE 6.20   If conventional PID control is used during heat-up (t_o), integral saturation causes temperature overshoot. Therefore, the conventional PID controller must be supplemented with added features to provide it with the required start-up characteristics. The added feature can be a "batch unit" or the "dual mode unit." If the proportional band is less than 50%, the batch unit will give good results, whereas if a proportional band wider than 50% is required, the dual-mode unit is likely to be more effective.

set-point temperature (T_r). This area (A) in Figure 6.20 is memorized by the integral mode of the PID controller as energy it owes to the process and, therefore, *after* the set point is reached, it continues to supply heat until the energy depth is compensated by an equal quantity of heat, which is added over the set point. In many processes, such overshoot cannot be tolerated and, therefore, the conventional PID algorithm cannot be used during the heat-up period.

An ideal reactor temperature controller will permit rapid automatic heat-up to reaction temperature without overshoot and then will accurately maintain that reaction temperature for several hours. This is a difficult goal to accomplish, because the dynamics of the controlled process will go through a substantial change as the heat load first drops to zero and then as the cooling load gradually evolves when the reaction is started.

1. Batch Unit

The purpose of the batch switch is to detect whether the output signal to the control valve or the slave set point (m or m_1 in Figure 6.21) has reached the high limit (HL), which corresponds to a fully open heating valve, or the maximum jacket temperature. When that is the case, the integral feedback is switched from the controller output, either to a gradually reduced or constant (preload = PL) signal. The preload is selected to correspond to the expected output (load) *after* the integral feedback (F) is switched back to the controller output.

The batch switch stays in the preload position as long as

$$\text{Error (e)} > (\text{HL} - \text{PL}) \, \text{PB}/100 \tag{13}$$

and during that period PL remains to be the constant of integration. Therefore, during this period, the batch switch holds the controller output at its trip point of HL, by biasing the controller in proportion to deviation. When the error(e) drops below (HL − PL)PB/100, the batch switch returns the integral feedback to the controller output and integration is resumed.

Without the batch unit, the controllers illustrated in Figure 6.21 would receive a feedback signal (F) equaling the output (m_1). Therefore, whenever there is an error (e), the output signal is driven continuously by the positive feedback through I (a first-order lag, having a time constant I) until bias (b) reaches the saturation limit. Once in this saturated state (the reset is wound up), the output signal "m_1" will equal "b" even if the error has returned to zero. This is the reason that in Figure 6.20 the temperature keeps rising even after it has reached set point (SP = c, e = 0). Without the

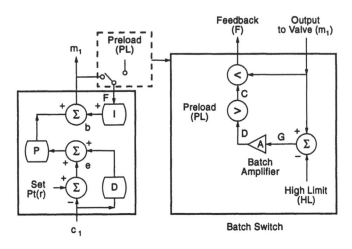

FIGURE 6.21   Inserting the batch switch into an interacting PID algorithm.

batch unit, therefore, control action cannot begin until an equal and opposite area of error is experienced.

Under normal operation, the output to the valve is below the high limit (HL). Therefore, G is positive and the amplifier (on the right of Figure 6.21) drives D upward, which causes m_1 to be less than C. In this state, the low selector selects m_1 as the feedback signal, and the controller behaves as a conventional PI controller. (The batch switch is closed as shown in Figure 6.21, to the left.)

When m_1 exceeds HL, the amplifier drives down D, C, F and b, and thereby limits m_1 from exceeding the HL setting.

If there is no preload setting or selector in the batch switch, it is possible that the feedback (F) will drive the bias (b) too far down, which can result in an undershoot, as shown in Figure 6.22. The purpose of the preload (PL) is to prevent this undershoot and thereby minimize the heat-up period with neither over- nor undershoot (Figure 6.22).

It can be seen from the above that a PID-type batch controller requires a total of five adjustments, because HL and PL must be set and the three control modes must be tuned. HL should be set at the maximum allowable jacket water temperature, which would then eliminate the need for a separate limit, such as the HIC in Figure 6.12.

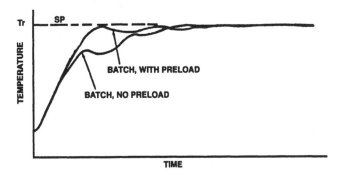

FIGURE 6.22   The preload provided by the batch unit, if properly adjusted, will eliminate both under- and overshoot.

The correct setting for PL is the master controller output at that time when reaction has started and a steady state has been reached between the generation and the removal of the heat of reaction. If, for example, the jacket water temperature during steady state is 90°F, this value could be selected as the preload setting, which will be the output of the master (and the set point of the slave) when the reaction temperature has been reached. Actually, the PL setting should be a few degrees lower than this value, say 87°F or 88°F, to allow for the contribution of the integral action of the controller from the time the proportional band is entered to the time when the set point is reached (Equation 13).

The effectiveness of the batch unit is reduced when the reactor requires a wide proportional band, say in excess of 50%. With a wide band (as shown in Figure 6.23) reset action begins earlier, which would lengthen the heat-up time. In such a case, one might consider using "dual mode control."

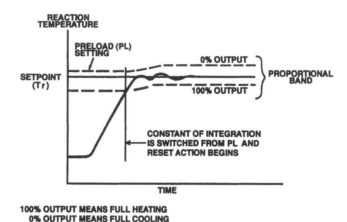

FIGURE 6.23 The wider the proportional band, the slower the heat-up, if batch switch is used.

2. Dual Mode Control

The name "dual mode" implies that two modes of control are applied, one for the heat-up period, the other when the reaction temperature (T_r) has been reached and the reaction has started. During this second mode, after a time delay control is transferred to a preloaded PID controller. To maximize productivity (minimize heat-up time) during the heat-up period, the maximum available rate of heating is applied. The application of full heat can be terminated on the basis of one of two criteria:

If *minimum heat-up time* is the design criteria, heating is continued in excess of the amount needed to reach T_r and this excess is then removed by a short period of full cooling (left side of Figure 6.24). If *minimum energy use* is the criterion, full heating is applied only until the reaction temperature has approached close enough (ΔE_{min2} in Figure 6.24) so that, if heating is stopped at that point, the temperature will coast up to T_r without overshoot.

In the dual-mode unit, which is set for *minimum heat-up time*, the preload is estimated as in the case of the batch unit, but it is not reduced for integral correction (not lowered from 90°F to 87°F or 88°F in our example), because in this case reset does not begin until the error is zero.

The sequence of operation is as follows:

1. Full heating is applied until the reactor is within 1 or 2% of its set point temperature. This margin is set by E_{min1}. During this state SS-1 and SS-2 are in position "A" in Figure 6.25.
2. When E drops to E_{min1}, time delays TD-1&2 are started, and full cooling is applied to the reactor for the duration of TD-1, which is a minute or so to remove the thermal

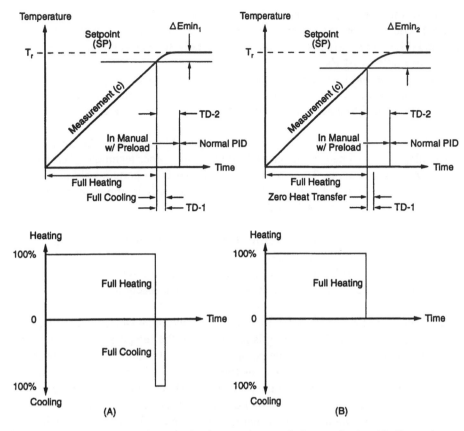

FIGURE 6.24 Dual-mode control can be implemented to maximize production (A, fastest heat-up) or to optimize energy use (B).

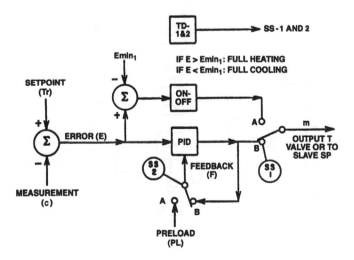

FIGURE 6.25 The dual-mode controller sequences full heating, full cooling, and finally preloaded PID for optimum start-up of potentially unstable batch reactors.

inertia of the heat-up phase. When TD-1 times out, SS-1 switches to position "B" and
the PID controller output is sent to the slave as its set point. This output is fixed at the
preload (PL) setting, which corresponds to the steady-state jacket temperature (estimated
in our example as 90°F).

3. When the error and its rate of change are both zero, estimated by the setting of TD-2,
   this time delay will switch SS-2 to position "B." This switching also transfers the PID
   loop from manual to automatic, and closes its external feedback loop.

If properly tuned, the dual-mode unit is the best possible controller, because by definition,
optimal switching is unmatched in the unsteady state by any other technique. On the other hand,
this loop requires seven settings. Three of these — P, I, and D — pertain only to the steady state
of the process; the other four — PL, E_{min1}, TD-1, and TD-2 — will determine start-up performance.
The effect of these adjustments is self-evident:

E_{min1} should be increased in case of overshoot and lowered if undershoot is experienced.
PL has the same effect as in Figure 6.22.
TD-1, if set too long, will bring the temperature down after the set point is reached.
TD-2 is not very critical, but should be set for a time when reaction has started.

If the dual-mode unit is configured to function in the *minimum energy use* mode, Figure 6.26
describes the operation. Here, full heating is applied until the temperature rises within E_{min2} of
T_r on the right side of Figure 6.24. At that point, no more heat is added (nor removed), and TD-
1 and TD-2 timers are started. When TD-1 has timed out, SS-1 in Figure 6.26 is switched from
zero heat transfer to position B, allowing the preloaded PID output to be the output of the master
controller. This condition is maintained until TD-2 also times out, switching SS-2 into position
B. At this point, the PID, which was in manual with its output held at PL, is switched to normal
automatic operation.

3. Model-Based Control

Model-based heat-up algorithms can be configured in two blocks, the "predictor block" and the
"corrector block."[4] The predictor block is a process model of the reactor and serves to determine

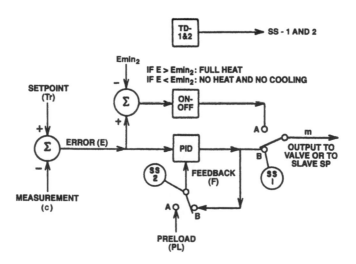

FIGURE 6.26 Dual-mode controller configured for minimum energy use.

the amount of heat input required (Qh) at any point in the batch heat-up cycle, so as to minimize the heat-up period while eliminating overshoot. The corrector block serves to refine the model by correcting the constants and calculation parameters in the process model by comparing the model to the actual process. Figure 6.27 illustrates the operation of such a model-based heat-up control system.

The variable manipulated by the "predictor" is the heat input rate (Qh), which initially is set to maximum, similar to the actions of the dual-mode control algorithms. The difference with the dual-mode strategy is that here, as the reactor temperature approaches its target set point, the manipulated variable is switched to a minimum rate of heating (Qh)min. The switching from (Qh)max to (Qh)min occurs at the time when the steady-state reactor model indicates that the new steady state temperature will match the set point.

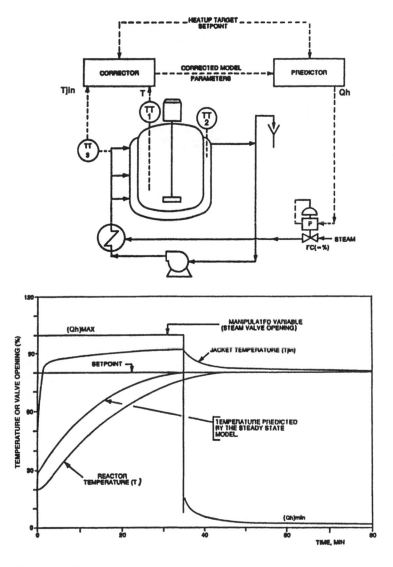

FIGURE 6.27 Model-based heat-up controls switch from maximum to minimum heating at the point where the model predicts that enough heat has been introduced for the reactor temperature to reach its set point, once the new steady state is reached.

4. Rate of Temperature Rise

In highly unstable, accident-prone reactors that have a history of runaway reactions, an added level of protection can be provided, based on the permissible rate of temperature rise during heat-up. This is usually superimposed on one of the previously discussed control systems on a selective basis as a safety backup. The protection is provided by calculating the actual rate of temperature rise in units of °F/minute and sending that signal as the measurement to RRIC in Figure 6.28. The set point of this controller is programmed as a function of the heat-up state.

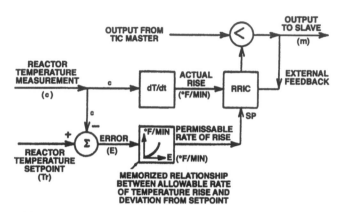

FIGURE 6.28   The rate of temperature rise constraint can be applied as added protection.

Toward the end of the heat-up period, when the reaction temperature has been almost reached (the error is near zero), the value of the permissible rate of rise is set to be much lower than at the beginning of heat-up. Doing so prevents the process from building up a thermal inertia as it approaches the region of potential instability.

5. Other Heat-Up Strategies

Several other strategies will speed heat-up without overshooting. One such scheme is to place the derivative mode between the transmitter and the controller. This way, the higher the rate of temperature rise, the higher the reactor temperature appears to be, and, therefore, the controller output drops before the set point is reached. The disadvantage of this strategy is its high noise sensitivity.

The various model-based control strategies such as IMC (internal model control), MPC (model predictive control), the Dahlin and Smith predictor controllers, and the PID algorithms with dead-time compensation have already been discussed in connection with Figure 6.10 and will also be examined in connection with continuous reactor controls.

III. BATCH STIRRED TANK REACTORS (BSTR)

The batch reaction cycle includes cleaning, purging, pressure testing, charging, heat-up, initiation, reaction, end-point determination, stripping, and discharging. With the exception of heat-up controls, which have already been covered, the control and automation of each of these phases will be discussed in the following paragraphs.

The nature of the batch reactor process is such that it never reaches a steady state. The process is in a continuous state of transformation and change; therefore, the control system parameters should also be in a continuous state of change. Some of these changes are rather drastic; for example, when the system switches from heating to cooling. When such changes occur, the controller parameters must be changed immediately and equally drastically, which is usually done by "programmed adaptation," meaning that the PID algorithm settings for the various stages of the reaction cycle are stored in a table and are recalled as needed.

Adaptation of tuning constants can also occur slowly using adaptive algorithms as a function of various measurements, for example, the rising of the level (mass) in the reactor or the reduction in the exothermic heat release as the reaction rate drops off. Adaptation can also be based on various rules of thumbs, such as utilizing the rate of temperature rise during heating up the reactor to predict the reactor time constants, which in turn can be used to tune the temperature controller.

In a batch process, where the same product is made in consecutive batches, preload is often determined empirically. This means that if in batch #19 the transition to reaction phase was stable and the controller output at the beginning of that phase was 52%, it is a fairly safe assumption that in batch #20 the same 52% preload will also provide stable transition. In some reactor heat-up applications the preload can also be predicted on the basis of the rate of temperature rise during heat-up.

A. BATCH CHARGING

Reactants and catalysts can be charged into the reactors either by weight or by volume, using flow meters, pumps, feeders, and weighing systems, as shown in Table 6.1. The main considerations in making the selection are accuracy and reliability of the data obtained. Because material balances are always based on the weight (not the volume) of the reactants, all things being equal, the mass or weight type sensors are preferable; using such sensors eliminates the need for density or temperature compensation. If they cannot be considered for some reason, the solid-state volumetric sensors (vortex) would be preferred over the ones with moving parts (turbines, pumps, positive displacement meters), because sensors with moving parts tend to be high-maintenance devices, with frequent need for recalibration.

TABLE 6.1
Performance Capabilities of Batch Charging Instruments

Detector	Type	Min. Error	Rangeability
Load cell weighing	Mass	±0.1%FS	NL
Turbine-type flowmeters	Volumetric	±0.25%AF	10:1
Positive displacement flowmeters	Volumetric	±0.25%AF	10:1
Metering pumps	Volumetric	±0.5%FS	20:1
Mass flow meters	Mass	±0.2%AF	20:1
Vortex shedding flow meters	Volumetric	±0.75%AF	10:1
Level	Volumetric	±0.25 in.	NL

AF — Actual flow,
FS — Full scale,
NL — No limitation

Therefore, it might be advisable to place the reactor on load cells and charge all ingredients that weight more than 10% of the total batch by weight. Reactants and additives that are needed in lesser quantities can be added under Coriolis mass flowmeter control with the load cells providing only backup. Catalysts or ingredients needed in extremely small quantities, representing less than 1% of the total batch, can be added by metering pumps or specialized feeders.

B. Sequencing Logic Controls

The sequence logic can be reduced to a time-ordered combination of a few basic actions, which are the components of the process and control states described in Figure 6.29.

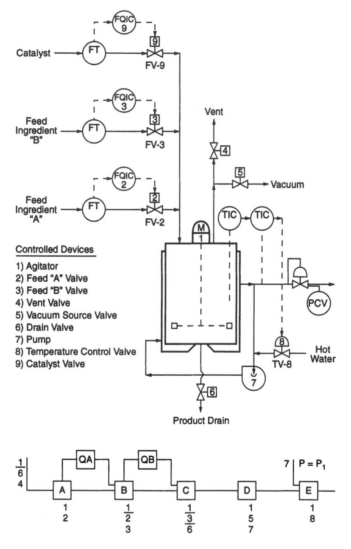

FIGURE 6.29 In this batch reactor example of sequencing logic, the upper portion shows the physical equipment involved, and the lower portion shows the sequenced process states.[3]

The upper portion of the illustration shows the physical equipment involved, including the nine controlled devices whose states are being logically sequenced. The lower portion of the illustration identifies the sequenced process states. During process state A the required quantity of raw material A is being charged. The prerequisites for the initiation of state A are that 1 and 4 be on and 6 be off. This means that the agitator must be on, the vent valve must be open, and the drain valve must be closed. When these prerequisites are satisfied, state A is initiated by turning 2 on. This means that the charge valve of reactant A is opened. The required quantity of this ingredient is set by the recipe as Qa. When this target is reached, process state A is terminated and process state B is initiated, which then performs its own tasks defined by the sequencing logic. In a more sophisticated reactor control system, the number of controlled devices and the number of logical sequence steps is much greater, but the basic concept is the same.

To minimize the software preparation costs, it is advisable to obtain the formal approval of all project participants before the design proceeds beyond the preliminary stage described in Figure 6.29. Once all the comments are incorporated and all required approvals are received, the more detailed stage of software development, the preparation of the "time sequence diagrams," can start. Figure 6.30 illustrates such a diagram, describing the steps in the process "A" in Figure 6.29, charging of ingredient A into the reactor.

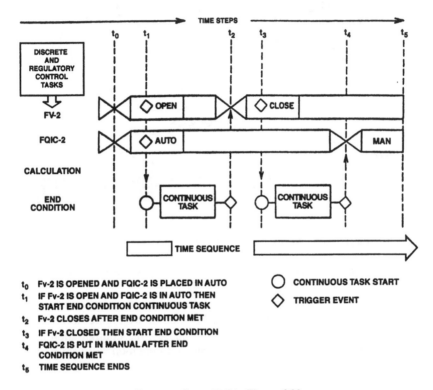

FIGURE 6.30 The time-sequence diagram of step "A" in Figure 6.29.

The time-sequence diagram shows the sequential changes in control actions and their time relationships. The sequence of events can be followed along dotted lines for vertical time coincidence and along the solid lines in a horizontal direction for sequence control.

The diamond symbol is used to indicate a trigger event, and a vertical dotted line indicates the time coincidence of trigger events. When two or more diamonds occur in the same relative time line, an "AND" logic condition is assumed.

The format of the time-sequence diagram proceeds from left to right in discrete time steps, with relative time being the horizontal coordinate.

Figure 6.30 illustrates a time-sequence diagram for a flow loop, such as the one shown on the top of Figure 6.54. First, the valve FV-2 is opened and the controller FQIC-2 is placed in automatic (t_0). When FV-2 is open and FQIC-2 is in "auto," the continuous task of flow integration is started (t_1). When the desired total is reached (t_2), FV-2 is closed. When FV-2 is closed (t_3), the integration of flow continues to check for leakage. If, at the end of a preset time interval, the total flow is still zero, FQIC-2 is switched back to manual (t_4) and the system is ready to start step "B" in Figure 6.29.

In batch process units, the batch sequence is subdivided into *process states*. Each state is given a unique name, such as *Charge, React, Heat, Cool, Hold, Discharge, Wash,* and *Empty*. Within

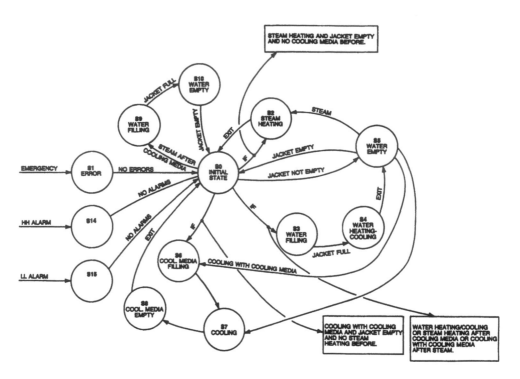

FIGURE 6.31   State diagram of a reactor jacket system.[5]

each state, discrete control functions, continuous-regulatory control functions, and safety and permissive interlocks are performed.

In each process state, a number of control modules can be in operation and each module can be in a number of states, as is illustrated for a jacket module by Figure 6.31. The totality of the recipe and tuning data, the time sequence steps and phases, the configuration of the regulatory functions, and the various safety interlocks all combine into the total software package of a batch-control system.

C. Model-Based Temperature Controls

In the previous paragraphs we have discussed the temperature controls used during heat-up and the conventional cascade controls used during reaction. In critical reactor control applications, one might also consider the use of model-based control.

When a reaction is highly exothermic and if runaway reactions can occur, the use of a model-based control system is justified. In order to develop the total model, three component models need to be developed:

1. *Equipment model*, describing the effect of heat-transfer media on reactor temperature (T_r)
2. *Kinetic model*, describing the effect of T_r on the chemical reaction
3. *Calorimetric model*, describing the effect of reaction rate on T_r[4]

When an exothermic reaction is taking place within the reactor, a feedback cascade loop such as that shown in Figure 6.13 is needed to provide stable temperature control by matching the rate of heat removal to the rate of heat generation. If the TRC in Figure 6.13 was tuned at a time when there was no reaction taking place within the reactor, it will not perform properly when exothermic heat is being generated. If the gain of the master TRC is fixed, the loop might become unstable if the reaction is autocatalytic (does not require a catalyst). Model-based controls can improve on such performance.

One such model-based approach is to use a heat-release-rate estimator. The rate of heat release (Q in Figure 6.32) can be estimated on the basis of reactor temperature (Tr) and jacket-inlet

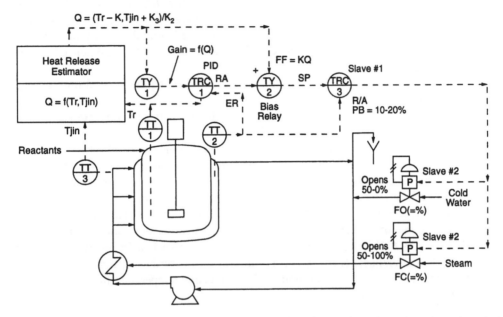

FIGURE 6.32 Heat-release estimator increases the gain of TRC-1 as the estimated total exothermic heat release rises and the process gain drops in a batch reactor.

temperature (Tjin) in accordance with

$$Q = (Tr - K_1\ Tjin + K_3)/K_2 \qquad (14)$$

In exothermic reactions Q is the major load variation that tends to upset the stability of the temperature controls. Once this disturbance load is estimated, it is possible to feedforward that estimate to the slave set point as a bias. This bias relay is TY-2 in Figure 6.32 and the feedforward bias (FF) it adds to the slave set point is a function of Q. The feedback master (TRC-1) naturally corrects the total effect on the jacket temperature set point and thereby overcomes the errors in the feedforward model.

The gain of the feedback master (TRC-1) is varied as a function of Q. As Q increases, the process gain drops, and therefore the gain of the nonlinear adaptive-gain feedback controller is increased in order to keep the gain product for the loop constant (Figure 10.4).

D. MAXIMIZING PRODUCTION

Optimization of a chemical batch reactor usually means that its throughput is maximized while the operating costs are minimized. These goals can be met by simple, traditional techniques such as by optimizing the utility distribution in such a way that no reactor is held up in its reaction sequence because it has to wait for the availability of a particular resource. Other obvious optimization techniques include the increasing of the reactant charging rates, the maximization of reactor heat-up or cool-down rates, and the maximized utilization of coolant availability during the exothermic reaction itself.

In a batch reactor, production rate increases with temperature, and therefore production can be maximized by maximizing temperature. This can be accomplished by a valve position controller (VPC) that raises the batch temperature set point whenever the coolant valve is less than 90% open. Unfortunately, the dynamic characteristics of this loop are undesirable because of the inverse response of the loop. When the coolant valve opens to more than 90%, the VPC will lower the set point of the temperature controller. This will temporarily *increase* the demand for coolant, although once the excess sensitive heat is removed, it will lower it. The longer the dead time introduced by this inverse response, the larger the margin needed for safety and stability, and hence the lower the set point of the VPC.

For such applications the VPC can have *only* integral mode and the integral time setting must be long to guarantee stability.

When attempting to maximize the heat removal capacity of a reactor, one can also precool the reactants prior to charging them to the reactor.[4]

The reaction rate can be measured in a reactor by detecting the exothermic heat release. Figure 6.32 has shown a method to estimate the rate of heat release. Whether an accurate measure or just an estimate of the heat release rate (Q) has been made, that information can be used for control. In Figure 6.3, the heat release (Q) was only used as the basis for adjusting the gain of the TRC master and as a feedforward anticipator; however, it can do more, because by converting the nonlinear reaction rate of the batch into a linear rate, the batch reaction time is much reduced. As shown in Figure 6.33, the batch reaction is completed much faster when the reaction rate is constant (linear) than when a nonlinear (first- or second-order) curve describes its time characteristics.[4] In order to keep the reaction rate from dropping off, the reaction temperature has to be increased (it is a nonisothermal batch reaction).

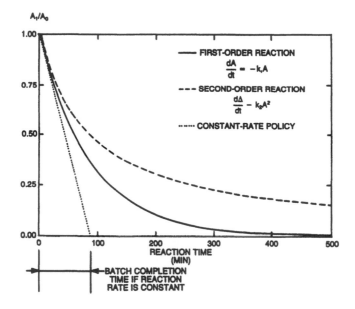

FIGURE 6.33 The batch reaction time is a function of the reaction rate within the reactor. If constant rate is maintained, the batch reaction time can be reduced drastically.[4]

Changing the operation of a batch reactor from a constant reaction temperature to a variable reaction temperature mode is not an easy proposition to carry out.[6] One of the disadvantages of a nonisothermal operation is that the formed products might change as the reaction temperature changes. Yet, in some processes, the economic advantages outweigh the required complexities in modeling and control. One possible configuration is to use the reactor shown in Figure 6.32, and use the heat-release estimator in that figure to modify the set point of TRC-1 in such a way that the reaction rate will be held constant. For some reactions the temperature settings required to provide constant reaction rates can be calculated, while for more complex reactions it is necessary to determine them experimentally.

Figure 6.34 shows a cascade configuration in which the reaction rate controller (XRC-4) is the master and it adjusts the batch temperature (TRC-1) set point so as to keep the reaction rate constant. The measurement to the reaction rate controller can come from a heat-release estimator (Figure 6.34) or from the actual measurement of the exothermic heat release to be discussed in the next paragraph.

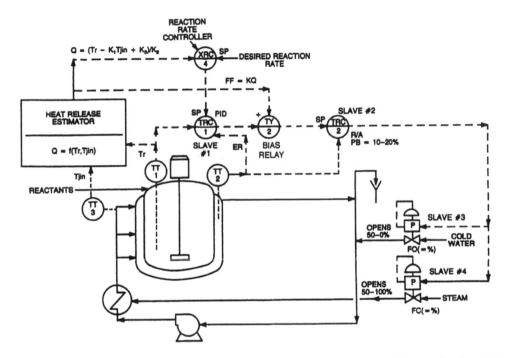

FIGURE 6.34 Nonisothermal (variable reaction temperature) batch reactor control shortens batch cycle time by keeping the reaction rate constant.

The temperature of the batch at any one time is a function of the balance between the exothermic heat that is generated and the quantity of heat that is being removed by the coolant. If the two are not in balance, the batch temperature changes. In the previously described control schemes, the exothermic heat was manipulated by changing the rate of reaction and the cascade loop controlled the batch temperature by adjusting the coolant flow to match that load. In some of the faster and more stable processes, this configuration can be reversed as shown in Figure 6.35. Here the rate of cooling is fixed (at a relatively high rate) and the temperature controller (TRC-4) is adjusting the reactant flow (exothermic heat generation) to maintain a balance.

At the beginning of the batch the concentrations of the precharged other reactants are high, the concentration of product in the batch is low, and therefore the process gain is very high. As the reaction progresses, product concentration rises, unreacted reactant concentrations drop, and as a consequence the gain of the process drops. By the end of the reaction, the process gain (Gp) can be reduced ten- or even one-hundred-fold. As is shown in Figure 10.4, good control requires that the gain product of the loop be constant. Therefore, as Gp drops it is necessary to increase the controller gain (Gc) in proportion. This is done by FY-1 in Figure 6.35, which measures the total amount of reactant charged and increases the gain of TRC-4 as that total rises.

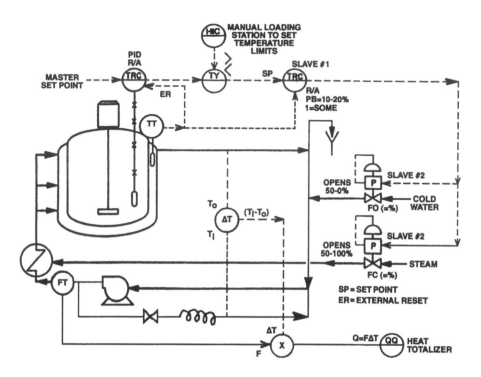

FIGURE 6.35 With constant coolant rate the reactant flow can be the cascade slave, and because process gain drops off as more and more reactants are converted to product, the controller (TRC-4) gain is automatically increased as a function of total reactant charged.

The function of the reaction rate (FY-1) is nonlinear and reaction-specific. TT-2 provides a feedforward signal so that if the coolant temperature would rise and thereby the cooling rate would drop off, this would also decrease the gain of TRC-4, thereby making it cause a smaller change in the slave set point.

The slave (FRC-1) set point is limited by the low signal selector TY-4. It compares two limits to the set point generated by TRC-4 and picks the lowest of the three signals. During start-up, safety is served by KT-4, which ramps the slave set point up at a safe rate to protect against overshooting. The slave (FRC-1) is preloaded to minimize start-up overshoot or transients. The amount of preload is based on experience and serves to stabilize the loop more quickly than if the initial value of the FRC-1 output was zero. The main advantage of such control systems is that they minimize the batch reaction time and thereby maximize plant productivity.

E. HEAT RELEASE DETERMINATION

By multiplying the jacket circulation rate by the difference in temperature between inlet and outlet, it is possible to determine Q, the amount of heat released or taken up by the reaction.

Under steady-state conditions, at high circulation rates the difference between T_o and T_i is about 5°F (3°C) or less. During load changes, dynamic compensation is needed. As more cooling is requested, the jacket inlet temperature will drop, but the jacket outlet temperature will stay unaffected until the jacket contents are displaced. If Q is to be used for control purposes, such dynamic errors must be removed by compensation. This is accomplished by delaying the jacket inlet temperature measurements into the ΔT transmitter by a time equal to the delay through the jacket. In Figure 6.36, this is accomplished by simulating the jacket using a length of tubing whose dead time is adjustable by changing the flow rate. A similar result can be obtained by placing an inverse

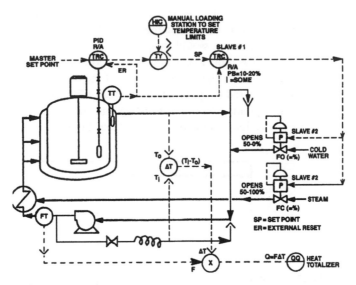

FIGURE 6.36 To correctly determine heat release, the jacket inlet temperature reading must be delayed by the residence time of the jacket, so that the ΔT transmitter will be dynamically compensated.

derivative relay, set for the jacket displacement time, into the transmitter output signal; in digital systems, the desired goal can also be reached by electronically simulating the transportation lag.

The correct selection of the ΔT transmitter is very important because in order to be able to detect Q within a $\pm 1\%$ error, the ΔT must be detected within $\pm 0.02°F$ ($\pm 0.01°C$). This can be accomplished only with the best RTD type transmitters with spans as narrow as -5 to 0 to $+5°F$. When the absolute value of the jacket temperature varies from, say, $50°F$ to $250°F$ throughout the reaction cycle, it is also necessary to correct the calibration of the ΔT transmitter as the absolute temperature changes.

Once the value of Q is accurately determined, it can be used for many purposes. The instantaneous value of Q signals the rate of heat removal or addition, and the time integral of Q gives the total heat that has been added or removed. A change in the value of ΔT under standard conditions can signal fouling and the need for cleaning the reactor heat transfer surfaces.

F. Endpoint Detection and Control

The reliable determination of batch cycle endpoint is important for two reasons:

1. One way to increase plant productivity is to reduce batch cycle time.
2. If the reaction continues beyond the optimal endpoint this can reduce process yield due to side reactions.

The completion of a batch reaction can be approximated on the basis of cycle time, total charge of reactants (level or weight), rise or fall of reaction pressure, total exothermic heat release (Figure 6.36), or a variety of analytical and physical measurements.

If one reactant is charged to the reactor in substantial excess to the others, this will usually guarantee the complete conversion of the other reactants, and end-point control is not required unless the product is also used as the recycle stream.

If the reactants are fed to the reactor in the same proportion in which they react, end-point control is required. If the reactants and product are all in the same phase (all liquids or all gases), the endpoint is usually determined through the use of an analyzer.

If some analytical property, for example, pH, is used in detecting the endpoint, it is important to realize that during most of the reaction the measurement will be away from set point. Therefore, integral action must not be used because it would saturate and an overshoot would result. This is one of the few processes in which the proper selection is a proportional plus derivative controller with zero bias, modulating an equal-percentage reagent valve. If the derivative time is set correctly, the valve will shut quickly as the desired endpoint pH is reached. With excessively long derivative time setting, the valve will close prematurely (Figure 6.37), but this is not harmful because the proportional mode will reopen it as long as the pH is away from the set point.

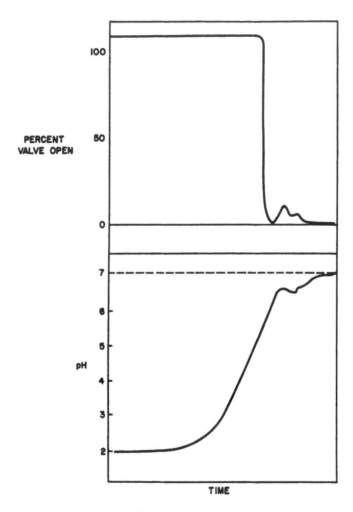

FIGURE 6.37   pH-based endpoint is controlled by proportional and derivative (PD) controllers.

For the measurement of pH, conductivity, and resistivity, on-line analyzers are available. The pH probe shown in Figure 6.38 can be used for on-line measurement in a glass-lined reactor. When maintenance or recalibration is needed or when the reactor is empty, the probe can be lifted out by a pneumatic actuator and immersed into buffer solution to protect it from getting dry.

When the reactants are in different phases, the endpoint can be controlled on the basis of level or pressure. If the product is gas, the endpoint can be maintained by the level controller, whereas if the product is liquid, the endpoint control can be based on pressure. When this pressure rises, that is an indication of excess gaseous reactant accumulation in the reactor. In response, the pressure controller (PC) will lower the gas flow; thereby, it will increase the liquid reactant concentration

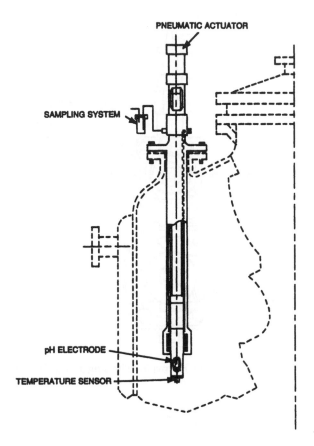

FIGURE 6.38 Immersion pH probe for reactors.

by increasing its relative flow. This in turn will increase gas consumption, and the balance will be reestablished and the correct endpoint maintained.

1. Probe-Type Analyzers

Probe-type analyzers do not require sampling systems, because the analyzer itself is inserted directly into the process. In the case of chemical reactors, analyzers can be inserted either into the reactor vessel or into the discharge piping. Regardless of their locations, probes will give accurate readings only when clean. Therefore, automatic cleaning of on-stream analyzer probes is a good idea. Figure 6.39 illustrates a flow-through, brush-type probe cleaner assembly that is provided with a sight glass allowing visual observation of the cleanliness of the probe. Such units are also available with spray nozzle attachments for water or chemical cleaning of the electrode. Side entry into a vessel with a retractable probe is the standard installation for fermentors, as illustrated in Figure 6.40.

One common method of endpoint detection is the measurement of viscosity either by monitoring the torque on the agitator or by using viscosity sensors. Such viscometers can be inserted in pipelines or directly into the reactor. It is essential that they be located in a representative area and be temperature-compensated.

In reactions in which the endpoint can be correlated to density, some of the more sensitive and less maintenance-prone densitometers can also be considered. The radiation-type design tends to meet these requirements if pipeline installation is acceptable. These units can operate with a minimum span of 0.05 specific gravity unit and an error of ±1% of that span, or an absolute error of ±0.0005 specific gravity.

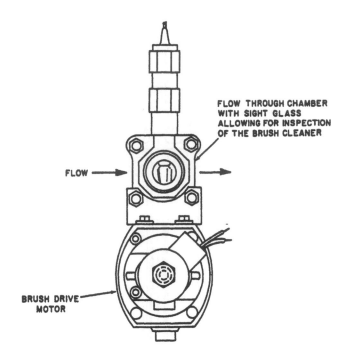

FIGURE 6.39 The flow-through probe cleaner is provided with a sight-glass for visual observation (Courtesy of Amico)

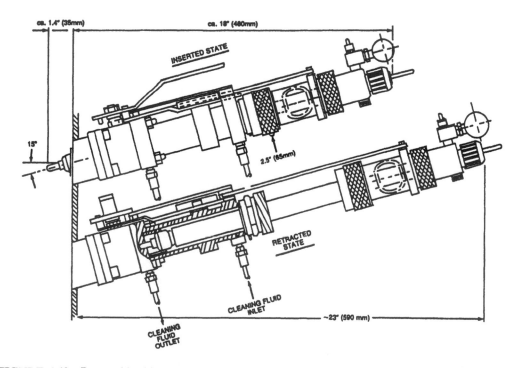

FIGURE 6.40 Retractable side entry probes are often used on fermentors. (Courtesy of Ingold).

In a number of reactions, refractometers have been found to be acceptable solutions to endpoint detection. When the RI analyzer is on-stream, human error tends to be reduced and analysis time is shortened, with a corresponding increase in productivity. On the other hand, an on-stream installation requires more maintenance attention, because the RI measurement must be made to high precision, such as ±0.0005 RI units. This requires accurate temperature compensation, plus protection through cooling of the heat-sensitive photocells, which are time-consuming and expensive to replace when ruined by high temperature.

If the endpoint of a reaction is detectable by such measurements as a change in color, opacity, or the concentration of suspended solids, the self-cleaning probe shown in Figure 6.41 can be considered for use. Here the reciprocating piston in the inner cavity of the probe not only serves to clean the optical surfaces through its wiping seals but also guarantees the replacement of the sample in the sampling chamber by fresh material upon each stroking of the piston.

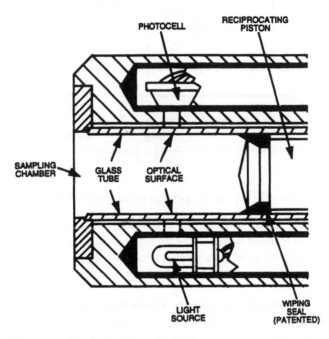

FIGURE 6.41 Probe-type suspended solids detector used on biological sludge applications. (Courtesy of Monitek Technologies Inc.).

G. STRIPPING CONTROLS

In most batch reactions, when the reaction is completed the batch is concentrated through the removal (stripping) of unreacted chemicals and solvents. In some plants, stripping is performed in the reactor vessel and, therefore, the speed of the stripping operation affects overall plant production.

During stripping or refluxing phases, the reactor can be controlled on the basis of heat input (Q). Therefore, the system shown in Figure 6.36 would be switched automatically from temperature control to heat input control (Figure 6.42) whenever the reactor enters a stripping or refluxing phase. The heat input during refluxing is usually set to be sufficient to maintain a state of slow boiling. During stripping, the heat input is usually set to complete the stripping in some empirically established time period; for example, an hour or two.

The time integral of Q represents the total reaction heat, which is an indicator of product concentration or percentage conversion. It can be used to introduce additives at predetermined conversions and to determine reaction endpoint. Through these automated steps, the need for taking grab samples can be eliminated, resulting in reduced overall cycle time and, therefore, increased production rate.

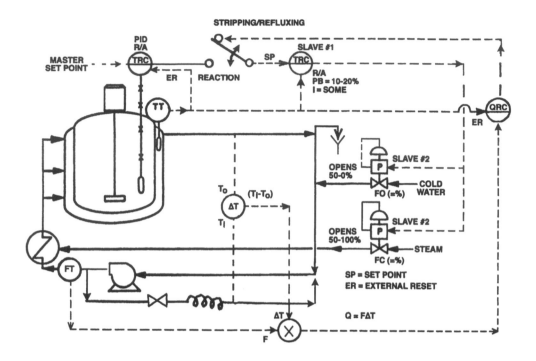

FIGURE 6.42 Control can be switched from temperature to heat input as the reaction cycle enters a stripping or refluxing phase.

If the endpoint of the reaction is detectable by infrared beam attenuation, such analyzers can also be put on-stream or can be inserted in pipelines and will detect the attenuated total reflectance of the IR beam. This type of analyzer is particularly suitable for the detection of water in hydrocarbons.

In addition to the analyzers dicussed above, a wide variety of computer-supported fiber-optic probes (FOPs) are also available for on-line analysis. Other analyzers, such as the gas chromatograph (GC), the high-pressure liquid chromatograph (HPLC), and the mass-spectrometer (MS) are also powerful analytical tools and are available both as laboratory or on-line devices.[7]

The analyzers are limited by sampling difficulties, by the intermittent nature of some models, and by incomplete mixing within the reactor. Because sampling and sample preparation result in dead time, sampling time should be minimized or eliminated altogether; this can be done by placing the sensor directly into the reactor, which results in tight control and fast response.

When stripping is done by direct steam injection, the controls shown in Figure 6.42 are not applicable. The controls of a vacuum stripper serve to remove the solvent and the unreacted monomers from the batch in minimum time but without causing foaming. The steam addition to the stripper (FIC in Figure 6.43) is usually programmed as a function of time (FY). As the batch gets more concentrated with polymer products and leaner in monomers or solvent, the more likely it is to foam and therefore the rate of steam addition is reduced. This is usually done by a hardware or software programmer (FY), which relates steam flow to time. This relationship is adjustable, and if foaming is experienced during a stripping cycle (LSA in Figure 6.43), the steam rate curve programmed in FY is lowered to avoid the repetition of foaming in the next batch.

The temperature of the batch is a function of (1) the balance between the amount of heat introduced by the steam and (2) the amount of cooling caused by the vaporization of the monomers. The vacuum level in the stripper is therefore controlled by temperature. This guarantees that the vacuum is low enough so that all the solvents and monomers are removed, but that it does not drop

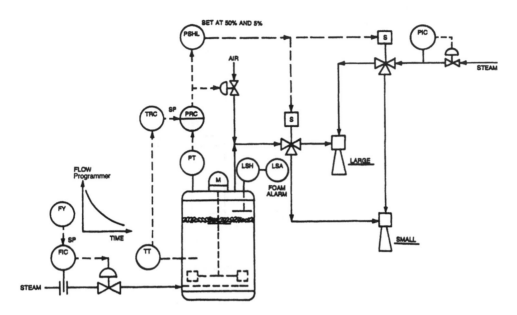

FIGURE 6.43   The controls of a vacuum stripper serve to remove the solvent and the unreacted monomers in minimum time but without causing foaming.

too low, to the point where the stripping steam would condense and would not be removed by the vacuum and therefore would cause dilution of the batch.

IV. CONTINUOUS STIRRED-TANK REACTORS (CSTR)

Ideally, once a continuous reactor is in operation, its operating conditions (feed rate, product withdrawal, cooling, etc.) should remain constant. In fact, the activity of catalysts varies, heat-transfer surfaces get dirty, and utility conditions or rates also vary. These changes affect the gain and time constants of the process, including its residence time and conversion. Most continuous reactors are operated under temperature control, except gas-liquid or gas-phase reactors, which are usually controlled by pressure or by residence time (level). In most cases, the product quality is only monitored, while in some the product analyzers automatically adjust the temperature or pressure set points.

A. GAS OR GAS-LIQUID REACTORS

In gas phase reactions, in oxidation and hydrogenation reactions, or in high-pressure polymerization, the reaction rate is also a function of pressure. If, in a batch reaction, the process gas is completely absorbed, the controls in Figure 6.44 would apply. Here, the concentration of process gas in the reactants is related to the partial pressure of the process gas in the vapor space. Therefore, pressure control results in the control of reaction rate. This loop is fast and easily controlled.

Certain reactions not only absorb the process gas feed but also generate by-product gases. Such a process might involve the formation of carbon dioxide in an oxidation reaction. Figure 6.45 illustrates the corresponding pressure control system. Here, the process gas feed to the reactor is on flow control, and reactor pressure is maintained by throttling a gas vent line. This particular illustration also shows a vent condenser, which is used to minimize the loss of reactor products through the vent.

In the case of continuous reactors, a system such as shown in Figure 6.46 is often employed. Here the reactor is liquid full, and both the reactor liquid and unreacted or resultant by-product

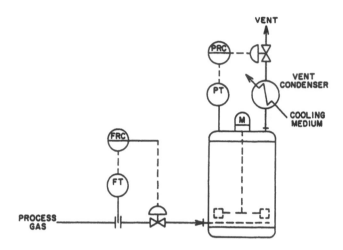

FIGURE 6.44   Reactor pressure can be controlled by gas make-up.

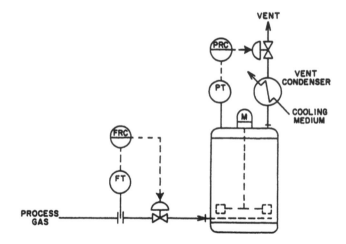

FIGURE 6.45   Reactor pressure can be controlled by vent throttling.

gases are relieved through the same outlet line. Reactor pressure is sensed, and the overflow from the reactor is throttled to maintain the desired operating pressure. Process gas feed and process liquid feed streams are on flow control.

It may be desirable to place one of the flow controllers on ratio in order to maintain a constant relationship between feed streams. If the reaction is hazardous, and there is the possibility of an explosion in the reactor (for example, oxidation of hydrocarbons), it may be desirable to add safety devices, such as a high-pressure switch, to stop the feed to the reactor automatically.

If one of the reactants differs in phase from both the other reactants and the product, inventory control can be applied (Figure 6.47). In the case of gaseous products (A), reactor level is controlled by modulating the liquid reactant; with liquid products, the gaseous reactant is modulated to keep reactor pressure constant (B). A purge is needed in both cases to rid the reactors of inerts either in the liquid or in the gas phase.

B. VACUUM CONTROL

Some reactions must be conducted under vacuum. The vacuum source is frequently a steam jet type ejector. Such units are essentially venturi nozzles with very little turndown. This constant

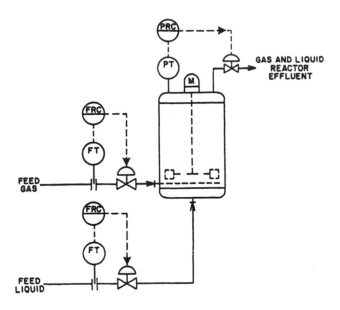

FIGURE 6.46 Arrangements such as this can be used to control pressure in continuous reactors.

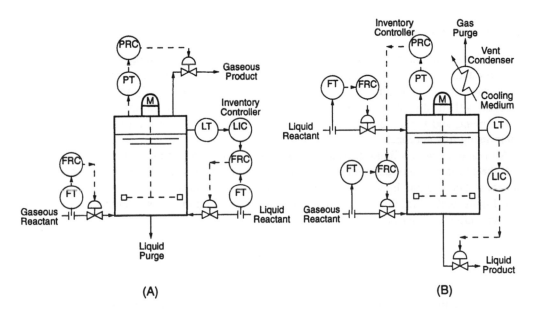

FIGURE 6.47 Inventory controllers can set production rate with both gaseous (A) and liquid (B) products.

capacity vacuum source is frequently matched to the variable capacity reactor by wasting the excess capacity of the ejector. This is done by creating an artificial load through the admission of ambient air. When the vapor generation rate in the reactor is low, the steam jet is still operating at full capacity, sucking in and ejecting ambient air. The corresponding waste of steam can be substantially reduced if the system shown in Figure 6.48 is applied. Here, a low load condition is detected by the fact that the air valve is more than 50% open, and this automatically transfers the system to the "small" jet. A later increase in load is detected as the air valve closes (at around 5%), at which point the system is switched back to the high capacity jet.

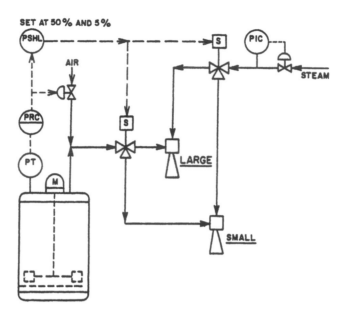

FIGURE 6.48   Cost-effective operation of vacuum pressure controls can reduce steam waste substantially.

C. RESIDENCE TIME CONTROL

In continuous reactors it is usually desirable to provide a constant residence time for the reaction. If the production rate (inflow to the reactor) varies, the most convenient way of keeping the residence time constant is to maintain a constant ratio between reactor volume (V) and reactor outflow (F). The ratio V/F is the residence time, where the volume (V) can be approximated by a level measurement, and F can be measured as outflow. Figure 6.49 illustrates a residence time control system using this approach.

All reactions with a liquid product can benefit from residence time control.

If the endpoint can be determined on the basis of reaction pressure (as in Figure 6.47), residence time control can be combined with end-point control by the addition of a pressure controller (PIC-01 in Figure 6.49), which will automatically cut the production rate set point to zero, when the reaction endpoint is reached.

V. OPTIMIZING CONTROLS

In batch reactions, one of the ways to increase productivity was to increase the rate of reaction by raising the reaction temperature (Figure 6.33). In continuous reactions, one method of maximizing production is to make sure that the potentially available cooling capacity is fully utilized.

In exothermic reactions, one of the critical safety constraints is coolant availability. In Figure 6.50 the optimizing controller (OIC) detects the opening of the coolant valve, and if it is less than 90% open, it admits more feed by increasing the set point of the FRC. Thus, production rate is always maximized, but only within the safe availability of the coolant. When the coolant valve opens beyond 90%, production rate is lowered so that the reactor will never be allowed to run out of coolant.

Another safety feature shown in Figure 6.50 is the reaction failure alarm loop (RA). This loop compares the reactant charging rate to the reactor with the heat removal rate and actuates an alarm if the charging rate is in substantial excess.

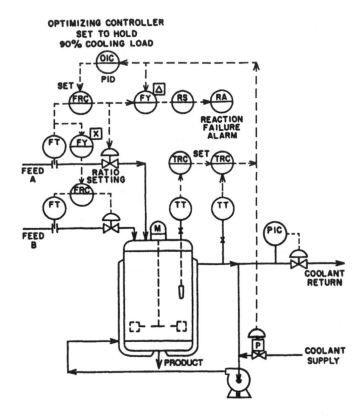

FIGURE 6.49   Residence time control is used in chemical reactors to maintain a constant ratio between reactor volume and reactor outflow.

In Figure 6.50, the valve position controller (OIC) maintains the feed rate as high as the cooling system can handle. This results in a variable production rate, because changes in coolant temperature or variations in the amount of fouling of the heat transfer areas will change production.

The dynamic response of the reactor temperature to changes in reactant flow is not favorable. A change in the feed flow must change the reactant concentration before a change in reaction rate (and therefore in heat evolution) can change the reaction temperature. This effect adds a large secondary lag to the valve position control loop. This necessitates the use of a three-mode controller, which in turn limits the use of this control configuration to stable loops only. The slower the OIC loop (the longer its lag), the lower should be the set point of OIC in Figure 6.50, to provide the required margin for stability.

When the concentration time constant is equal to or larger than the thermal time constant, temperature control through feed flow manipulation is no longer practical.[2] In such reactors, only the manipulation of the coolant will give stable operation. The strategy of valve position control is still useful in such applications, but in these cases it must manipulate supplementary cooling.

VI. REACTOR SAFETY

Safety problems can arise in chemical reactors as a result of many causes, including equipment failure, human error, loss of utilities, or instrument failures. Depending on the nature of the problem, the proper response can be to "hold" the reaction sequence until the problem has been cleared or to initiate an orderly emergency "shutdown" sequence. Such actions can be taken manually or automatically (Figure 6.31).

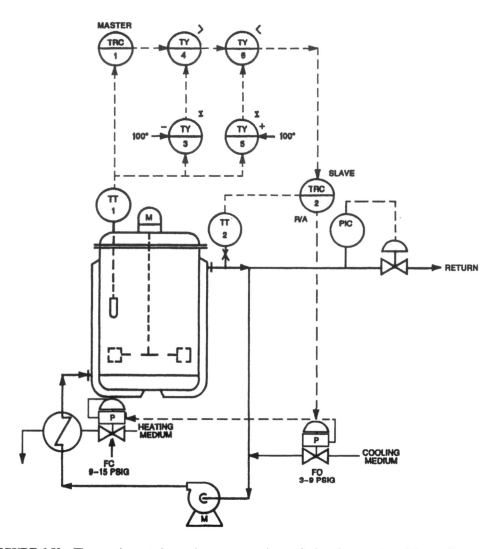

FIGURE 6.50   The reaction rate in continuous reactor is matched to the capacity of the cooling system.

After a "hold" or "emergency" condition has been cleared, an orderly sequence of transitional logic is required to return the reactor to normal operation. It makes no difference whether the emergency was caused by a pump or valve failure or whether the reactor was put on hold to allow for manual sampling and laboratory analysis of the product: a return sequence is still required. The reentry logic determines the process state when the interruption occurred and then decides whether to return to that process state or the previous one in order to reestablish the conditions that existed at the time of the interruption.

A reactor control system should provide the following features:

- The ability to maximize production
- The ability to minimize shutdowns
- The ability to maximize on-line availability of the reactor
- The ability to minimize the variations in utility and raw material demand
- The ability to provide smooth operation in terms of constant conversion, yield, and product distribution
- Easy start-up and shutdown

However, the overriding, primary design objective is that the reactor must be safe for both the operating personnel and the environment.

Safety is guaranteed by monitoring both the potential causes of emergencies and their consequences. If the cause is detected and responded to in time, the symptoms of an emergency will never develop; in that sense, the monitoring of potential causes of safety hazards is preferred. Some of the potential causes of emergencies and the methods used to sense them are listed below:

Causes	Methods of Detection
Electric power failure	Volt or watt meters
Instrument air failure	Pressure switch
Coolant failure	Pressure or flow switches
Reactant flow failure	Pressure or flow switches
Catalyst flow failure	Pressure or flow switches
Loss of vacuum	Vacuum switch
Agitator failure	Torque or RPM detector
Pump failure	Pressure or flow switch
Valve failure or mispositioning	Position sensing limit switches
Measuring device failures	Validity checks

Each of the above failures can actuate an annunciator point to warn the operator and can also initiate a "hold" or "shutdown" sequence, if the condition cannot be corrected automatically. Automatic correction will require different responses in each case.

For example, in the case of electric power failure, a diesel generator or steam turbine might be started. If instrument air failure is detected, an alternate compressor might be started, or an alternate gas, such as dry nitrogen, might be used. In addition, the safe failure positions for all control valves must be predetermined, as in Figure 6.13. In the case of loss of coolant, an alternate source might be used, or the type of automatic constraint control that was shown in Figure 6.50 might be applied.

It is also desirable to provide intelligent man-machine interfaces which will protect against unsafe operator actions. The simplest example of such limits is to prevent the operator from moving set points outside safe limits.

If reactant or catalyst flow fails, automatic ratio loops (Figure 6.50) can cut back the flow rates of related streams; alternatively, if there are alternate storage tanks from which the reactant can be drawn, these tanks might be accessed. A loss of vacuum can also be corrected by switching to an alternate vacuum source if such redundancy is economically justifiable. Critical pumps and valves can be provided with spare backup units, and agitators can be furnished with multiple drives. Failed sensors and defective instruments can be automatically replaced if redundant or voting systems are furnished.

Symptoms of an emergency might involve temperature, pressure, composition, or other conditions as listed below:

Symptom	Detectors
High reactor temperature	High temperature switch
High jacket temperature	High temperature switch
Low jacket temperature	Low temperature switch
High reaction rate	High heat release detector
Reaction failure	Low heat release detector
High rate of temperature rise	Rate of rise detector
High pressure	High pressure switch and relief valve
Abnormal level	Level and weight switches
Abnormal composition	On-stream analyzers

The control system will respond to the symptom of an abnormal condition as it did to failure causes, but here the probability of an emergency shutdown is higher. For example, if, in a potentially unstable reactor, the temperature or pressure has exceeded its high limit, it is possible to switch the reactor to full cooling and to shut off the reactant feeds, but after that, only such drastic action might be taken as depressurization or even the transferring of the reactor contents into a blowdown tank.

The consequences of high jacket temperature are less serious, and the limit shown in Figure 6.12 is usually sufficient to prevent the formation of hot spots. In some processes, the critical consideration is low jacket temperature because that can cause "frosting" or "freezing" of the reactants onto the inner wall of the reactor. As the frozen layer thickens, heat transfer is blocked and the reactor must be shut down. If used on such a process, the control system in Figure 6.12 would also be provided with a low limit on the slave set point.

When the reaction rate suddenly rises or when it approaches the capacity of the cooling system, it is necessary either to increase the heat removal capacity quickly (Figure 6.15) or to cut back on the reactant flows (Figure 6.50). It is also necessary to detect if the reaction has failed by comparing the reactant feed rate with the heat removal rate (Figure 6.50) or by monitoring heat release (Figure 6.36). In yet other processes, it is desirable to limit the rate of temperature rise (Figure 6.28).

A. Glass Lining Protection

Some of the chemical reactors are glass-lined to resist corrosion or to protect against contamination or discoloration of the product. When a reactor is glass-lined, the integrity of the lining can be damaged, because the thermal expansion characteristics of glass and metals are different. Therefore, it is necessary to provide control systems that serve to protect the lining against cracking (Figure 6.51) and that are able to detect if such damage has occurred (Figure 6.52).

The integrities of various glass-lining materials are guaranteed for different temperature differentials between the metal and the glass. The lining protection controls serve to keep the difference between the temperature of the lining (batch temperature detected by TT-1 in Figure 6.51) and the temperature of the metal (jacket temperature detected by TT-2) under the guaranteed limit. Figure 6.51 illustrates the protection loop for a situation in which the maximum differential is 100°. The goal of the controls is to keep the set point of the slave (TRC-2) within 100° of the batch temperature (TT-1). The purpose of TY-3 and TY-4 is to make sure that the slave set point does not drop too low. Therefore, TY-3 subtracts 100° from the TT-1 signal, while TY-4 selects the higher between that and the master (TRC-1) output. Similarly, the purpose of TY-5 and TY-6 is to prevent the slave set point from rising too high. TY-5 adds 100° to the TT-1 signal, while TY-6 selects the lower between that signal and the TRC-1 output.

Figure 6.52 shows the control system used to monitor the integrity of the glass lining, when the batch is electrically conductive. The fault test is made by inserting a probe into the batch and checking the resistance between that and the metallic tank wall. If the glass lining is damaged, the resistance drops as the metal is exposed to the conductive batch. The fault monitor shown in Figure 6.52 is designed for the scanning of 10 glass-lined reactors.

B. Multiple Sensors

Multiple temperature sensors can be used for safety or maintenance reasons. In a fixed-bed reactor, for example, in which the location of the maximum temperature might shift as a function of flow rate or catalyst age, multiple sensors would be installed and the highest reading selected for control, as shown in Figure 6.53. If analyzers are considered to be unreliable, a pair of redundant analyzers in combination with a high signal selector (Figure 6.54) can also increase reliability.

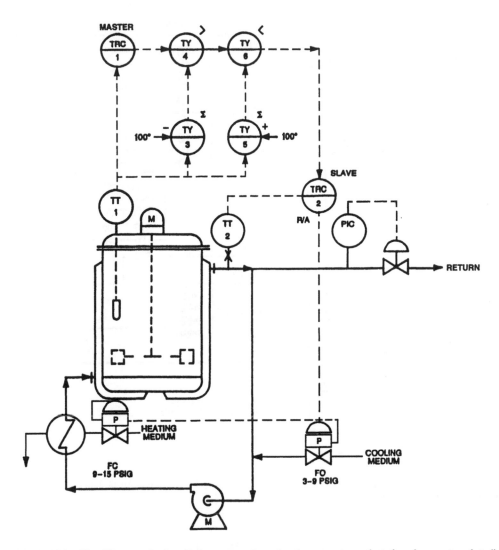

FIGURE 6.51 The illustrated glass-lining protection circuit guarantees that the slave set point (jacket temperature) will never be more than 100° away from the batch temperature.

On the other hand, if the reason for the use of multiple sensors is to increase the reliability of the measurement, the use of median selectors is the proper choice (Figure 6.54). A median selector rejects both the highest and the lowest signals and transmits the third. This type of redundancy protects against the consequences of sensor misoperation, while also filtering out noise and transients that are not common to two of the three signals.

Another method of increasing sensor reliability is the use of voting systems. These also consist of at least three sensors. Reliability is gained, as the voting system disregards any measurement that disagrees with the "majority view."

C. Instrument Reliability

A risk analysis requires data on the failure rate for the components of the control system. Such data has been collected from user reports; the findings of one such survey are shown in Table 6.2. When the mean time between failure (MTBF) of the individual instrument components has been estimated, the MTBF of the various control loops can be established.[8] An example of this

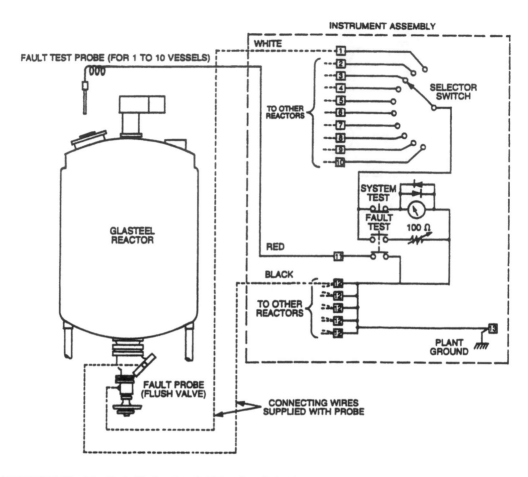

FIGURE 6.52 The Fault Finder signals if the glass lining in a glass-lined reactor is damaged. (Courtesy of Factory Mechanical Systems.)

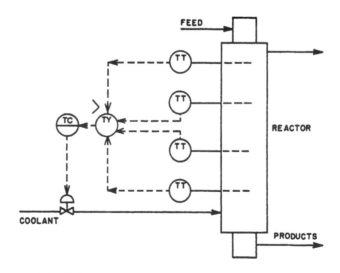

FIGURE 6.53 To control the highest temperature in a fixed-bed reactor, this method can be used.[3]

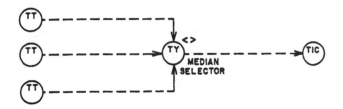

FIGURE 6.54 A median selector protects from sensor error or failure.

technique is illustrated in Figure 6.55. Once the MTBF of each loop has been established, the reliability of the whole reactor control system can be increased by increasing the MTBF of the highest risk loops. This usually is done through self-diagnostics; preventive maintenance; the use of backup devices; and, in critical instances, the technique of "voting," described in connection with Figure 6.54.

TABLE 6.2
Failure Rate for Control System
Components

Variable	Instrument	Mean Time Between Failures
Level	Bubbler	1–2 yrs.
	d/p transmitter	1–5 yrs.
	Float and cable	0.2–2 yrs.
	Optical	0.1–5 yrs.
Flow	Flume and weir	0.5–5 yrs.
	Venturi, etc.	2 mo.–5 yrs.
	Propellers	1 mo.–1 yr.
	Positive displacement	1 mo.–1 yr.
	Magnetic	0.5–10 yrs.
Density	Nuclear	1–3 yrs.
	Mechanical	1–6 mos.
Analysis	pH and ORP	1–4 mos.
	Dissolved O_2	1–9 mos.
	Turbidity	1–6 mos.
	Conductivity	1–4 mos.
	Chlorine gas	0.5–1 yr.
	Explosive gas	0.2–1 yr.
	TOC	0.1–1 mo.
Miscellaneous	Temperature	0.5–2 yrs.
	Pressure	0.1–5 yrs.
	Speed	0.6–5 yrs.
	Weight	0.6–2 yrs.
	Position	0.1–1 yr.
	Sampling	0.1–1 yr.

While risk analysis is complex and time consuming, a plant can be designed for a particular level of safety, just as it can be designed for a particular level of production. The reliability of the result is as good as the data used in the analysis. For this reason data collected by users, testing laboratories, or insurance companies should be used, instead of manufacturers' estimates. For

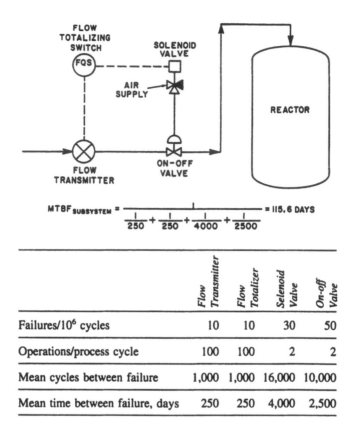

$$MTBF_{SUBSYSTEM} = \frac{1}{\frac{1}{250} + \frac{1}{250} + \frac{1}{4000} + \frac{1}{2500}} = 115.6 \text{ DAYS}$$

	Flow Transmitter	Flow Totalizer	Solenoid Valve	On-off Valve
Failures/10^6 cycles	10	10	30	50
Operations/process cycle	100	100	2	2
Mean cycles between failure	1,000	1,000	16,000	10,000
Mean time between failure, days	250	250	4,000	2,500

FIGURE 6.55 The mean time between failure (MTBF) for a control loop can be calculated based on the MTBF of its components.[8]

instrument reliability and performance data, one good source is the International Association of Instrument Users, SIREP-WIB.

REFERENCES

1. Luyben, W.L., Batch reactor control, *Instrum. Technol.*, August 1975.
2. Shinskey, F.G., *Process Control Systems*, 4th ed., McGraw-Hill, New York, 1996.
3. Nagy, I., *Introduction to Chemical Process Instrumentation*, Akadémiai Kiadó, Budapest, 1992.
4. Szeifert, F., Chovan, T., and Nagy, L., Dynamic Simulation and Control of Flexible Chemical Technologies, Automation '92, Budapest, 1992.
5. Chemiflex Reactor User's Manual, Chinoin Pharmaceutical and Chemical Works, Budapest, 1991.
6. Klado, M., The case of a real engineering design problem, *Chemtech*, 1971.
7. Lipták, B.G., *Instrument Engineers' Handbook, Process Measurement and Analysis*, Chilton Book Co., 1995.
8. Buckley, P.S., Protective controls for a chemical reactor, *Chem. Eng.*, April 4, 1975.
9. Molvar, A.C., Instrumentation and Automation Experiences, 600/2-76-198, U.S. EPA.

BIBLIOGRAPHY

Alsup, M., Recipe Configuration for Flexible Batch Systems, ISA/93 Technical Conference, Chicago, September 19-24, 1993.

Coughhanowr, D.R., *Process Systems Analysis and Control*, 2nd ed., McGraw-Hill, New York, 1991.

Gamache, M.E., *Statistical Process Control Handbook*, McGraw-Hill, New York, 1995.

Houpis, C.H., and Lamont, G.B., *Digital Control Systems*, 2nd ed., McGraw-Hill, New York, 1992.

Kelly, K.P., Training operators with simulation software, *Control*, June 1992.

Lipták, B.G., *Instrument Engineers' Handbook, Process Control*, Chilton Book Co., 1995.

Longwell, E.S., Dynamic Modeling for Process Control, ISA Conference paper, Chicago, September 1993.

Luyben, W.L., *Process Modeling, Simulation and Control for Chemical Engineers*, McGraw-Hill, New York, 1990.

Marlin, E.T., *Process Control*, McGraw-Hill, New York, 1995.

McCabe, W.L., Smith, J.C. et al., *Unit Operations in Chemical Engineering*, 5th ed., McGraw-Hill, New York, 1993.

Molnár, F. and Nagy, T., Flexible reactor controller (in Hungarian), *Mérés Automatika*, 39, 1991.

Nagy, I., *Introduction to Chemcial Process Instrumentation*, Akadémiai Kiadó, Budapest, 1992.

Nowicki, P.L., OOP & expert systems transform batch recipes, *InTech*, September 1992.

Procyk, L.M., Batch process automation, *Chem. Eng.*, May 1991.

Shinskey, F.G., Controlling temperature in batch reactors, *InTech*, April 1992.

Shinskey, F.G., *Process Control Systems*, 4th ed., McGraw-Hill, New York, 1996.

Szeifert, F., Chovan, T., and Navy, L., Dynamic Simulation and Control of Flexible Chemical Technologies, Automation '92, Budapest, 1992.

Tom, T.H., DCS Selection for Flexible Batch Automation, ISA/93 Technical Conference, Chicago, September 19-24, 1993.

Watts, J.C., Model Based Control Using Nonlinear Models, ISA Technical Conference, Chicago, September 1993.

7 Dryers

The equipment discussed in this section serves to dry solids. The process involves the removal of liquids, such as water or solvents, by adding heat to vaporize them. The types of dryers can be grouped into batch designs and continuous designs. The continuous dryers can be further subdivided on the basis of (1) whether the dryer itself is heated (*nonadiabatic* contact drying) or the dryer is unheated and the heat of vaporization is provided by preheated air (*adiabatic* convective drying); (2) for adiabatic dryers, whether the dryer is *co-current* (the solids and the drying air move in the same direction), *countercurrent* (hot air and solids move through the dryer in opposite directions), or *fluidized bed*; and (3) whether the dryer is operated under positive pressure or under vacuum. The control of each of these designs will be briefly discussed in this chapter.

The subject of dryers is so extensive that some limitations must be imposed on the discussion. Therefore,

1. The dryers and drying principles discussed will be limited to those related to the removal of a volatile solvent from a solid material. Such processes as the removal of a solvent from an air stream (air drying) or the removal of one solvent from another (drying of an organic solvent) will not be covered.
2. The solvent to be removed will be taken as water, although an occasional reference may be made to other solvents.

I. THE PROCESS

While the latent heat for evaporating the moisture from the solid particle can be provided by the surface heat of directly heated conductive contact dryers (nonadiabatic) or by the hot air introduced as the heat source of the convective (adiabatic) dryers, the solid particle is still dried in a four-phase sequence. (1) The solid particle is heated up to the drying temperature. (2) The moisture from the completely wet surface of the solid particle is being evaporated. (3) The surface is partially dry. (4) The surface is dry. Drying occurs completely through internal evaporation and diffusion.

In Figure 7.1 section A-B represents the period of product entry into the dryer. Since some heat is necessary to bring the material to the initial drying temperature, evaporation during this phase is slow. The next section, B-C, is a period of evaporation of surface moisture. The temperature near the surface of the solids during this time is the wet-bulb temperature of the air which is directly in contact with the product. After the surface moisture has evaporated, the rate of evaporation drops off (section C-D). This phenomenon is due in part to a "case-hardening" of the surface and in part to the long path necessary for the water to migrate to the surface. If the solid is one that has water of crystallization, or bound water, the rate will then drop off even more, as the bound moisture diffuses to the surface as shown in section D-E of the curve.

The transition between the "constant" rate drying and the "falling" rate drying zone (when dry areas appear on the solid particle surface) is identified as the critical moisture W_c. The rate of

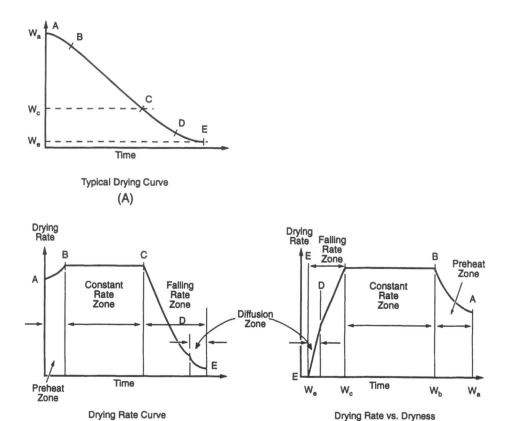

FIGURE 7.1 The four phases of drying. A-B: Preheating rising rate. B-C: Wet surface, constant rate. C-D: Partially dry surface, falling rate. D-E: Dry surface, diffusion. W_a = initial, W_c = critical, W_e = final moisture content of the dried product.

drying in the C-D falling rate zone is proportional to W/W_c. In the D-E zone, where bound water has to diffuse to the surface of the particle, the rate of drying falls even further.

A. AIR AS A DRYING MEDIUM

Adiabatic dryers obtain their heat from the hot air, which enters them and uses the sensitive heat to vaporize the moisture from the process solids. In this process the air is cooled, but its energy content (enthalpy) does not change, because the sensitive heat loss is compensated by an equal rise in enthalpy due to increased moisture content (adiabatic process). Figure 7.2 shows the constant enthalpy lines (total heat lines with units of BTU/lb of dry air). These sloping constant enthalpy lines are parallel with the wet-bulb temperature lines in the psychrometric chart and represent the path which the drying air follows. Air enters at a high temperature and a low relative humidity (such as at point A, 115°F and 30% RH) and leaves at a lower temperature but with an increased humidity content (such as point B, 91°F and 80% RH).

If one assumes that the dryer is operated in the falling-rate zone and if in that zone the rate of drying is proportional to the actual-to-critical moisture content ratio (W_P/W_c), then product moisture content (W_P), according to Shinskey, can be calculated as

$$W_p = \frac{W_c F_a C_p}{A H_w \zeta} \ln \frac{\left(T_i - T_w\right)}{\left(T_o - T_w\right)} = K\, W_c \ln\left(\frac{T_i - T_w}{T_o - T_w}\right) \tag{1}$$

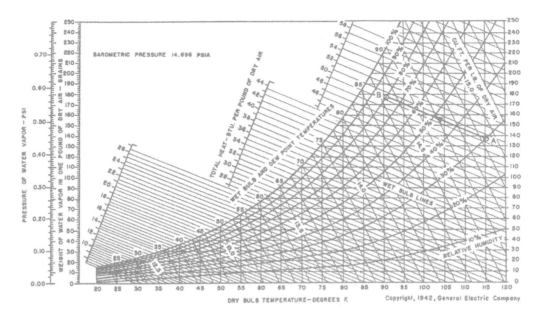

FIGURE 7.2 Psychrometric chart.

where A = solids surface area, C_p = specific heat of the air, F_a = air flow, H_w = latent heat of water, $K\text{-}F_aC_p/AH_w\zeta$, T_i = air inlet temperature, T_o = air outlet temperature, T_w = air wet bulb temperature, and ζ = mass transfer coefficient.

Equation 1 shows that the product moisture will be constant if the ratios before and after the logarithm remain constant. The terms before the logarighm are usually constant, with the exception of air flow (F_a) and particle size (A), but even these variations are cancelled by proportional changes in the mass transfer coefficient (ζ) and the critical moisture content (W_c), leaving the F_a/ζ and the W_c/A ratios constant.

Therefore, if the control objective is to hold the moisture content of the product (W_p) constant, this can be achieved by maintaining the ratio of the two temperature differences under the logarithm constant. In an adiabatic dryer, the wet-bulb temperature (T_w) remains constant throughout the dryer, as no heat is added or removed; only the sensible heat is exchanged for latent heat. Therefore, if a drier product is desired, it can be obtained by increasing the outlet temperature (T_o), while keeping the inlet (T_i) constant.

The amount of heat transferred from the air to the solids is increased if the temperature difference between them is increased. If the solids are at constant temperature (constant rate zone), the enthalpy of the air will also be constant and the air will only exchange its sensible heat for the latent heat of the water which joins it. Therefore in this zone (from point B to C in Figure 7.1), the wet-bulb temperature is also constant. The heating air enters a dryer at some high temperature (Ti) and leaves at an outlet temperature (To) which is low at the beginning of a batch drying cycle (Figure 7.3) and rises as the solids heat up and dry. The dry-bulb temperature of the outlet air must always be higher than its wet-bulb temperature (Twb), because at 100% relative humidity (Twb = dew point) the air would no longer be able to pick up moisture from the solids. As shown in Figure 7.3, the drier the solids get, the more motivation is needed to drive off the remaining moisture. This requires an increase in the temperature of the outlet air and might also necessitate an increase in the difference between the temperature of the outlet air and of the solids.

In continuous dryers, the temperature profiles differ from that shown for batch dryers in Figure 7.3. The majority of the industrial dryer units are co-current longitudinal (adiabatic) dryers (Figure 7.4), which can be used on thermally sensitive products. Air enters at a high temperature (Ti),

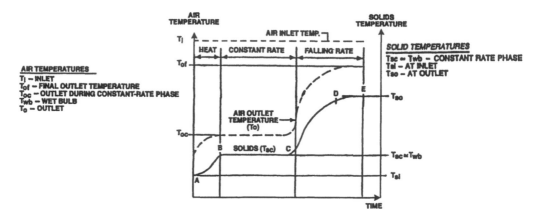

FIGURE 7.3 Product temperature rises in a batch dryer from Tsi to Tso as it is heated by air, which enters at Ti and leaves at To.

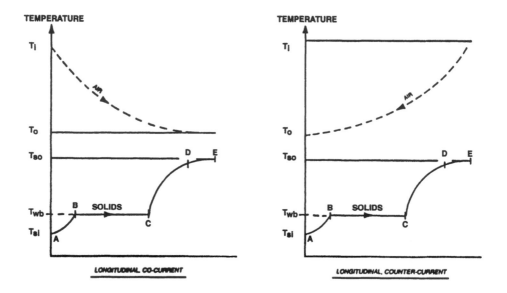

FIGURE 7.4 Temperature profiles in longitudinal adiabatic dryers.

which speeds the drying process but is usually acceptable even with temperature-sensitive products, because the solids temperature is limited to the wet-bulb temperature (Twb) as long as the particle surface is wet. Counter-current longitudinal dryers are used on thermally stable solids (cement, carbon black), which must be dried to very low moisture levels (Figure 7.4).

In continuous adiabatic fluidized bed dryers the average moisture content of the solids in the bed is almost the same as the moisture content of the product. Most fluid bed dryers are generating products with less than the critical moisture content (W_c in Figure 7.1) and are operated in the falling drying-rate zone. Under these conditions the air outlet temperature (To) determines product dryness, if feed flow and feed moisture content are constant.

B. DRYER TYPES

Each material has some unique drying characteristics depending upon its substance, the solvent, the affinity of these for each other, and the surroundings of the particular dryer.

The theoretical drying rate depends upon feed rate and driving force. The latter is a combination of temperature difference between the product and the surroundings and the moisture content of the dryer atmosphere. Other factors of equal importance are the intimacy of material surface contact with the heating medium and the degree of agitation or surface renewal.

The three design features necessary to a drying operation are a source of heat, a means of removing the solvent from the vapor space surrounding the product, and a mechanism to provide agitation or surface renewal. The extensive assortment of dryer types available serves as testimony to the fact that drying is among the most difficult processes. A few representative batch and continuous dryer types are listed in Table 7.1. Control schemes for them will be presented in the following paragraphs.

TABLE 7.1
Dryer Characteristics

Type	Examples	Heat Source	Method of Moisture Removal	Method of Agitation	Typical Feed	Product Dead Times
			Batch Dryers			
Atmospheric	Tray	Hot air	Air flow	Manual	Granular or powder	Min. to hours
Vacuum	Blender tumbler	Hot surface	Vacuum	Tumbling	Granular or powder	Min. to hours
	Tray	Hot surface	Vacuum	None	Solid or liquid	Min. to hours
	Rotating Blade	Hot surface	Vacuum	Mixing blade	Solid or liquid	Min. to hours
Specialty	Fluid bed	Hot air	Air flow	Air	Granular or powder	Minutes
			Continuous Dryers			
Heated cylinder	Double-drum	Hot surface	Air flow	None	Liquid or slurry	Min. to hours
Tumbling	Rotary	Hot air	Air flow	Tumbling	Granular	Min. to hours
	Turbo	Hot air	Air flow	Wipe to successive shelves	Granular	Min. to hours
Air stream	Spray	Hot air or combustion gas	Air flow	Not required	Liquid or slurry	Seconds
	Flash	Hot air	Air flow	Air	Granular or powder	Seconds
	Fluid bed	Hot air	Air flow	Air	Granular	Minutes

C. Moisture Measurement and Control

Moisture analyzers and other detectors capable of measuring the moisture content of solids are discussed in the *Instrument Engineers' Handbook*, Process Measurement Volume, Section 8.33. Because this is not an easy measurement, particularly in continuous dryers, indirect measurements utilizing the relationship stated in Equation 1 are also used.

In batch systems, one might take samples for moisture analysis at various periods. Once approximate drying conditions have been established, checking of moisture is required only near the estimated end point.

A similar procedure can be followed for continuous systems. Empirical runs can be made to establish dryness of a product with given feed and dryer characteristics so that operating curves can be developed. The control scheme is designed to maintain these conditions with occasional feedback and with corrections from grab samples. Analyzers are also available for moisture analysis that can be adapted for automatic, closed-loop dryer control. The most successful of these units rely on infrared, microwave, or capacitance detection. Care must be exercised in the selection to allow for changes in bulk density or for void spaces, which will introduce an error. Some of these sensors do not detect bound moisture.

Even if a fully reliable solids moisture analyzer is available, some prefer to control the dryer on the basis of a temperature-based inferential system and use the analyzer only to update the inferential model. This is because most dryer processes are dominated by dead time (Table 7.1) due to the slow transportation of solids, and information on load changes can be obtained sooner on the basis of the air temperatures. If direct analyzer control is used, by the time the off-spec product reaches the analyzer, without correction for air temperature, and the change in product moisture is detected, there will be additional off-spec material in the dryer, for which the feedback correction will come too late. Therefore, the more advanced dryer controls use the analyzers for feedback trimming of the faster (and partially feedforward) inferential temperature models.

Although spray and flash dryers have a retention period of less than a second, holdup in the majority of continuous dryers ranges from 30 minutes to an hour. This fact makes automatic feed control difficult. It is also often difficult to measure the variation in moisture content of the feed. The feed rate is metered by a screw conveyor or another similar device.

Dryer controllers are usually proportional and reset only, because dryer dynamics do not generally warrant the use of the rate response.

Some dryer control systems are still relatively unsophisticated. The drying rate curve levels off in the area where most product specifications lie (Figure 7.1), therefore the control of final conditions is not critical. In addition, many moisture specifications are somewhat liberal in recognition of the difficulties in handling both moist and dry material. The specifications also allow for a possible change in moisture content after drying as most dried materials are affected by humidity.

II. BASIC CONTROLS (BATCH DRYERS)

Batch dryers are used when the nature of the product (lumber) requires it, when the knowledge of the identity of a batch is of importance (pharmaceutical or food industries), or when the quantities being dried are small. The disadvantage of batch drying is that the process has no steady state and, therefore, the operating conditions are continuously changing (Figure 7.3). In the following paragraphs, the control systems for batch type tray, vacuum, fluid bed, and freeze dryers and kilns will be described.

A. Atmospheric Tray Dryers

The term "atmospheric dryer" designates a batch dryer that operates at close to atmospheric pressure and obtains the heat for drying from the hot air supply. The same medium is relied upon to remove the generated moisture. The two most common forms are the tray dryers, in which trays covered with material to be dried are loaded into racks, and the truck dryers, which are similar except that the racks of trays are mounted on trucks.

Most often, the wet material is placed on trays that are placed on the rack of a carriage and moved into the atmospheric dryer, which includes a fan and a heat exchanger (Figure 7.5). Steam flow is throttled by inlet temperature, inlet air flow is set manually, and outlet air flow (blowdown) is throttled to maintain the dew point of the wet-bulb temperature within the dryer chamber. This reduces the cost of drying. When air humidity is measured, that information can also be used to determine end point. In the case of uniform products, the dryer cycle is often terminated on a time cycle. When solvent (instead of water) is being driven off, controls are also needed to make sure that explosive air concentrations are not reached.

In tunnel dryers, the carriage is not stationary, but moves through the tunnel, while the individual drying zones follow each other in series and are controlled as in Figure 7.5.

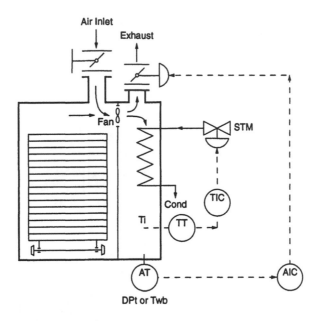

FIGURE 7.5 Small atmospheric tray dryer.

In all tray dryers, the available control parameters are the air velocity and distribution, temperature, and humidity. The velocity and distribution of the air are often set manually by dampers and are not changed. Humidity can also be a manual setting by virtue of adjusting the damper in the recirculation line. If recirculation is not used and if product requirements warrant, dehumidification is used on the inlet air. Both of these provisions are shown (dotted) in Figure 7.6.

Control of the atmospheric temperature within the dryer is accomplished by regulation of the steam to the heating coil. The thermal bulb can be placed in the dryer or in the inlet or outlet air. The sizing of the steam valve is determined by the dryer airflow and temperature requirements as analyzed on psychrometric charts (See Figure 7.2), and by the pressure of steam used. It is necessary during the latter portion of the drying cycle to reduce steam flow to a level sufficient only to heat the air to the dryer temperature. The load at this time is limited to the dryer heat loss. A temperature override (TIC) is frequently included in the heater discharge to limit this maximum inlet temperature.

It is not uncommon to use a program controller for temperature control of a batch tray dryer, particularly on solvent service. In this way the rate of evaporation is controlled and the concentration of solvent in the air is *held below the explosive range*.

B. VACUUM AND FREEZE DRYERS

Most dryers can be operated under vacuum. Parameters that can be controlled in batch vacuum dryers are absolute pressure (vacuum), rate of tumbling, or agitation, and temperature of the heat-transfer medium. The general rule is to control vacuum and rate of tumbling manually and concentrate on the transfer fluid temperature for automatic control. A tray dryer, for example, becomes a vacuum dryer if the trays are placed in a vacuum chamber and a heat-transfer medium is circulated through the hollow shelves (Figure 7.7). Here, hot water temperature is controlled by steam heating on a selective basis. Under normal conditions, the air temperature controller (TIC-01) is in control, but if the water temperature reaches a high limit (which might discolor or otherwise harm the product), the high-temperature controller (TIC-02) takes over. The operating vacuum is controlled either manually or by PIC-03.

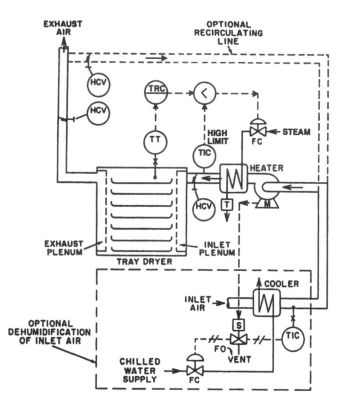

FIGURE 7.6 Larger atmospheric tray dryer.

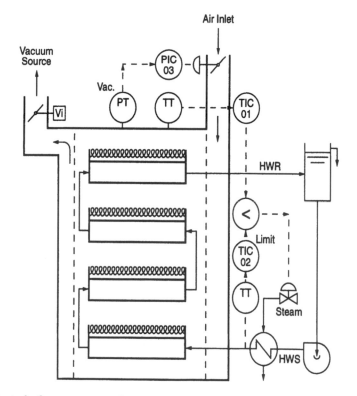

FIGURE 7.7 Control of a vacuum tray dryer.

Some of the vacuum tray dryers operate at very low temperatures and are known as "freeze dryers." Refrigeration is required for the heat-transfer fluid in the first part of the cycle to freeze the dryer and its contents. The circulating fluid is then gradually warmed to provide heat for evaporation. In these nonadiabatic dryers, too, the latent heat of evaporation is provided by the heating media (electricity or brine) to the trays, which heat the product by conduction (Figure 7.8). For low vacuum operation, Pirani or ionization-type vacuum sensors are used.

Moisture content is often detected by electrical conductivity[1] or by equilibrium vapor pressure (EVP). EVP is determined by closing V1 in Figure 7.8 and thereby periodically isolating the vacuum chamber from condenser #2. In less than a minute, PIC-03 will detect a rise in absolute pressure (see bottom of Figure 7.8) that will reach the EVP corresponding to the ice temperature

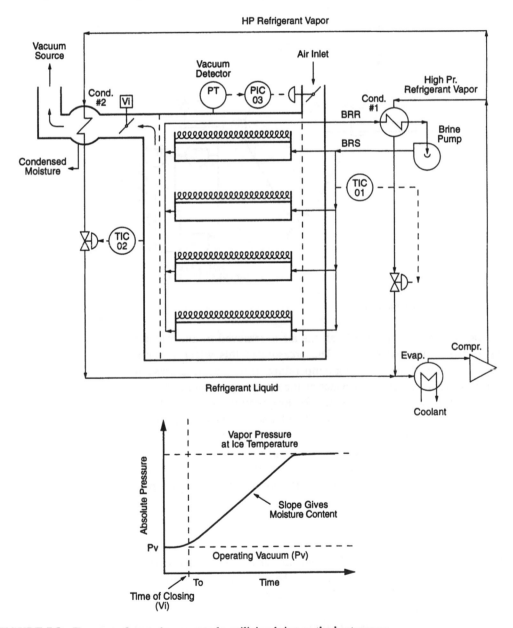

FIGURE 7.8 Tray-type freeze dryer controls, utilizing brine as the heat source.

in the product. The slope of this curve (the speed of pressure rise) indicates the moisture content of the product.

The heating controls (TIC-01 in Figure 7.8) in freeze dryers can be limited by electric resistivity sensors. When resistivity drops (conductivity rises), that can be an indication that the ice is melting on the trays and therefore the temperature of the heating medium should be reduced. A sudden rise in electrical resistance can be used as an indication of batch completion.

Other vacuum dryer designs include the double-cone blender-dryer (Figure 7.9), which tumbles the product in a jacketed vessel. An alternate design is the rotary dryer, which is also jacketed, but here the vessel is stationary and the product is agitated by a hollow mixing blade, which is provided with a cavity for the passage of the heat-transfer fluid.

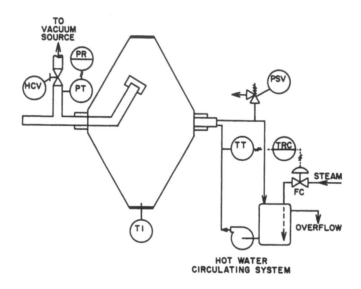

FIGURE 7.9   Vacuum blender-dryer.

In vacuum dryers practically all the air is removed by the vacuum, and therefore the vapor space is filled with water vapor. The pressure of this water vapor is a function of the vapor pressure of water at the operating temperature. If the absolute pressure inside the vacuum dryer is p and the vapor pressure of water at the operating temperature is pv, the pressure difference (pv — p) will be the driving force for evaporation. This differential can be measured by a differential vapor pressure transmitter[1] by placing the water-filled bulb into the solids being dried and connecting the other side of the d/p cell to the vapor space in the vacuum dryer (Figure 7.10). When the solids are wet, the (pv — p) reading is likely to be small, and as the product is dried the difference is likely to increase as more and more driving force is required to evaporate the remaining water. Therefore, the monitoring of the (pv — p) signal can be used to signal the end point of the batch cycle.

C. KILNS

Kilns are rotary dryers that can be adiabatic (dried by hot air) or jacketed. The dried product is usually intermixed by the slow rotation of the kiln, but the overall design is very much a function of the nature of the product being dried. For example, if a batch of lumber is being dried, tumbling by rotating the kiln could damage the product. In such cases, the product is not intermixed and uniform drying is achieved by periodically (once every couple of hours) reversing the direction of the hot air through the kiln. Also, the nature of wood is such that large moisture gradients can cause cracking. One method of protecting against cracking is to limit the air inlet temperature (T_i)

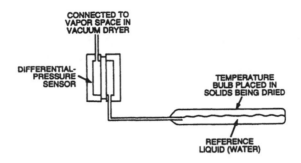

FIGURE 7.10 Differential vapor pressure transmitter[1] can detect the differential pressure which is the driving force for evaporation.

so that the temperature difference $(T_i - T_o)$, which determines the rate of drying, will not exceed a safe limit (Figure 7.11). At the beginning of the batch, while the lumber is wet, ΔTIC-02 is likely to be in control, but after that phase, TIC-01 is likely to take over.

To obtain uniform product quality, it is desirable to maintain the wet-bulb temperature constant in the kiln. A lumber kiln provides a relatively clean environment, so a conventional wet-bulb thermometer can be used as the measurement of TIC-03, which throttles the recirculation damper. This damper is likely to be closed at the beginning of the batch and will start opening only when the wood has been partially dried. The continuous measurement of T_w allows the use of Equation 1 to predict the time when the batch is done. (The constant K in the equation is estimated on the basis of previous batches, with each variety of lumber requiring a different value.) When the batch is done, the actual moisture content of the wood is measured by conductivity or capacitance gauges.[1]

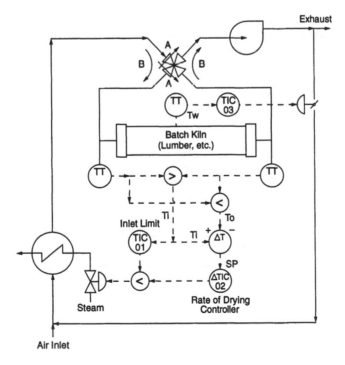

FIGURE 7.11 Batch kiln dryer for lumber.

D. Fluid-Bed Dryers

In a fluid-bed dryer the product is held in a portable cart with a perforated bottom plate. Heated air, blown (pressure) or sucked (vacuum) through the plate, fluidizes the product to cause drying. The airflow is adjusted to get proper bed fluidization, and a timer can be set for the drying period. In the simplest designs, control is restricted to temperature control of the inlet air, as indicated in Figure 7.12.

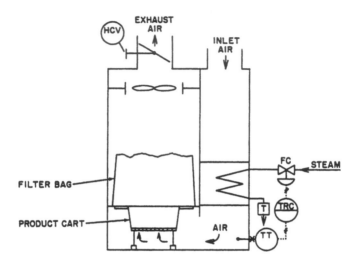

FIGURE 7.12   Batch fluid-bed dryer.

A duplex control scheme is sometimes used to take advantage of the high heat transfer properties of the fluid bed. The bed is sufficiently well mixed so that a high inlet temperature can be used at the beginning of the cycle without danger of damaging the product as long as the temperature is reduced during the latter stages of drying (during the falling-rate phase in Figure 7.1). This is accomplished by the use of a controller on the outlet air that adjusts the set point on the inlet air controller. As the bed temperature rises during drying, the set point of the inlet air temperature controller is reduced (Figure 7.13).

III. OPTIMIZING CONTROLS (BATCH)

When drying pharmaceutical powders, the container is charged with the dry powder, fitted with a filter bag or with fine screens at the top and bottom, and steam heated air is drawn upward through the bed. The time-temperature relationship for both the air and the powder has been given in Figure 7.3. While moisture covers the product particles, the rate of drying is maximum (Figure 7.1) and constant. During this period, the solids are at the wet-bulb temperature (T_w), the outlet air temperature is a little higher (Figure 7.3), and Equation 1 gives the moisture content of the dried solids.

A more advanced batch dryer control scheme is one in which the dew point of the incoming air is controlled automatically and the termination of drying is not based on time but rather on the measurement of outlet temperature (To). Figure 7.14 illustrates these features. The use of a dew-point controller which automatically modulates the amount of recycled air contributes to both stability and energy conservation. It provides for stability by eliminating one uncontrolled variable (ambient humidity), and it contributes to energy conservation by reusing some of the heat content of the exhaust air. During the constant-rate drying phase, the dew-point controller (AIC in Figure 7.14) keeps the amount of recycled air near zero, because at that time the air outlet temperature (To) is relatively low and its humidity is high. On the other hand, during the falling-rate phase of drying, when To is high

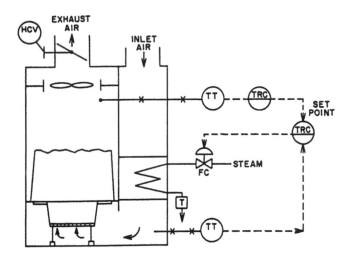

FIGURE 7.13 Batch fluid-bed dryer with cascade control.

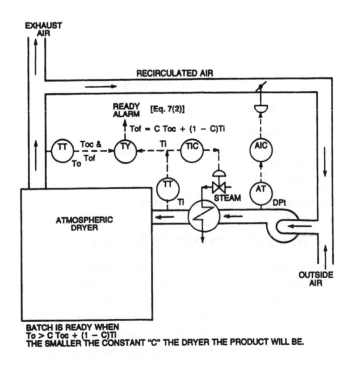

FIGURE 7.14 A more advanced batch dryer control system.

and humidity is low, the control loop will recover a fair portion of the exhaust heat and thereby will automatically reduce the opening of the steam valve.

The other control loop in Figure 7.14 serves to determine the outlet air temperature at which the drying has been completed (Tof in Figure 7.3). Shinskey has found that Tof is a function of the air inlet temperature (Ti) and of the air outlet temperature during the constant-rate drying phase (Toc). TY in Figure 7.14 memorizes T_{oc} during the constant-rate drying phase and continuously calculates T_{of}, using the following equation:

$$Tof = (C)Toc + (1 - C)Ti = Toc + C (Ti - Toc) \qquad (2)$$

The constant C is a function of the desired dryness of the product. Its value ranges from 0 to 1.0, and the smaller it is, the drier the product will be. Therefore, C can be adjusted empirically: increased if the product is too dry or decreased if the product is too wet.

The advanced control features are incorporated into the batch fluid-bed dryer controls shown in Figure 7.15. The moisture content of the product (W_p) is continuously calculated by Equation 1, while the end point of the batch is signaled when the value of T_o reaches the final value (T_{of}) predicted by Equation 2.

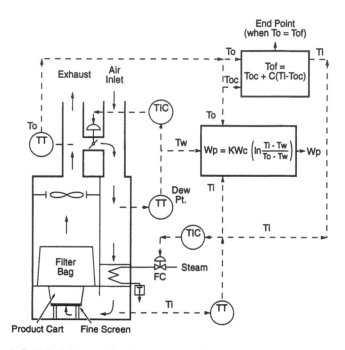

FIGURE 7.15 Batch fluid-bed dryer with advanced controls.

IV. BASIC CONTROLS (CONTINUOUS)

Most large-scale industrial processes producing dry solid materials use continuous dryers. The control of a dryer is a challenge, because the actual dryness (moisture content) of the product is difficult to measure, and therefore control generally depends on an inferential model of secondary variables such as temperature. These variables can be measured more reliably and faster (less affected by the transportation lag in the dryer) than are analyzers. Therefore, when analyzers are used, they are often applied to update or trim the inferential model, instead of directly controlling the process.

The majority of continuous dryers require control of the flow and temperature of an air stream. Temperature detection of an air stream is inherently sluggish in response. The holdup in the majority of dryers is on the order of 30 minutes to an hour (Table 7.1). During this period, the effects of input changes are blended, and fast response is therefore not required. Notable exceptions to this are flash and spray dryers with very short holdup periods.

Dryer controllers generally have proportional and reset modes only. Setting them in the field requires a good deal of patience, since the full effect of a change in a variable may take half an hour or more to be realized. Some variables, notably humidity of the inlet air, may defy regulation if exhaust recycling is not available.

A. Drum and Double-Drum Contact Dryers

The heated cylinder dryer is one of the oldest dryer designs. In these directly heated, nonadiabatic dryers, conductive heat transfer is taking place. The drum dryer can be steam jacketed, while, in the double-drum dryer (Figure 7.16), the steam is fed through the hollow shaft and condenses on

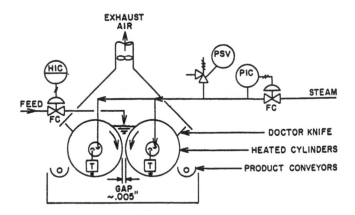

FIGURE 7.16   Double-drum dryer.

the inner surface of the drum. In the double-drum dryer, liquid is fed into the "valley" between the heated cylinders. The drums, rotating downward at the center, receive a coating of the liquid with the thickness depending upon the spacing between the rolls. The material must be dry by the time it rotates to the doctor knife, where it is cut off the roll.

The variables available for control are the speed of the cylinders, the spacing between them, the liquid level in the valley, and the steam pressure in the cylinders. The first two (speed and spacing) are usually adjusted manually. The liquid level is maintained by throttling the feed stream. Some attempts have been made to control this level automatically, but they have been thwarted by three basic difficulties: The height of the level is only 6 to 9 in. (150 to 225 mm); the liquid is constantly in a state of extreme agitation, bubbling, and boiling; and the liquid is highly concentrated and tends to plug conventional level sensors that depend on physical contact for measurement. See Chapter 3 in the *Process Measurement* volume[1] for non-contact-level detectors. The control of feed can be manual, indicated in the diagram by a manual loading station (HIC) or on flow control. In the past all of the other controls were on manual, and only the steam pressure was controlled automatically. Today packages are available with higher levels of automation.

To lower the drying temperature, heat-sensitive materials such as food products or soap are dried in steam-jacketed vacuum dryers. Here, the vacuum removes practically all the air, so the vapor space is essentially filled with water vapor. The driving force for evaporation is the difference between the partial pressure of the water in equilibrium with the solids at the dryer temperature (p_p) and the absolute pressure in the dryer (p). In emulsions and suspensions the equilibrium pressure (p_p) equals the full vapor pressure of the water (p_v), while in solutions (such as soap) it is reduced in proportion to the mole fraction (x) of water in the mixture ($p_p = p_v x$). Shinskey[2] reports that the moisture content (x) in the product of a continuous soap dryer can be controlled by maintaining the driving force for the drying process ($p_v - p$) constant. This measurement is a natural task for the differential vapor pressure transmitter (Figure 7.10), with its bulb filled with water and inserted into the dried product (Figure 7.17).

The main target of advanced controls is better steam controls. Because drying is performed by placing the process materials (paper, flake food products, soaps) onto a heated surface (a nonadi-

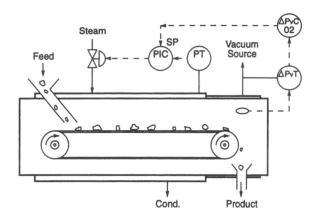

FIGURE 7.17 Nonadiabatic drum-type vacuum dryer under differential vapor-pressure control.

abatic process), the overall heat transfer is a function of the surface temperature (steam pressure) and of the moisture content (thermal conductivity) of the process material. Because the heat transfer area is fixed, the best method to respond to load variations is to change the steam pressure, which in turn will change the dryer surface temperature. Load variations can be detected by measuring the orifice differential (h in Figure 7.18). An increase in h indicates an increase in load, which means that the steam supply pressure (P) to the dryer should be increased.

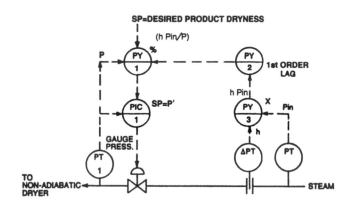

FIGURE 7.18 The ratio of the gauge pressure of the steam into the dryer (P) and the orifice differential corrected for supply pressure variations (hPin) can be adjusted to obtain the desired product dryness.

Shinskey[2] has determined that if the ratio of h to P is held constant, the product moisture content will also be constant. In Figure 7.18, this ratio control loop is also corrected for supply pressure variations and is provided with dynamic compensation. The dynamic compensation is needed (PY-2) so that when an increase in load (h) causes an increase in the PIC-1 set point (P′) and thereby the steam valve opens, this opening will *not* result in another increase in P′ until the dryer has had time to respond fully by increasing its surface temperature and heat transfer coefficient. In this way the first-order lag in PY-2 will protect against positive feedback.

B. Rotary Dryers

The term "tumbling dryers" is used here to designate equipment in which material is tumbled, or mechanically turned over, during the drying process. Two examples in this class are the rotary dryer and the turbo dryer.

The cross section of a rotary dryer is shown in Figure 7.19. As the shell is rotated, the material is lifted by the flights and then dropped through an air stream. The speed of rotation, the angle of elevation, and the air velocity determine the material holdup time. Variable-speed drives are sometimes incorporated to change the rotation rate, but they are usually manually controlled. Airflow is not varied as a control parameter.

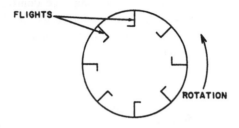

FIGURE 7.19   Cross section of rotary dryer.

A typical control scheme for a rotary, counterflow dryers (material direction opposite air direction) is shown in Figure 7.20. Primary controls maintain the airflow and inlet air temperature. Secondary considerations are pressure control to maintain pressure within the dryer and temperature and level alarms at the dry product outlet. A temperature alarm is also provided at the air outlet

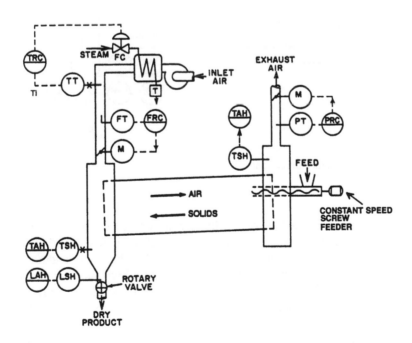

FIGURE 7.20   Rotary dryer controls.

to prevent dryer overheating when feed is stopped. Direct setting of the air inlet temperature is based on the supposition that the product temperature approaches it before the solids reach the discharge end. Better control of moisture is obtained if the control is based directly on the temperature of the product, but there are practical difficulties in providing a material holdup for measurement with consistent renewal. The feed rate is also manually controlled, since the long holdup (dead time) makes automatic adjustments impractical.

Figure 7.21 describes the conventional controls of co-current rotary dryers. In such a configuration, the optimum inlet air temperature is the maximum that the product can tolerate and the utility can produce. The high-temperature alarm (TAH) is needed if high temperature can degrade the product. Alarms are also provided on the steam and air valve positions (VPA) because, if either is fully open, the product moisture content can no longer be controlled. One limitation on the controls shown is the temperature-flow cascade loop, due to the dead time (transport lag) of the air in the dryer. Another limitation is that the outlet temperature set point should not be a constant but should rise (a small amount) with air flow.

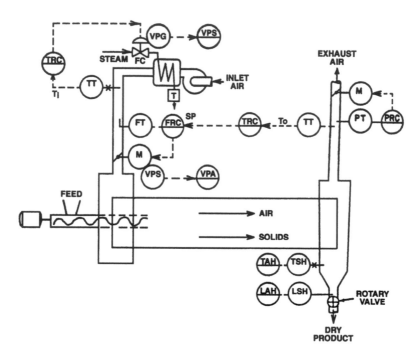

FIGURE 7.21   Co-current rotary dryer controls.

C. Turbo Dryers

In a turbo dryer the material is dried on rotating horizontal shelves (Figure 7.22). They are arranged in a vertical stack, and the product is wiped from each shelf through a slot after a little less than one revolution. A leveler bar then spreads it evenly on the shelf below. In addition to the general countercurrent of hot air, fan blades on a central shaft impart a horizontal velocity pattern.

Since the internal fan provides consistent air circulation, it is feasible to control the air throughout. Motorized dampers are provided for the lower air input and for the combined flow to the upper sections. The division between the upper inlets is adjusted by manual dampers. The inlet air temperature is controlled by throttling of the steam valve, and the motorized dampers are adjusted by the corresponding temperatures. Circulation rates of the center fan and the shelves and the feed rate of the material are manually controlled.

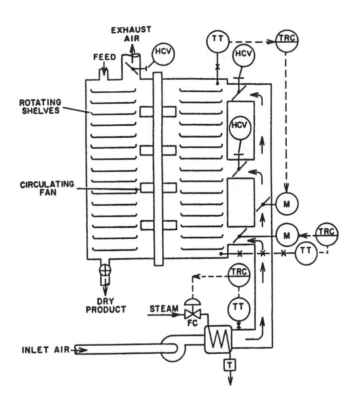

FIGURE 7.22 Turbo dryer.

D. SPRAY DRYERS

These dryers are used in applications where the product is a slurry produced by a product that is suspended in water, such as in the case of the prilling towers in nitrogen fertilizer plants. As a function of the spraying method applied, they can be nozzle or rotating plate types. In the nozzle variety, the product is sprayed under pressure, while in the plate type the spraying effect is produced by the feed stream's impact on a plate that is rotating at high speed. One of the control objectives usually is to keep the size of the droplets uniform.

The spray dryer is an exception to the generalization of large transportation lag times in dryers, because its holdup is normally on the order of tenths of a second. A liquid or thin slurry feed is atomized into a chamber in which there is a large flow of hot air (Figure 7.23). The inlet air temperature must be quite high, and therefore direct-fired heaters are used whenever possible.

The feed is introduced by a high-pressure, manually controlled pump. Airflow is also manually regulated and balanced. Process conditions are maintained by temperature control near the outlet end. The temperature controller regulates the firing rate of the fuel-air mixture through a gas control unit. A temperature switch is provided to shut off both the feed and the fuel in the case of fire or other abnormally high temperature conditions.

An improved control configuration is illustrated in Figure 7.24. The roles of TRC-1 and TSH-3 are unchanged, but here TRC-2 is added, which guarantees uniform spraying conditions by holding the temperature of the sprayed fluid constant. Another added feature is the flow ratio controller between the liquid and air flows (FRC-4). The measurement of this controller is the liquid flow into the dryer,[4] while the set point is the sum of primary and secondary air flows. To better solidify the sprayed droplets in fertilizer applications, the secondary air is usually cooled. Alarms might detect the potentials for explosion, product plugging, flooding, high temperature, and loss of air flow.

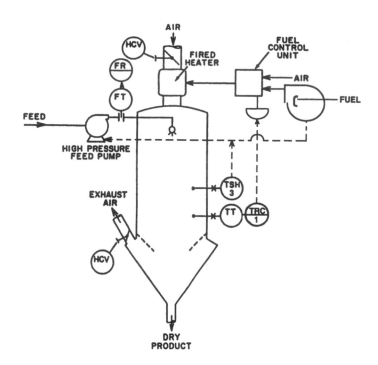

FIGURE 7.23   Spray dryer.

Some of the more advanced control concepts that are explained in connection with Figure 7.28 can also be implemented on spray dryers. When direct-fired air heaters are used, the inferential control model (Equation 4) needs to be modified, partially because moisture is being added to the air by the products of combustion. Another factor is that the supply air in direct-fired dryers is usually at 1,000°F (538°C) or more, and at such temperatures the variations in ambient humidity have practically no effect, and therefore the term "a" in Equation 4 can be disregarded.

E. FLUID-BED DRYERS

The fluid-bed dryer has a fluidized bed of material maintained by an airflow upward through a perforated plate (Figure 7.25). Feed is controlled by a variable-speed screw, and discharge is by overflow of the bed through a side arm.

The bed is maintained at the desired product moisture content by a temperature controller that is detecting the temperature of the fluidized product or of the air space above.

Since the system of Figure 7.25 responds to outlet temperature but not to the absolute humidity of the air stream, it is limited by "reverse action": an increase in humidity of the entering air can cause a reduction in inlet air temperature instead of increasing it. This is because the effect of increased humidity is to reduce the drying rate, which represents less heat loss from the air, and therefore raises the outlet temperature. The controller (TRC) in Figure 7.25 compensates for this rise in temperature by lowering the inlet temperature, which further reduces the drying capacity of the air.

Two alternative control schemes are available that do not exhibit this effect. One is an adaptation of the system of Figure 7.20, in which airflow rate and inlet temperature are directly controlled. The other is to apply the type of advanced controls that are described in Figure 7.28 in connection with the countercurrent longitudinal dryer.

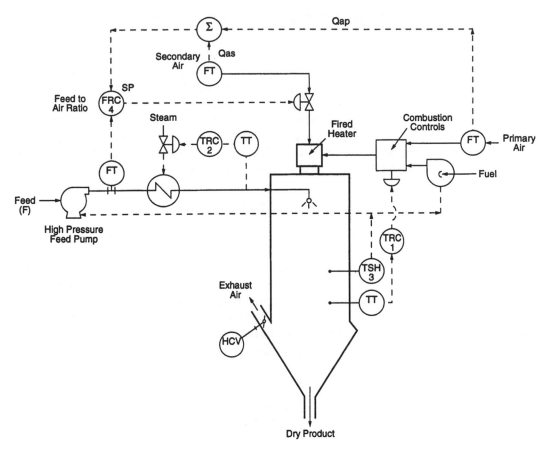

FIGURE 7.24 Spray dryer with gas to liquid flow ratio.

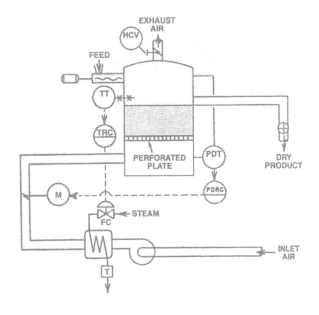

FIGURE 7.25 Fluid-bed dryer.

V. OPTIMIZING CONTROLS (CONTINUOUS)

A. FLUID-BED DRYERS

In an adiabatic fluidized-bed dryer both the bed temperature and moisture content can be considered uniform and close to the temperature and dryness of the product. Therefore, the solids temperature profile in a fluid-bed dryer is not like the curves in Figures 7.3 and 7.4; rather, it is more like a horizontal line. When the load increases, both inlet and outlet air temperatures (Ti and To) must rise based on the same type of relationship as is given in Equation 4 and also on the top of Figure 7.28. The drier the product, the higher the slope (b) of that line. If the feed flow (F) and moisture content are constant, the descriptive equation for the outlet temperature set point (T_o') is simplified into

$$T_o' = a + bTi \qquad (3)$$

where a relates to the dew point of the entering air, and b relates to the desired moisture content of the product.

For a condition in which the product moisture content required is in the "falling-rate region," meaning that the product moisture content (W_p) is less than the critical moisture content (W_c). In the falling rate region ($W_p < W_c$), the product moisture content is expressed by Equation 1. The constant (K) combines air and solids flows and air and solids moisture content, and the equation shows that the product moisture content (W_p) will be held constant if the ratio of the two temperature differences under the logarithm $[(T_i - T_w)/(T_o - T_w)]$ is constant (Figure 7.26).

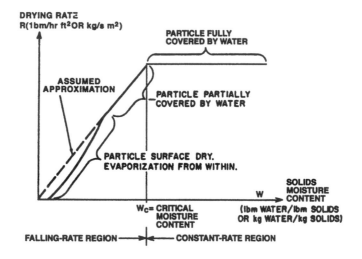

FIGURE 7.26   Drying rate versus moisture content.

To control moisture, temperatures at *both* ends of the dryer *must* change, and the difference between T_i and T_o must increase as the dryer load rises.

If the product is heat-sensitive, then a high limit is set on T_i. If the inlet temperature (T_i) is too high for convenient wet-bulb (T_w) measurement, this signal can also be obtained by measuring air humidity *upstream* to the air heater.

Figure 7.27 describes an advanced control system, which automatically programs the proper relationship between the air outlet temperature (T_o) and the corresponding desired inlet temperature (T_i'). This relationship is graphically described on the bottom of Figure 7.27, where one straight line corresponds to a particular feed condition (cF = Const.) with "a" being the intercept and "b" the slope of the straight line. These values (a and b) are feedback controlled and field calibrated. Raising either will increase the dryness of the product.

For any particular load condition, the desired air outlet temperature T_o' is calculated by the DCS system, which is solving Equation 4 ($T_o' = a + bTi + cf$). Some of the terms in this equation might not be detected continuously. For example, the feed moisture content (c) can be periodically measured and manually introduced as a constant. Similarly, the moisture content of the product (b) might not be detected continuously either. In that case (Figure 7.27), the value "b" is manually set (HIC-3) as the desired product dryness, and the actual value (AT-3) is tested only periodically.

As the desired air outlet temperature (T_o') is calculated, a positive feedback loop is formed, because an increase in T_i results in an increase in T_o', which increases the inlet air temperature controller (TRC-2) set point (T_i'), which further increases T_i. For proper operation of the system, this positive feedback must be compensated by the slower-acting negative feedback loop provided by the measurement of TRC-1 (T_o). Therefore, for the negative feedback to be effective, it must be made relatively faster, which necessitates the slowing of the set point response to a change in

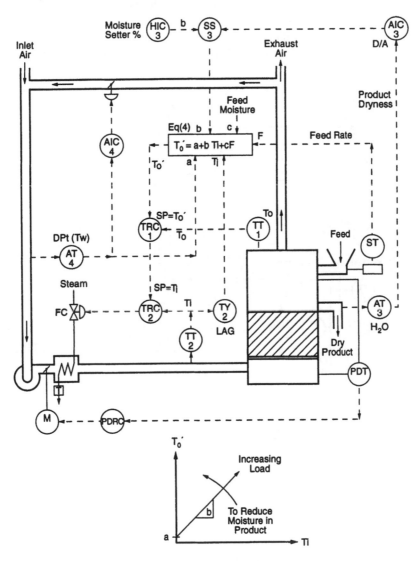

FIGURE 7.27 Advanced controls of a continuous fluid-bed dryer.

T_i measurement and therefore T_i must be lagged (TY-2). The overall controls will be stable if both the steady-state and the dynamic gain of the positive feedback loop are below that on the negative feedback loop. Setting the lag (TY-2) to exceed the integral time of TRC-1 guarantees the proper dynamic gain relationship, while the proper setting of "b" will assure steady-state stability.

If the dryer operates at low temperatures, it is also desirable to control and/or compensate for ambient humidity variations (AT-4). In Figure 7.27, the dew point (or wet-bulb temperature) of the incoming air is both measured ("a") and controlled (AIC-4). In high-temperature dryers (over 400°F or 200°C), the effect of ambient humidity variations becomes insignificant. The heat source in high-temperature dryers is often combustion, with the combustion products also entering the dryer. In such cases, the hydrogen in the fuel is also combusted into moisture, and the resulting effect on wet-bulb temperature must also be compensated.

B. ROTARY DRYERS

An improved control system can be obtained if some of the heat content of the exhaust air is reused and if the load variations are automatically controlled. The heat recovery is achieved and the dew point of the inlet air is controlled the same way as was discussed in connection with Figure 7.14, by the dew point AIC-4 shown in Figure 7.28. The matching of load variations with an appropriately

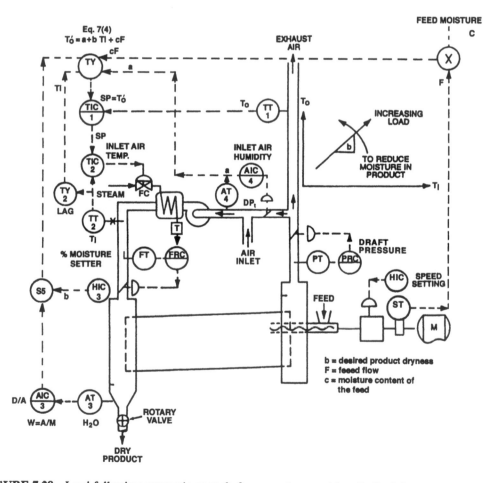

FIGURE 7.28 Load-following automatic controls for a countercurrent longitudinal dryer.

selected combination of inlet and outlet air temperatures (Ti and To) is similar to the controls in Figures 7.15 and 7.16, but needs further explanation.

Figure 7.4 illustrated the temperature profiles of both the solids and the air as they pass through a countercurrent longitudinal dryer. When the load on the dryer increases, it is necessary to increase the temperature difference (the distance between these curves), in order to transfer more heat and thereby drive off more moisture. Load can increase for one of three reasons: (1) the mass flow of the feed (F) can increase, (2) the moisture content of the feed can increase, or (3) the dew point of the drying air (a) can rise. When the load rises, it is not sufficient to only increase the air inlet temperature, because that would not necessarily increase the temperature difference (Figure 7.4) between the solids and the air throughout the dryer. What is needed is to increase *both* Ti and To in a way which is consistent with the characteristics of the process.

As it is shown on the top right of Figure 7.28, an increase in load requires an increase in both Ti and To. The slope of b of the line relating Ti to To is a function of the product dryness desired. The drier the product (measured by AIC-3 at the dry product outlet), the higher the value of b and therefore the larger increase in T_o is required for a unit increase in Ti. Therefore, the control system in Figure 7.28 is configured in such a way that the desired air outlet temperature (To') is calculated as the set point of TIC-1, which is the cascade master of the Ti controller, TIC-2. For calculating the set point for TIC-1, Shinskey[2] suggests the following relationship:

$$\acute{T}o = a + bTi + cF \tag{4}$$

where a = the dew point of the inlet air detected by AT-4, b = the desired product, which can be feedback-trimmed by AIC-3, c = the feed moisture content, which determines the impact of feed flow variations on the total load, and F = the mass flow of wet feed.

If the dew point (a) is controlled and the feed flow (F) is constant, this control system is simplified into a cascade loop where TIC-1 is the master of TIC-2, while the set point of TIC-1 is calculated as the product (bTi). The dynamics of this type of configuration must be carefully evaluated. Firstly, the response of To to a change in Ti is not instantaneous, and therefore the measurement signal (Ti) must be *lagged* (TY-2) before it is used in Equation 4. Secondly, the control system contains a positive feedback loop which has to be carefully configured so that it will not destabilize the process.

After a load increase, To drops, and therefore TIC-1 increases the set point of TIC-2, which results in an increase in Ti. As Ti rises, Equation 4 causes an increase in the set point of TIC-1 and the cycle is repeated. If the slope (b) on the top right of Figure 7.28 is properly set and the loop is properly tuned, the system comes to rest matching the higher load with a higher heat input according to the Ti-To relationship. On the other hand, if $\acute{T}o$ (the TIC-1 set point) can change faster than the speed at which To can respond to a change in Ti, instability and cycling can occur. The purpose of the first-order lag (TY-2) is to prevent that.

VI. CONCLUSIONS

In this chapter, a variety of traditional and advanced control systems have been described for several batch and continuous dryer designs. In all cases, the dual control goals are improved product quality and improved energy efficiency. The control strategies that were shown are not unique solutions and can often be interchanged or combined. Some improvements in dryer controls became available because of the development of better sensors and analyzers,[1] while others are by-products of new, more sophisticated, computer-based control techniques.

In dryers, the main limitation of feedback control is the large dead time on the process side (Table 7.1), which would make such loops as AIC-3 in Figure 7.28 ineffective if it were used as the direct cascade master of TIC-1 or TIC-2, instead of just adjusting "b" in the inferential process model. This feedback trimming of the model is fast enough to correct for changes in product particle size, for changes in the moisture content of the feed ("c"), or even for changes in atmospheric humidity ("a") if it is not otherwise controlled (AIC-4 in Figure 7.28). If product moisture is measured and controlled, it is essential to protect the controller (AIC-3 in Figure 7.28) from reset wind-up during dryer equipment breakdowns or feed interruptions, which are not uncommon. For that reason, auto-manual switching (A/M) is provided in addition to external feedback.

In dryers with short dead times (such as spray, flash, or fluid-bed dryers), successful feedback control based on continuous product moisture measurement is more feasible (AIC-3 in Figure 7.28), but, even in these installations, it is advisable to only trim the model and to provide manual backup (HIC-3) to allow for analyzer maintenance.

Moisture content can be tested by detecting the weight loss of grab samples after drying or continuously by conductivity, dielectric constant, IR reflection, neutron dispersion, nuclear magnetic resonance, or microwave absorption measurement.[1,5] When using infrared or microwave analyzers, the variations in product particle size, density, or bed thickness can result in noisy measurement signals. Because filtering always deteriorates the quality of control, the filter time constant should be set as low as displays and proper derivative mode response will allow.

REFERENCES

1. Lipták, B.G., ed., *Instrument Engineers' Handbook*, Measurement Volume, 3rd ed., 1995.
2. Shinskey, F.G., *Energy Conservation Through Control*, Academic Press, 1978.
3. Nagy, I., *Introduction to Chemical Process Instrumentation*, Akadémiai Kiadó, Budapest.
4. Miller, R.W., *Flow Measurement Engineering Handbook*, 3rd ed., McGraw-Hill, New York, 1997.
5. Carr-Brion, K., *Moisture Sensors in Process Control*, Elsevier, New York, 1987.

BIBLIOGRAPHY

DeThomas, F.A., Control of Moisture in Multi-Stage Dryer, ISA/93 Technical Conference, Chicago, September 19-24, 1993.
Kinney, T.B., Advanced Control with Optimization of Kiln-Dried Lumber, ISA Conference, 1985.
Krigman, A. Moisture and humidity, *InTech*, March 1985.
Lipták, B.G., "Optimizing Dryer Performance," *Chemical Engineering*, February and March, 1998.
Lipták, B.G., ed., *Instrument Engineers' Handbook*, Measurement Volume, 3rd ed., 1995.
Mettes, J., et al., Multipoint ppm Moisture Measurement Using Electrolytic Cells, ISA/92 Technical Conference, Houston, October 1992.
Miller, R.W., *Flow Measurement Engineering Handbook*, 3rd ed., McGraw-Hill, New York, 1997.
Nagy, I., *Introduction to Chemical Process Instrumentation*, Akadémiai Kiadó, Budapest.
Schultz, G., Relative humidity measurement with an analog transmitter, *Measurements & Control*, December 1992.
Shinskey, F.G., *Process Control Systems*, 4th ed., McGraw-Hill, New York, 1996.
Shinskey, F.G., *Energy Conservation Through Control*, Academic Press, 1978.
Wiederhold, P.R., Cycling chilled-mirror hygrometer, *Measurements & Control*, 1993.

8 Evaporators

I. THE PROCESS

The process of evaporation is as old as the process of solar evaporation of sea water to obtain salt. In the process of evaporation, usually a weak solution of a nonvolatile material is concentrated by the evaporation and removal of the solvent, which most often is water. The product can either be the clean solvent (sea water desalinization, boiler feedwater preparation) or the concentrated solution (corn or maple syrup, etc.). In some specialized applications (nuclear wastes), there is no marketable product.

The heat source is usually steam. The amount of water removed per unit of steam used increases with the number of evaporator stages (effects) and with design variations such as vapor recompression. The addition of trim heaters improves control quality at the price of efficiency. In all cases, the energy content of the steam is eventually lost to the environment.

A. TERMINOLOGY

Boiling Point Rise   This term expresses the difference (usually in °F) between the boiling point of a constant composition solution and the boiling point of pure water at the same pressure. For example, pure water boils at 212°F (100°C) at 1 atmosphere, and a 35% sodium hydroxide solution boils at about 250°F (121°C) at 1 atmosphere. The boiling point rise is therefore 38°F (21°C). Figure 8.1 illustrates the features of a Dühring plot

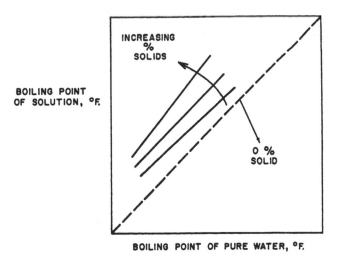

FIGURE 8.1   Dühring plot of boiling point rise.

in which the boiling point of a given composition solution is plotted as a function of the boiling point of pure water.

Capacity   The capacity for an evaporator is measured in terms of its evaporating capability, viz., pounds of vapor produced per unit time. The steam requirements for an evaporating train may be determined by dividing the capacity by the economy.

Co-Current Operation   The feed and steam follow parallel paths through the evaporator train.

Countercurrent Operation   The feed and steam enter the evaporator train at opposite ends.

Economy   This term is a measure of steam use and is expressed in pounds of vapor produced per pound of steam supplied to the evaporator train. For a well-designed evaporator system the economy will be about 10% less than the number of effects; thus, for a triple-effect evaporator the economy will be roughly 2.7.

Multiple-Effect Evaporation   Multiple-effect evaporations use the vapor generated in one effect as the energy source to an adjacent effect (Figure 8.2). Double- and triple-effect evaporators are the most common; however, six-effect evaporation can be found in the paper industry where kraft liquor is concentrated, and as many as 20 effects can be found in desalinization plants.

Single-Effect Evaporation   Single-effect evaporation occurs when a dilute solution is contacted only once with a heat source to produce a concentrated solution and an essentially pure water vapor discharge.

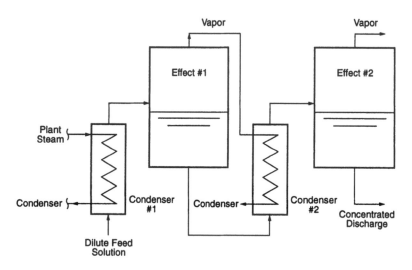

FIGURE 8.2   Multiple-effect evaporator.

B. THE PROCESS MODEL

In some cases neither vapor nor concentrated discharge has any market value, as in the case of nuclear wastes. Evaporators can be arranged in forward-feed, reverse-feed, or parallel-feed configurations with each stage being heated by the vapors of the previous stage.

In multiple-effect evaporation, the temperature in the last effect is the boiling point corresponding to the vacuum available from the vacuum source. The temperature difference between effects is a function of the pressure difference required to move the product forward. Also, in parallel flow (co-current) applications, the liquid phase arriving from the upstream effect will be above its boiling point, due to the drop in pressure, while the vapor removed from each effect is superheated by the boiling point elevation. Typical temperature differences between effects are about 30°F (16.7°C), the per-effect efficiency is 95% to 98%, and steam is usually available at 175 PSIA (12.1 bar) or more. Additional effects provide diminishing returns because (in addition to the efficiency of each effect) more and more heat is required to vaporize the same amount of water as the pressure drops.

For one set of conditions (2 PSIA vacuum source, 33°F difference between effects, 97% efficiency), Shinskey[1] calculates the economy as a function of the number of effects as follows: 1 (0.95), 2 (1.85), 3 (2.71), 4 (3.52), 5 (4.28), 6 (4.98), 7 (5.62), and 8 (6.21).

One effective way to control evaporators is to establish the required material and energy balances in the steady state (Figure 8.3) and then correct these balances for the dynamics (dead times) of the process. When solids are suspended in water (or other liquids), one might need to obtain percent solids information by measuring specific gravity (or density) and vice versa. Figure 8.4 illustrates the conversion.

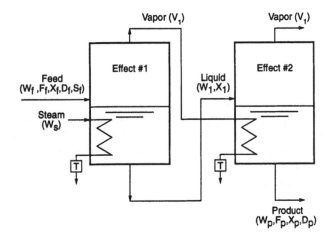

FIGURE 8.3 Terminology for evaporator material balance.

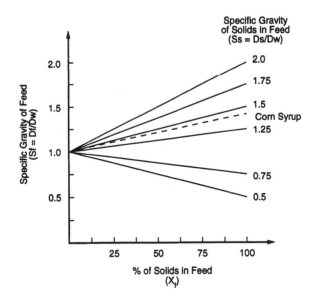

FIGURE 8.4 Conversion between % solids and density or specific quantity.

Where D_f = feed density (lb/gal or kg/l), D_p = product density (lb/gal or kg/l), D_w = nominal water density (8.33 lb/gal or 1.0 kg/l), E = economy (mass of total vapor per unit of steam), F_f = volumetric feed flow (GPM or l/m), S_f = specific gravity of feed (D_f/D_w), S_s = specific gravity of solids, V_1, V_2 = mass flow of vapor from each effect, V_t = total mass flow of vapors from all effects, W_f = total mass flow of feed (lbs or kgs per unit of time), W_1, W_2 = mass flow of liquid from each effect, W_p = mass flow of product (lbs or kgs per unit of time), W_s = mass flow of steam (lbs or kgs per unit of time), x_f = weight fraction of solids in feed, x_1, x_2 = weight fraction of solids in liquid from each effect, and x_p = weight fraction of solids in product.

In an evaporation process, the load variables are the flow and solids concentration of the feed (F_f and x_f), the manipulated variable is the mass flow of steam (W_s), and the controlled variable is the solids concentration of the product (x_p). Assuming a heat loss to the environment of 5%, Myron[2] gives the relationship between the manipulated and controlled variables in Equation 1.

$$0.95\ F_f D_f(1 - x_f/x_p) = F_f D_w S_f(1 - x_f/x_p) = 0.95\ W_s E \qquad (1)$$

For an in-depth discussion relating to methods of computing the economy (E) of a particular evaporator system, see References 2 and 4.

For each feed material, a relationship between the density of the feed material (D_f) and its solids weight fraction (x_f) has usually been empirically determined by the plant or is available in the literature. See Reference 4 where the density-equals-percent-solids relationship of many inorganic compounds is available.

The $S_f(1 - x_f/x_p)$ portion of Equation 1 is a function of the feed density, $f(D_f)$, i.e.,

$$f(D_f) = S_f(1 - x_f/x_p) \qquad (2)$$

Assume, for example, that a feed material (to be concentrated) has the solids—specific-gravity relationship shown in Table 8.1. If this feed material were to be concentrated so as to produce a product having a weight fraction of 50% ($x_p = 0.50$), the $f(D_f)$ relationship of Equation 2 could be generated as shown in Table 8.1.

TABLE 8.1
Density to Weight Fraction Relationship

Solids Weight Fraction (x_f)	Specific Gravity ($S_f=D_f/D_w$)	$(1 - x_f/x_p)$	$f(D_f) = S_f$ $(1 - x_f/x_p)$
0	1.0000	1.0	1.0000
0.08	1.0297	0.840	0.865
0.16	1.0633	0.680	0.723
0.24	1.0982	0.520	0.571

This body of data is plotted in Figure 8.5. In all the cases investigated, the $f(D_f) = S_f$ relationship is a straight line having an intercept of 1.0, 1.0. The $f(D_f)$ relationship can then be written in terms of the equation of a straight line: $y = mx + b$, i.e.,

$$f(D_f) = 1.0 + m\ (S_f - 1.0) \qquad (3)$$

where m = slope of line.

Using the data of Table 8.1, the value of m is determined as

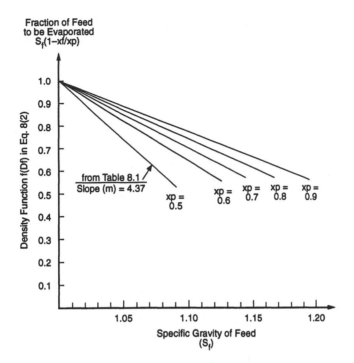

FIGURE 8.5 The fraction of the feed that must be evaporated depends on product concentration and feed specific gravity or density. The slope of these lines (m) is usually throttled by the feedback quality controller (DIC-9 in Figures 8.27 an 8.28).

$$m = \frac{1.000 - 0.571}{1.0000 - 1.0982} = -4.37 \tag{4}$$

Therefore,

$$f(D_f) = 1.0 - 4.37 \, (S_f - 1.0) = 1 - 4.37 \, (D_f/D_w - 1) \tag{5}$$

Substituting Equation 5 into Equation 1 and solving for the manipulated variable, steam flow (W_s),

$$W_s = \frac{500 \, F_f\left[1 - m\left(D_f - 1\right)\right]}{nE} \tag{6}$$

If feed rate is manipulated in response to a variable steam flow and feed density, Equation 6 becomes

$$F_f = \frac{W_s E n}{500\left(1 - m\left(D_f - 1\right)\right)} \tag{7}$$

The factor 500 is used to convert gallons per minute of water to pounds per hour of steam. Equations 6 and 7 can be used to provide steady-state feedforward control models. In these, the product concentration (x_p) controller provides feedback trimming by adjusting the slope "m" in the model. If steam flow (W_s) is the manipulated variable, the feedforward model is based on feed flow (F_f) and on the amount by which feed density exceeds that of water ($D_f - 1$). If the manipulated variable is feed flow (F_f), the feedforward model is based on the measurements of W_s and ($D_f - 1$). To convert the steady-state control model into a dynamically compensated one, lead/lag compensation has to be added.

C. TYPES OF EVAPORATORS

Six types of evaporators are used for most applications. The length and orientation of the heating surfaces determine the name of the evaporator.

Horizontal-tube evaporators (Figure 8.6) were among the earliest types. Today, their use is limited to preparation of boiler feed water, and to small-volume evaporation of severely scaling liquids, such as hard water, but this requires special construction at high cost. In their standard form they are not suited to scaling or salting liquids and are best used to applications requiring low throughputs.

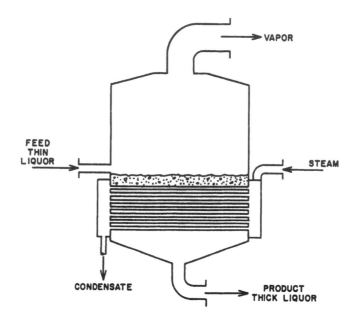

FIGURE 8.6   Horizontal-tube evaporator used in boiler feedwater applicatons.

Figure 8.7 illustrates a forced (pumped) circulation evaporator with an integral flash chamber. If salting or erosion is a potential problem, the evaporator can be located outside the flash chamber so that the tubes are easier to repair or clean and boiling does not occur in them. Compared with the horizontal-tube design, these evaporators are more expensive, have a higher residence time and cost more to operate.

The short-tube vertical evaporator (calandria or Robert type) is used in the sugar industry for cane sugar juice concentration (Figure 8.8). Liquor circulation through the heating element (tube bundle) is by natural circulation (thermal convection). Since the mother liquor flows through the tubes, they are much easier to clean than those shown in Figure 8.6, in which the liquor is outside the tubes. Thus, this evaporator is suitable for mildly scaling applications where low cost and convenient cleaning or descaling are desired. Level control is important—if the level drops below the tube ends, excessive scaling results. Ordinarily, the feed rate is controlled by evaporator level to keep the tubes full. The disadvantage of high residence time in the evaporator is compensated for by the low cost of the unit for a given evaporator load.

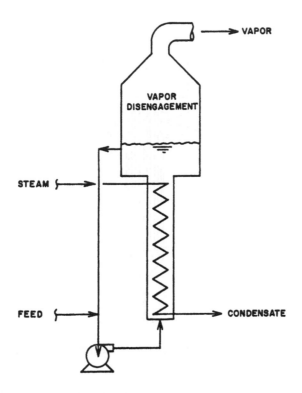

FIGURE 8.7   Forced-circulation evaporator.

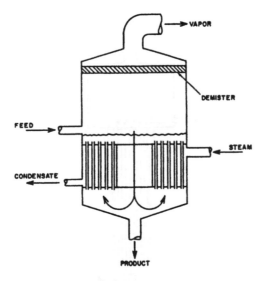

FIGURE 8.8   Short-tube vertical (calandria) evaporator, used in sugar juice concentration.

The rising film concentrator (RFC) or long-tube, vertical evaporator is illustrated in Figure 8.9. Due to its low cost-per-unit capacity, it is a popular design. It is used in the pulp and paper industry back liquor and in the food industry to concentrate corn syrup. These units have low residence (holdup) times because they have little internal circulation and are of the single-pass design. They require effective level control to maintain the liquid seal in the flash tank, and they are difficult to

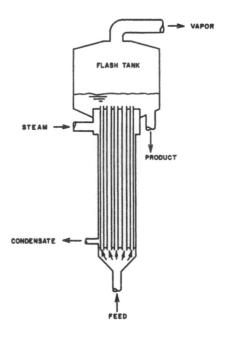

FIGURE 8.9   Long-tube vertical (Kestler type) evaporator, used in corn syrup and black liquor concentration applications.

control because of their sensitivity to operating conditions. These evaporators are tall (20 to 50 ft or 6 to 15 meters) and therefore require more head room than the other designs.

A falling-film evaporator (Figure 8.10) is commonly used with heat-sensitive materials. Phys-

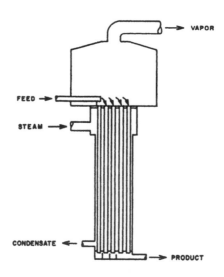

FIGURE 8.10   Falling-film vertical evaporators are used to concentrate heat-sensitive materials.

ically, the evaporator looks like a long-tube vertical evaporator, except that the feed material descends by gravity along the inside of the heated tubes, which have large inside diameters (2 to 10 inches, or 50 to 250 mm).

An agitated film evaporator, like the falling-film evaporator, is commonly used for heat-sensitive and highly viscous materials. It consists of a single large-diameter tube. The material that is to be concentrated is spread by mechanical wipers over the inside surface of the tube. As a result, high heat transfer coefficients are obtained while the usually highly viscous film is sinking downward.

Evaporator design variations include the use of external condensers (Figure 8.2), which serve to minimize liquid inventory and thereby reduce dead time.

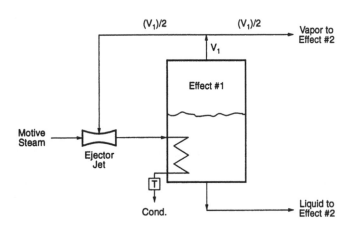

FIGURE 8.11   Thermal compression doubles economy of evaporation.

The use of thermal vapor compression (Figure 8.11) is a potential method of improving energy economy. It uses high-pressure steam to compress part of the vapor, which is generated by the first effect so that the condensing temperature of the compressed vapors will be high enough to boil the liquid in the first effect. Because the heat content of the motive steam also causes boiling, not all the generated vapors can be compressed. With this system, each unit of steam generates about twice as much vapor as it would have without an ejector. These designs are not well suited to variable load conditions because ejectors operate best under constant pressure and flow conditions and are not easily throttled. For increased rangeability, multiple jets can be considered.

Efficiency of evaporation can also be increased by the use of mechanical vapor recompression (Figure 8.12). In that case, all the generated vapor is recondensed and the rate of evaporation is throttled by adjusting the speed of the compressor. In a brine concentration application, the power consumption is reported[1] to be 72 KWhr/1000 gallons, which corresponds to a work-to-heat ratio of about 3%. The pressure in the evaporator can be controlled by a separate cooler or can be left uncontrolled, in which case it will float as a function of feed temperature. Noncondensables must be removed by purging and stripping.

D. Sensors, Valves, and Vacuum Sources

Good level control is a prerequisite of maintaining material balance and residence time. High levels can cause entrainment, and low levels can cause burn-on, which might result in discoloration, decomposition, or even explosions. In such a boiling environment, the extended diaphragm type or the water-purged d/p cells are often considered, although radiation-type level sensors are also common.[5]

Level is maintained by feed or outflow control. In sizing and selecting the control valve, one must consider that only small pressure drops are available in the last effect (usually only a couple

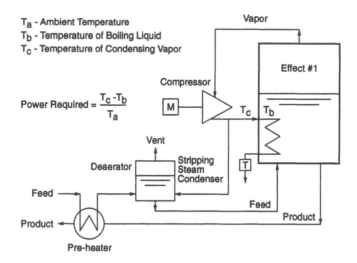

FIGURE 8.12 Mechanical recompression-based evaporators require steam only during start-up.

of PSID or 0.1 bar) and that the throttled liquid is often a scaling slurry, which is near its boiling point. Therefore, low-pressure drop slurry valves (saunders or pinch types[6]) are often considered.

The last few effects of an evaporator train are usually operated under vacuum. Barometrically compensated absolute pressure transmitters and ionization or heat-conductivity type vacuum gauge (Penning and Pirani) sensors[5] provide the measurement signals.

The vacuum is usually controlled either by air bleed into the vapor line (Figure 8.13B) or by throttling the cooling water flow to surface or barometric mixing condensers (Figure 8.13C and D). Air bleed is uneconomical because the expense of pumping the water offsets the savings realized by using a smaller valve on the air line. Direct throttling of the evaporator vapor flow is seldom used (Figure 8.13A) because the valve would have to be large, the vapor line may be 24 or 30 in. (600 or 750 mm), would usually have to be made of stainless steel, and would waste valuable pressure drop.

Similarly, the direct throttling of the vacuum source is seldom practiced. This is because, if the vacuum is generated by steam jets, they are essentially constant steam-flow devices, and multiple jets would be needed for modulation. If the vacuum source is a vacuum pump, its speed can be throttled, but the cost might not be justifiable (Figure 8.13E).

E. PRODUCT QUALITY MEASUREMENT

Some of the measurements used in controlling the product quality include (1) a boiling point elevation, (2) density, (3) refractive index, and (4) conductivity detection. The choice is a function of the product characteristics. It is desirable that the analyzer be specific and sensitive to the product concentration changes. For example, if the product is corn syrup (Figure 8.4) and its concentration is to be controlled at 75% solids, one should make sure that the densitometer is sensitive enough to detect a density change corresponding to a 0.5% change in solids.

The product quality detector must also be located in an accessible and representative location where interference from air bubbles or vibration have been eliminated. The analyzer should also be close to the last effect so that dead time, caused by transportation lag, be minimized in the control loop.

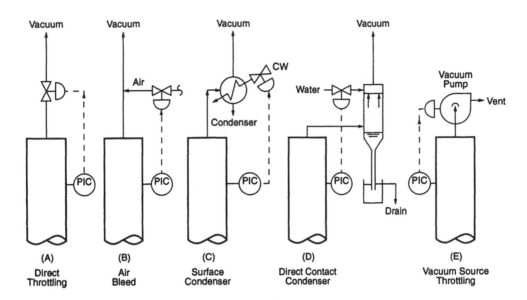

FIGURE 8.13 Methods of evaporator pressure modulation.

1. Boiling Point Elevation (BPE)

According to Dühring's rule, the boiling point of a solvent, such as water, *increases* nearly linearly with rising solution concentration. Therefore, the boiling point temperature of the solution will indicate its concentration if the pressure is constant or if the pressure variations are fully compensated. The water vapor that is boiled off from a solution is therefore superheated by the BPE.

Figure 8.14 gives the relationship between solids concentration and boiling point elevation for atmospheric sucrose solutions.[1] It can be seen that, if the product is to have high concentrations, the sensitivity of this measurement will exceed 2°F/(wt%), which will provide good control. On the other hand, at low concentrations, the sensitivity drops below 0.2°F/(wt%) and control sensitivity will suffer. The boiling point elevation (BPE) is a function of mole fraction (not weight fraction) and, therefore, solutes of lower molecular weight will give higher BPE values at the same concentration:

$$BPE = x(RT^2/\lambda) \tag{8}$$

where BPE = boiling point elevation, x = mol fraction of solute, R = universal const. (1.986 BTU/lb-mole°R or cal/gm-mole°K), T = solvent boiling point (absolute temperature units), and λ = latent heat of vaporization.

If the evaporator pressure is constant, the BPE can be detected by a single temperature measurement made at just below the liquid surface, usually right above the tube bundle of a calandria-type evaporator (Figure 8.15A). Some install the sensing bulb near the bottom of the pan, so that it will always be covered by liquid. This creates a head error, as the boiling temperature rises with hydrostatic head and therefore this error must be compensated.

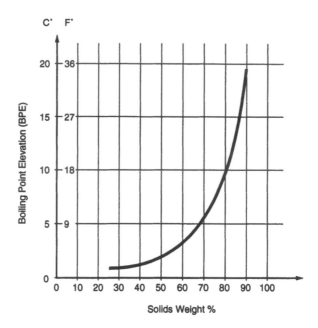

FIGURE 8.14 Boiling point rise of atmospheric sucrose solution.

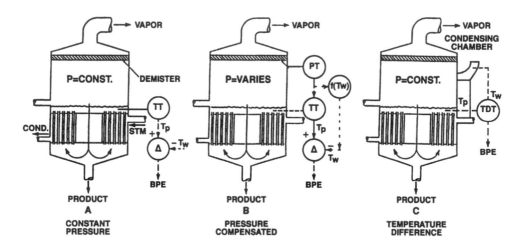

FIGURE 8.15 BPE measurement methods.

BPE is determined as the difference between the liquid product temperature (T_p) and the boiling temperature of water at the same pressure (T_w). If the pressure is not accurately controlled at a constant value, pressure compensation (Figure 8.15B) is recommended.

In case of temperature difference sensing, the boiling point of the water (T_w) is detected in the vapor space of a condensing chamber (Figure 8.15C), where hot condensate is forming and flashing. To obtain BPE, this detected T_w is subtracted from T_p. Because T_w responds much faster to a pressure change than does T_p, stable pressure control is preferred, although properly tuned dynamic compensation is also feasible. Temperature difference control has been found to be most effective when the evaporator pressure is controlled by water flow throttling to a direct contact barometric condenser (Figure 8.13C).

2. Density and Analyzers[5]

The hydrostatic head produced by a liquid column of known height can be detected as density. This *differential pressure* can be measured by bubbler tubes, purged d/p cells, or by flanged, extended diaphragm-type transmitters. These designs are more resistant to coating and plugging than are the small connecting tubes of regular d/p cells. Evaporator feed density is often detected by these devices.

On clean fluids, another density sensor to consider is the *buoyancy float*. It is sensitive to the drag forces of the flowing fluid, which can be minimized by placing it in a flow chamber on a minimum-flow bypass and by Teflon-coating the float.

U-tube weight and vibrating tube densitometers are used to detect product density if they can be protected from coating, which can result in calibration shifts and plugging. If *oscillating Coriolis* mass flowmeters are used to detect feed flow, they can also furnish a density reading. Their error is from 0.002 to 0.005 g/cc.

In the food industry, the *gamma radiation gauges* are popular because they are sensitive and are not subject to plugging. When the process fluid is viscous, the densitometer should be located at a flooded low point to protect against air entrainment.

Concentration can also be detected by continuously measuring the *refractive index* of some liquids. Their fast response is an advantage, while the need for calibration, temperature compensation, and for a sampling line, and the resulting dead time are their main disadvantages.

The electrolytic *conductivity* of fluids such as caustic soda changes with concentration and hence can be used to detect it. The electrodes are usually placed in a minimum-flow bypass. They too require calibration, temperature compensation, and are subject to cell plugging.

F. STEAM SUPPLY

Evaporators can balance and economize plant-wide steam management. If the evaporator feed storage capacity is sufficient, the level of evaporator operation can be adjusted to match the availability of waste steam. Similarly, if there is a shortage of low-pressure steam in the plant, the evaporators can supply it.

When waste steam is used as the energy supply and its availability varies, the goal is to maintain product quality by matching feed flow to steam, according to Equation 7, and allowing the feed tank capacity to absorb the variations. Figure 8.16 shows a scheme for letting feed tank level adjust steam flow set point, while the actual steam flow sets feed rate. The PI (proportional and integral) level controller has a set point of only 25%, so 75% of the tank volume is available to absorb upsets. The proportional band is set to match the set point, thereby providing sensitive response and guaranteeing that the steam set point will be reduced to zero before the tank is emptied.

In case of multi-effect evaporators, the steam generated by the individual effects can be used as plant steam supply or steam can be added to any of the vapor streams generated by the individual effects. This ability will substantially increase plantwide steam management flexibility. All that is needed to balance such flexible systems is to measure the entering and leaving steam flows and apply the proper multipliers, calculated on the basis of the individual effect economies listed in Figure 8.17.

So, if in a five-effect evaporator train, steam is extracted from the top of the second, the total steam loss is the ratio between the flow *not* passing through the last three effects (2.71) versus that passing all five (4.28) or 2.71/4.28 = −0.63.

Similarly, when steam is added after the third effect, the steam will pass through only two effects and therefore the total steam gain to the train is 1.85, while, if it passed all five, it would have been 4.28. Therefore the effective steam gain is 1.85/4.28 = +0.43.

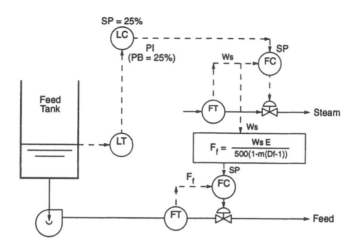

FIGURE 8.16 Variable waste steam availability.

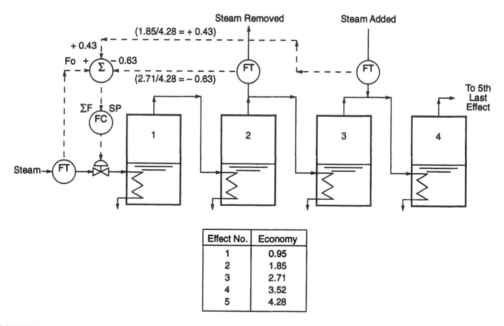

Effect No.	Economy
1	0.95
2	1.85
3	2.71
4	3.52
5	4.28

FIGURE 8.17 Steam extraction and addition to a five-effect evaporator train.

1. Steam Enthalpy

One load variable that, if allowed to pass through the process, would upset the controlled variable, i.e., product density, is steam enthalpy. In some applications the steam supply may be carefully controlled so that its energy content is uniform; in other applications substantial variations in steam enthalpy may be experienced. The objective is to consider the factors that influence the energy level of the steam supplied and to design a system that will protect the process from these load variables.

For saturated steam the energy content per unit weight is a function of the absolute pressure of the steam. If the flow of steam to the process is measured with an orifice meter, the mass flow of steam is

$$W_s = K_1 \sqrt{\frac{h}{v}} \tag{9}$$

where W_s = steam flow in lbs per hour (kg/hr), h = differential head measurement in feet (m), v = specific volume in cu. ft per lbm (m³/kg), and K_1 = orifice coefficient dependent on the physical characteristics of the orifice.

The total energy to the system is

$$Q = W_s H_s \tag{10}$$

where Q = energy to system per BTU per hour (J/h), W_s = steam flow in lbs per hour (kg/h), and H_s = heat of condensation in BTU per lb, or J/kg (enthalpy of saturated vapor minus enthalpy of saturated liquid).

Substituting Equation 9 into Equation 10 gives

$$Q = K_1 \sqrt{\frac{(H_s)^2}{v} h} \tag{11}$$

For evaporator applications, it is the heat content of the steam that needs to be accurately measured. Because evaporators are often operated on waste steam, and because the steam pressure, superheat and enthalpy often vary, it is desirable to pressure compensate all orifice-type flow measurements, as shown in Figure 8.18.[1] In later figures, this configuration is not shown in detail, but is represented by the symbol FT(h).

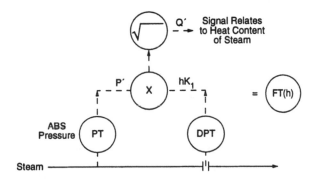

FIGURE 8.18   Steam heat-flow measurement [FT(h)].

2. Scaling the Heat-Flow Signal

For any particular application the steam pressure will vary within limits around a normal operating pressure. To demonstrate the design of a control system to compensate for variations in energy input to the process, assume that the steam pressure varies between 18 and 22 PSIA (124 and 152 kPa), with a normal operating value of 19.7 PSIA (135.9 kPa). The pressure transmitter has a range of 0 to 25 PSIA (0 to 172.5 kPa). These values of pressure variation and operating pressure are typical for evaporator operation. The value of the $(H_s)^2/v$ term appearing in Equation 11 will vary, depending on the steam pressure. Over a reasonably narrow range of pressures, the value of $(H_s)^2/v$ can be approximated by a straight line with the general form of

$$(H_s)^2/v = bP + a \tag{12}$$

where P = absolute pressure in PSIA (Pa), a and b = constants, and P' = scaled value of P (P' = P/P_{max}).

Table 8.2 shows the typical values of H_s, v, $[(H_s)^2/v]'$ and P and P' selected for the specified range of pressures and for a case in which the steam is condensed at 1.5 PSIG.

$$\left(H_s^2/v\right) \text{ normal} = \frac{(970.7)^2}{20.4} = 46,208$$

Rewriting Equation 12 in scaled form:

$$\left(\frac{(H_s)^2}{v}\right)' = bP' + a \tag{13}$$

The designer can either linearize the data — using any two points from Table 8.2 — or use at least square computation to find the best straight line.

TABLE 8.2
Specific Volume-Enthalpy Data

(P) Steam Supply Pressure PSIA (kPa)	(Hs) Heat of Condensation BTU/lbm† (MJ/ka)	(v) Specific Volume ft³/lbm (m³/kg)	$\left((H_s)^2/v\right) = \dfrac{(H_s)^2/v}{\left((H_s)^2/v\right)_{normal}}$	$P' = \dfrac{P}{P_{max}} = \dfrac{P}{25}$
18 (124.2)	969.1 (225.4)	22.2 (1.38)	0.916	0.720
19 (131.1)	970.2 (225.7)	22.1 (1.37)	0.965	0.760
19.7* (135.9)	970.7 (225.8)	20.4 (1.26)	1.00	0.788
20 (138)	971.2 (225.9)	20.1 (1.25)	1.02	0.800
21 (144.9)	972.1 (226.1)	19.2 (1.19)	1.07	0.840
22 (151.8)	973.0 (226.3)	18.4 (1.14)	1.11	0.880

* Normal operation.
† Assuming that the steam is condensed at 1.5 PSIG.

In a least square computation the following values of a and b constants of Equation 13 were obtained:

$$\left(\frac{(H_s)^2}{v}\right)' = 0.085 + 1.161\ P' \tag{14}$$

where a = 0.085 and b = 1.161.

Squaring Equation 11 and substituting in Equation 14:

$$(Q^2)' = (1.161\ P' + 0.085)\ hk_1 \tag{15}$$

The parenthetical portion of Equation 15 can be written so as to make the sum of the two coefficients equal 1.0, which simplifies its implementation using conventional analog hardware.

This is done by multiplying and dividing each term in the parentheses by the sum of the two coefficients.

$$\left(Q^2\right)' = 1.246\left(\frac{1.161\,P'}{1.246} + \frac{0.085}{1.246}\right)hk_1 \tag{16}$$

$$(Q^2)' = 1.246\,(0.932\,P' + 0.068)\,hk_1 \tag{17}$$

The instrumentation to implement Equation 17 is shown in Figure 8.19.

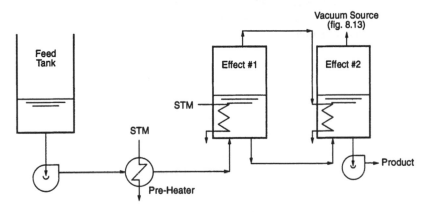

A) Co-Current (forward feed)

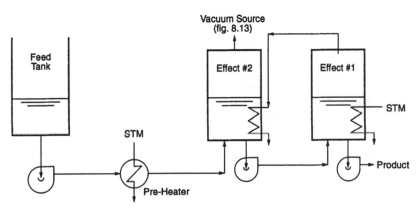

B) Counter Current (reverse feed)

FIGURE 8.19   Forward and reverse feed configurations.

II. BASIC CONTROLS

The product of an evaporation process can be the distilled solvent vapors (seawater desalinization) or the concentrated liquid product. If the product is a heat-sensitive concentrated liquid (food industry), co-current, forward-feed equipment configuration is used so that the boiling temperature in the effects will drop as the concentration rises (Figure 8.19A). When the product is heat sensitive, the evaporator should be designed with short residence times (small hold-up volumes).

If the product is highly viscous (and not heat sensitive), the countercurrent, reverse feed configuration is used because this results in a higher temperature and therefore a less-viscous product (Figure 8.19B). In this configuration, the liquid has to move from lower to higher pressure stages and therefore pumps are required between effects.

The evaporator process is characterized by large material and energy storages and by transportation lags (dead times) that usually exceed 30 seconds. The major disturbance variables are the load variables (feed rate, feed concentration) and steam availability or enthalpy, while lesser disturbances can be caused by pressure variations, variations in ambient heat losses, and tube fouling.

A. Single Effect Evaporator Controls

Figure 8.20 illustrates a single-effect evaporator, which is provided with simple feedback controls for the type of application where the production rate is determined by the heat input (FIC-1) and where steam availability is guaranteed.

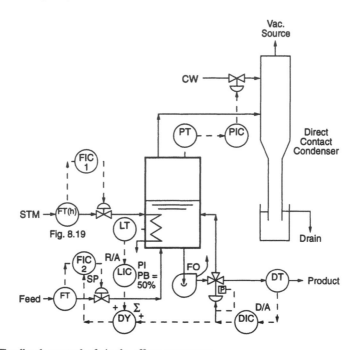

FIGURE 8.20   Feedback control of single-effect evaporator.

The feed rate is adjusted by level control (LIC), while product quality is maintained by throttling the rate of discharge from the evaporator (DIC). Such LIC could usually be tuned for a narrow proportional band, but here, because of the resonant nature of this loop, it is usually selected as a proportional-integral controller, with a PB = 50% or so. The use of a valve positioner or a complete feed flow control loop (FIC-2) is also recommended to overcome the limit cycle that tends to develop in such integrating processes and also to correct for the hysteresis and nonlinearity of the valve.

When the evaporator is operated under vacuum, a discharge pump is needed and the outlet valve is located on its discharge side. If the pump is the positive displacement type, the valve must be of the three-way design. One of the functions of the discharge valve is to prevent the outflow of weak solution during start-up.

During start-up, the LIC should admit only as much feed as is needed to match the rate of evaporation, while, when the desired product concentration is reached, the LIC must admit the equivalent sum of evaporation plus product withdrawal.

Evaporator levels are prone to oscillate due to interactions between level and product concentration control loops or between level and pressure loops. Figure 8.20 shows a simple (one-way) decoupler between DIC and LIC loops, which operates by adding the characterized equivalent of the outflow to the LIC output, thereby avoiding a level upset when the outflow rate is changed. Level oscillations can also be caused by pressure cycling. Therefore, it is important to provide stable and tight pressure control to all evaporators.

B. Multiple-Effect Evaporator Controls

Figure 8.21 illustrates a double-effect evaporator with level controllers (LICs) manipulating the inflows and the product composition controller (DIC) throttling the product flow. Such feedback controls can be considered when the steam supply is not restricted and the steam flow controller (FIC-2) is used to set the production rate. In this control system, if the product concentration drops (possibly because of a drop in feed concentration), the DIC responds by reducing the product withdrawal rate and therefore recirculates more liquid to the first effect. Because the feed flow set point (FIC-1) is the sum of the DIC output (which has been reduced) and the LIC-3 output (which also drops when the level rises), the net effect is an overall reduction in feed rate.

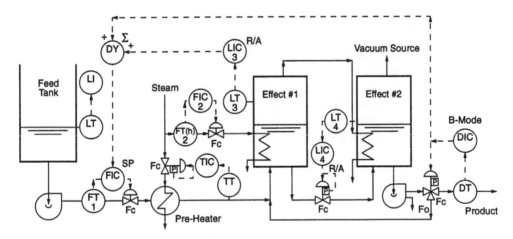

FIGURE 8.21 Feedback control of multiple-effect evaporator.

If the summer (DY) is properly calibrated, the resulting new steam-to-feed ratio will bring the product concentration back to specifications. On the other hand, this type of feedback control is not particularly stable because, after such a correction, the concentration in each effect changes successively and because it gets upset easily by changes in steam rate, which result from changes in production rate. It is for this reason that for heat-sensitive food product evaporators, for example (having short residence times), material-balance-based feedforward controls have been developed.

Figure 8.22 illustrates another feedback control configuration. Here, the feed flowrate (FIC-1) sets the production rate and, in response to load variations, the steam flow (FIC-2) is adjusted by the product quality controller (DIC). Similar to the controls shown in Figure 8.21, this configuration is also limited to applications where the major load variables (feed flow and feed concentration) are stable and steam availability is not limited or unpredictable.

Here, a three-mode density controller (DIC) is adjusting the steam flow to the first effect, while the level controllers (LIC-3 and 4) maintain the internal material balance of the effects. The steam flow [FT(h)] is detected on a heat input basis, as explained in Figure 8.18 and the DIC acts as the cascade master of FIC-2. Such cascade adjustment of a steam (heat)flow set point is particularly effective and recommended when steam pressure varies.

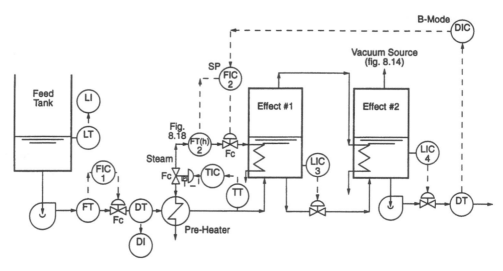

FIGURE 8.22   Feedback control of co-current dual-effect evaporator.

1. Selective Control

Figure 8.23 adds an automatic protection feature against the possibility of running out of steam. A valve position controller (VPC) detects the steam valve opening, and, when it reaches 90% (this on an equal percentage valve corresponds to 70% flow), it starts lowering its output signal and thereby the low signal selector (FY) takes control away from FIC-1. Therefore, during periods of steam shortage or when the feed is temporarily diluted due to some upset, the VPC sets the feed rate to correspond to the allowable maximum opening of the steam valve.

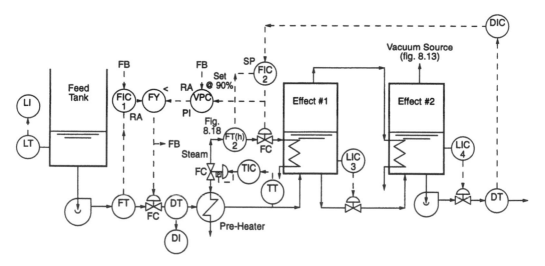

FIGURE 8.23   Selective valve position control provides automatic feed cutback during times of steam shortage.

The VPC is a PI controller with the integral mode dominating (as is the case in most valve position controllers). This allows it to operate smoothly even if the measurement signal (the steam valve opening) is noisy, as its output signal responds not to the error, but to the total area under the past error curve. Both controllers (FIC and VPC) are provided by external feedback taken from the output of the low selector FY. This way, whichever controller is in control will have its output signal and its feedback signal at the same values and therefore will operate as a normal PI controller. The controller which is not selected will operate as a proportional-only controller with a bias, where the bias is the external feedback signal. This guarantees that at the time of switchover the two outputs will be identical, and therefore the switchover will occur bumplessly (at the time of switchover the error in the idle controller has just reached zero, and therefore its output equals its external feedback signal).

2. Feedforward Control

In most evaporator applications the control of product density is constantly affected by variations in feed rate and feed density to the evaporator. In order to counter these load variations, the manipulated variable (steam flow) must attain a new operating level. In the pure feedback or cascade arrangements this new level was achieved by trial and error as performed by the feedback (final density) controller.

A control system able to react to these load variations when they occur (feed rate and feed density) rather than wait for them to pass through the process before initiating a corrective action is termed *feedforward control*. Figure 8.24 illustrates in block diagram form the features of a

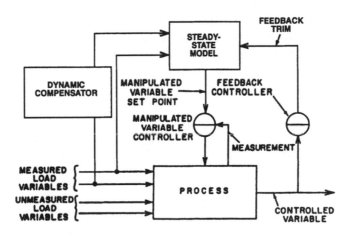

FIGURE 8.24   Feedforward system.

feedforward system. There are two types or classes of load variations—measured and unmeasured. The measured load signals are inputs to the feedforward control system, where they compute the set point of the manipulated variable control loop as a function of the measured load variables. The unmeasured load variables pass through the process, undetected by the feedforward system, and cause an upset in the controlled variable. The output of the feedback loop then trims the calculated value of set point to the correct operating level. In the limit, feedforward control would be capable of perfect control if all load variables could be defined, measured, and incorporated in the forward set point computation. Practically speaking, the expense of accomplishing this goal is usually not justified.

At a practical level, then, the load variables are classified as either major or minor, and the effort is directed at developing a relationship that incorporates the major load variables, the manipulated variables, and the controlled variable. Such a relationship is termed the steady-state model of the process. Minor load variables are usually very slow to materialize and are hard to measure. In terms of evaporators, minor load variations might be heat losses and tube fouling. Load variables such as these are easily handled by a feedback loop. The purpose of the feedback loop is to trim the forward calculation to compensate for the minor or unmeasured load variations.

Without this feature the controlled variable could be off the set point. Thus far we have discussed two of the three ingredients of a feedforward control system: the steady-state model and feedback trim.

The third ingredient of a feedforward system is dynamic compensation. A change in one of the major loads to the process also modifies the operating level of the manipulated variable. If these two inputs to the process enter at different locations, there usually exists an imbalance or inequality between the effect of the load variable and the effect of the manipulated variable on the controlled variable, i.e.,

$$\frac{\Delta \text{ controlled variable}}{\Delta \text{ load variable}} \neq \frac{\Delta \text{ controlled variable}}{\Delta \text{ manipulated variable}} \tag{18}$$

This imbalance manifests itself as a *transient* excursion of the controlled variable from set point. If the forward calculation is accurate, the controlled variable returns to set point once the new steady-state operating level is reached. In terms of a co-current flow evaporator, an increase in feed rate will call for an increase in steam flow. Assuming that the level controls on each effect are properly tuned, the increased feed rate will rapidly appear at the other end of the train while the increased steam flow is still overcoming the thermal inertia of the process. This sequence results in a transient decrease of the controlled variable (density), and the load variable passes through the process faster than the manipulated variable. This behavior is shown in Figure 8.25.

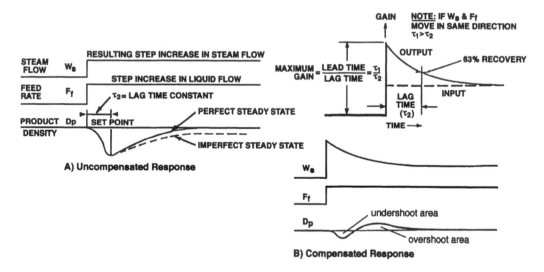

FIGURE 8.25 Load variable (F_f) faster than manipulated variable Ws. Applies to co-current evaporators.

The same sequence is seen in Figure 8.26, except that the manipulated variable passes through the process faster than the load variable. Such behavior may occur in a countercurrent evaporator operation. This dynamic imbalance is normally corrected by inserting a dynamic element (lag, lead-lag, or a combination thereof) in at least one of the load measurements of the feedforward control system. Usually, dynamic compensation of that major load variable, which can change in the severest manner (usually a step change), is all that is required. For evaporators this is usually the feed flow rate to the evaporator. Feed density changes, although frequent, are usually more gradual, and the inclusion of a dynamic element for this variable is not warranted. In summary, the three ingredients of a feedforward system are (1) the steady-state model, (2) process dynamics, and (3) addition of feedback trim.

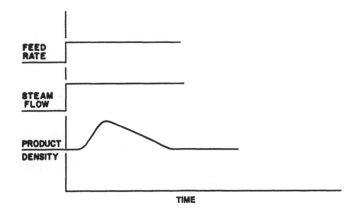

FIGURE 8.26   Manipulated variable faster than load variable. (Counter-current evaporator.)

III. OPTIMIZING CONTROLS

A. MATERIAL-BALANCE BASED FEEDFORWARD

Figure 8.5 has shown the relationship between feed-specific gravity (S_f), product concentration set point (x_p), and the fraction of the feed that must be evaporated. Equations 6 and 7 provided steady-state material-balance models for the evaporation process. The slope of the operating line in Figure 8.5 ("m") varies and is usually feedback trimmed by the product quality controller to correct for minor variable (which are neglected by the model) and for sensor errors. Equation 6 is used when the dominant load variables are feed flow and feed density (F_f and D_f), while the steam flow is the manipulated variable (Figure 8.27) and Equation 7 is used, when the heat source is an unpredictable waste steam stream and the feed rate is therefore adjusted to match steam availability.

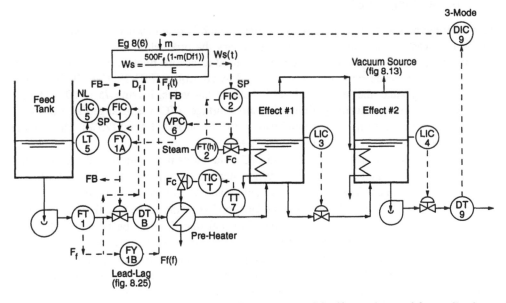

FIGURE 8.27   Material balance-based, dynamically compensated feedforward control for applications with variable feed and reasonably reliable steam supply.

1. Steam as the Manipulated Variable

In Figure 8.27, the feed flow is detected by FT-1 and, because a change in this load variable effects the product density *faster* than does a change in steam flow (manipulated variable), lead-lag compensation is needed to dynamically synchronize the two. As was shown in Figure 8.27, such dynamic compensation results in forcing the steam flow (W_s) to initially "overrespond" to a change in feed flow, and this excess steam is gradually removed over the set lag time (T_2), returning W_s to the new steady-state value.

The required lag time (T_2) is found by measuring the time it takes for the uncompensated response (Figure 8.25A) to drop to its minimum. The lead time (T_1) is initially set to two lag times ($2T_2$). If in the resulting response (Figure 8.25B) the undershoot area exceeds the overshoot area, T_1 is increased; if it is less, the lead time is decreased, until the areas are equal. Once the proper T_1/T_2 ratio is found, one can check if the areas of error are reduced by increasing or decreasing both settings, while maintaining their ratio.

The dynamically compensated feed flow signal [$F_f(t)$] is used in the material-balance calculation, while the uncompensated signal (F_f) is sent as measurement to the feed flow controller (FIC-1). The set point of FIC-1 is cascade-adjusted by a nonlinear feed tank level controller, which brings evaporator operation in line with feed availability. The material-balance calculation also looks at the feed density (D_f) which usually does not change very fast and therefore does not require dynamic compensation, and the feedback trim (m) from the product quality controller (DIC-9). The output (W_s) of this calculation is used to set the steam flow controller (FIC-2).

If for any reason the steam supply would become insufficient, the steam valve would open to 90% and the valve position controller (VPC-6) would take the control away from FIC-1, through the low signal selector (FY-1A), which then selects the VPC output to reach the feed valve. In such periods of operation, only as much feed is admitted as will not require a higher than 90% steam valve opening. To protect against reset wind-up, the actual feed valve opening is sent as external feedback (FB) to both FIC-1 and VPC-6.

2. Feed as the Manipulated Variable

Figure 8.28 illustrates a feedforward control system which is usable when the heat source for the evaporator is a waste steam and it is acceptable to vary the production rate so as to match steam availability. In this control system, the feed tank is used to absorb fluctuations in steam supply as LIC-5 provides the cascade set point for the steam flow controller (FIC-2). The LIC is set at 20 or 25%, and, at steady state, the feed tank level is held at that value, leaving 75 to 80% of the tank volume to accumulate feed during periods of steam shortage. LIC-5 is arranged as a proportional and integral controller, with its proportional band set to equal its set point (25%). This guarantees that its output will drop to zero before the tank is emptied.

As steam availability drops, level rises and LIC-5 increases the steam flow controller set point to its maximum limit, which is set by FY-2. When steam supply is restored, LIC-5 will continue to demand maximum steam flow until the feed tank level is returned to set point.

The material balance around the evaporator (Equation 7) sets the actual feed flow rate by adjusting the FIC-1 set point (F_f) in a feedforward manner. F_f is calculated on the basis of feed density (D_f), actual steam flow (W_s), and the feedback trimming signal (m) from the product quality controller, DIC-9. When the steam supply is lost, feed flow drops to zero, the level controllers LIC-3 and 4 close their respective valves, and the feed accumulates in the feed tank. When the steam supply is restored, it is utilized to its maximum until the feed tank level is restored to its set point.

Because the effect of a steam flow change travels slower through the evaporators than does the effect of a change in feed flow, if accurate dynamic matching is desired, the steam flow signal should enter the feed flow calculation (Equation 7) with a delay. In Figure 8.28, this first-order time lag can be set on FY-2B and its output is the dynamically compensated steam flow signal (W_s).

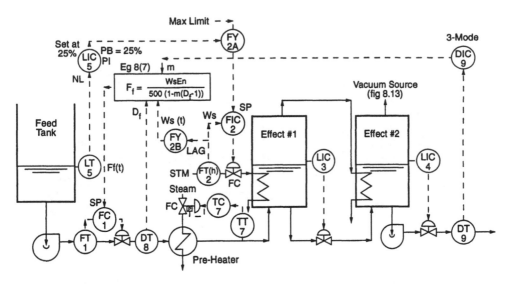

FIGURE 8.28 Feedforward controls for waste steam-operated evaporator application.

B. TRIMMING CONTROLS

Low-holdup evaporators (Figure 8.10) are difficult to control because they lack stabilizing capacity and their dead times (T_d) are large relative to their time constants (T_1). This problem can be mitigated by partial recirculation of the product (Figure 8.21), but this cannot be used with many products. It was for these applications that the otherwise expensive trim heaters have been introduced to achieve finer control than previously possible. With the introduction of material-balance based and dynamically compensated feedforward controls, the need for trim heaters has been substantially reduced, but not completely eliminated.

Figure 8.29 illustrates the addition of trim controls to the falling-film type process, which is shown in Figure 8.27. The main difference is that, in Figure 8.29, the product-density-control function is split in two: DIC-9A adjusts the feedforward model as a PI type feedback controller, while DIC-9B controls the trim heater. This second controller is a three-mode (PID) controller that is provided with an external feedback received from the main steam flow setpoint. This causes it to act like a PD controller, which is so biased that the two steam valves will always move in the same direction and be at the same opening if their controllers are on set point. The external feedback (FB) signal to DIC-9B (the main steam valve opening) biases it to follow and respond to both the feedforward inputs and to changes in steam pressure.

Product quality responds faster to this trim heater than to material-balance control, but such installations require not only an extra heater but also an increased operating cost. It is more expensive to add steam at the trim heater because that steam will pass through only that one heater, instead of passing through all.

Similar to the use of trim heaters, it is also possible to introduce a small stream of feed directly to the last effect. The use of this technique also improves the response of the system to variations in product quality but is also inefficient, as that stream of feed is exposed only to that one effect instead of to all. Trim controls were used prior to the introduction of feedforward controls, and in many cases the use of trim controls has been discontinued after the installation of feedforward controls.

C. Scaling the Material Balance

If the material balance (Equations 6 and 7) is implemented by analog hardware (Figures 8.30 and 8.31), the scaling of the analog computing instruments is necessary to ensure the compatibility of

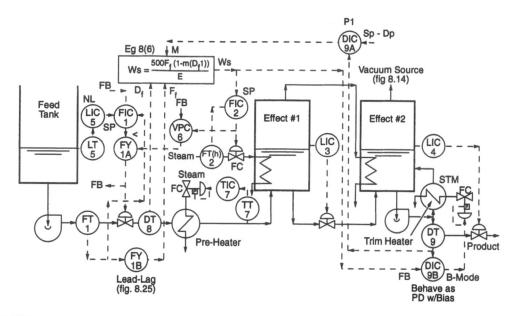

FIGURE 8.29 Closer control and faster response is provided by the addition of a trimming heater.

input and output signals. This is most effectively achieved by normalizing, i.e., assigning values of 0.0 to 1.0 to all inputs and outputs. Scaling or normalizing is a three-step procedure:

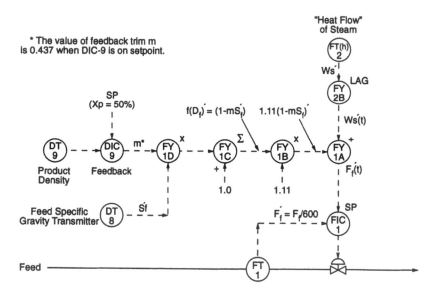

FIGURE 8.30 Scaled analog implementation of the dynamically compensated, material balance-based feedforward control, if feed flow (F_f) is the manipulated variable.

1. The engineering equation is defined.
2. A normalized equivalent is written for each term in the engineering equation.
3. The normalized equivalents are substituted into the engineering equations.

The first step — defining the engineering equations—has already been done for a sample process (Table 8.1). The fraction of feed to be evaporated (the feed-density function) has already been defined in Equation 2 and has already been graphically illustrated in Figure 8.5 for a product

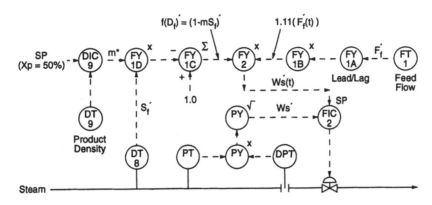

* The value of feedback trim m
is 0.437 when DIC-9 is on setpoint.

FIGURE 8.31 Feedforward control with steam flow (W_s) being the manipulated variable.

concentration (x_p) of 50%. The slope of that line has been calculated in Equation 4 as –4.37. Using that calculation, the density function [$f(D_f)$] for the sample process in Table 8.1 has been given in Equation 5 as

$$f(D_f) = 1 - 4.37(S_f - 1) = 1 - 4.37(D_f/D_w - 1) \qquad (18)$$

If D_w is 8.33 lbs/gal (1 kg/l), then

$$f(D_f) = 1 - [(4.37/8.33)(D_f - 1)] = 1.525 - 0.525 \, D_f \qquad (19)$$

1. Scaling Example

Assume that the process properties of the material being concentrated are as described in Table 8.1 and the other constants or transmitter ranges are as follows:

Feed flow (F_f): 0 to 600 GPH (0 to 2.3 m³/hr)
Feed specific gravity (S_f): 1.000 to 1.100
Feed density (D_f): 8.33 to 9.163 lbs/gallon (1 to 1.1 kg/l)
Steam flow (W_s): 0 to 2500 lbs/hr (0 to 1125 kg/h)
Evaporator train economy (nE): 1.8 (lbs of vapor per lb of steam)

The scaled equivalent of a variable in an engineering equation is the *sum* of the zero value of the transmitter plus its range value multiplied with the scaled variable. The scaled (normalized) values are given as 0 to 1.0 or as 0 to 100% and are distinguished from the engineering values by the addition of a prime ('). Therefore, the scaled values of feed flow, steam flow, and feed density or specific gravity are denoted as F_f', W_s', and D_f and S_f'. Therefore,

$$F_f = 600F_f'$$

$$W_s = 2500W_s'$$

$$D_f = 8.33 + 0.833D_f'$$

$$S_f = 1.00 + 0.10S_f'$$

The values of D_w (8.33) and nE (1.8) need not be scaled, as they are constants. Because the density function $f(D_f)$ already has a range of values between 0 and 1.0, its scaled value is identical:

$$f(D_f) = f(D_f)'$$

Assuming that the feed concentration is detected by a specific gravity sensor having a range of 1.0 to 1.1, we can substitute the scaled equivalent of the engineering value S_f into Equation 18 as follows:

$$f(D_f) = 1.0 - 4.37(S_f - 1) = 1.0 - 4.37(1.000 + 0.100S_f' - 1.0)$$

$$f(D_f) = 1.0 - 0.437S_f' \tag{20}$$

The relationship between W_s and F_f was derived by Myron[2] as

$$W_s = F_f D_w f(D_f)/nE \tag{21}$$

Substituting the scaled values of the engineering terms in Equation 21 gives:

$$2500W_s' = (600F_f')8.33(1.0 - 0.437S_f')/1.8$$

Therefore,

$$W_s' = 1.11F_f'(1.0 - 0.437S_f') = 1.11\ F_f'f(D_f)' \tag{22}$$

If, instead of steam flow, the manipulated variable is feed flow, the scaled relationship becomes

$$F_f' = W_s'/1.11\ (1.0 - 0.437S_f') \tag{23}$$

Figure 8.30 illustrates the analog components required to implement Equation 23. When the product-density controller (DIC-9) is on set point, its output signal ("m", which is the feedback trim of the feedforward model) has the value of 0.437. When DIC-9 is off set point, it changes the slope "m" as needed. FY-1D multiplies "m" with the scaled value of the feed-specific gravity (S_f), FY-1C subtracts that product from 1, and FY-1B multiplies it with 1.11 to arrive at the denominator of Equation 23. FY-1A takes the scaled and dynamically compensated value of the steam flow [$W_s'(t)$] and divides it with the denominator signal from FY-1B to generate the dynamically compensated set point for FIC-1.

Figure 8.31 illustrates the scaled analog implementation of the controls when steam flow is the manipulated variable, based on Equation 22. The only major difference is that the dynamic compensator (FY-1A) is a lead-lag device in this case.

IV. CONCLUSIONS

In the case of evaporators, the saying "One can control a process only if one fully understands it," is particularly true. Some of the most important factors include the temperature sensitivity and handling characteristics of the product, acceptability of product recycling, the pressure and availability of the steam-heat source, the availability of feed storage, the residence times in the individual effects, the nature of the vacuum source, etc. Because these factors do vary widely, this chapter provided a wide variety of control solutions.

With many exceptions, the following generalizations can be made. The trend is to eliminate the use of trim heaters, product recycling, and single-loop feedback controls. The most favored systems are multi-effect with a direct contact condenser-type vacuum source or with vapor recompression. The most favored controls are dynamically compensated, material balance-based feedforward systems, which are often cascaded to nonlinear feed tank level controls.

Multi-effect evaporators can also be used to provide plantwide steam management flexibility by letting the evaporator act as either a generator of low-pressure steam or a user of waste steam.

REFERENCES

1. Shinskey, F.G., *Energy Conservation Through Control*, Academic Press, 1978.
2. Myron, T.J. and Lipták, B.G., *Instrument Engineers' Handbook, Process Control*, Section 8.16, Chilton Book Co., 1995.
3. McCabe, W.L., and Smith, J.C., *Unit Operations of Chemical Engineering*, 5th ed., McGraw-Hill, New York, 1993.
4. Perry, R.H. and Green, D.W., *Perry's Chemical Engineers' Handbook*, 7th ed.
5. Lipták, B.G., *Instrument Engineers' Handbook, Measurement and Analysis Volume*, Chilton Book Co., 1995.
6. Lipták, B.G., *Instrument Engineers' Handbook, Process Control Volume*, Chilton Book Co., 1995.

BIBLIOGRAPHY

Balázs, T. and Nagy, I., Relative Gain Array for an Evaporator Control System, CHISA 84 Conference, Prague, 1984.
Chen, C.S., Computer controlled evaporation process, *Instrum. Control Syst.*, June 1982.
Farin, W.G., Control of triple-effect evaporators, *Instrum. Technol.*, April 1976.
Hoyle, D.L. and Nisenfeld, A.E., Dynamic feedforward control of multi-effect evaporators, *Instrum. Technol.*, February 1980.
Myron, T.J. and Lipták, B.G., *Instrument Engineers' Handbook, Process Control*, Section 8.16, Chilton Book Co., 1995.
Nagy, I., *Introduction to Chemical Process Instrumentation*, Akadémiai Kiadó, Budapest, 1992.
Perry, R.H. and Green, D.W., *Perry's Chemical Engineers' Handbook*, 7th ed.
Shinskey, F.G., *Energy Conservation Through Control*, Academic Press, 1978.

9 Fans

I. THE PROCESS

Vapor and gas transportation is part of most industrial activity, and the reduction of this transportation cost contributes to plant optimization. This chapter describes the automatic control of fans and describes strategies for minimizing their costs of operation.

Fans transport large volumes of low-pressure gases (Figure 4.2). Their discharge pressures are usually in inches or in millimeters of water column, while their capacities can approach a million CFM. Air conditioning applications of fans is discussed in connection with Figure 11.5. Here, the emphasis will be on industrial fan applications.

A. Process Dynamics and Tuning

The pressure controller is usually a proportional-only controller; integral action is unnecessary. When proportional-only controllers are used, the gain generally can be as high as 20 (PB = 5%), although when the controlled pressure is low (inches or centimeters of water), the gain is usually lower.

When two-mode (PI) controllers must be used for some reason, the typical gain setting is 1.0 (PB = 100%, although, when they are fine tuned, it can be anywhere between 50 and 500%). The typical integral time setting is 1 minute per repeat (reset setting of one repeat per minute, although if fine tuned it might end up anywhere between 0.5 and 10). When implemented digitally, the typical scan time is 0.2 seconds, which can vary as a function of application from 0.02 to 1.0. When controlled by suction or discharge dampers, the ideal characteristic is equal percentage, while if the capacity controls are in a bypass around the fan station, the ideal damper characteristics are linear.

The pressure control loop is self-regulating and lag dominated where the lag is a function of the ratio between the process volume and the flow delivered by the fan station. The process dead time is usually nearly zero (0.05 to 0.5 seconds); the process time constant ranges from 5 seconds to 5 minutes as a function of process volume, while the open loop gain ranges from 10 to 100.

B. Fan Curves and Types

Figure 9.1 describes the discharge-pressure characteristics (curve #1), efficiency (curve #2), and power consumption (curve #3) of a constant speed fan, together with a mostly friction system curve (curve #4). Often the discharge pressure is expressed in units of head, so that a 1-in. water pressure is the equivalent of 69.4 feet of air if the air density is 0.075 lbm/ft³. Fan efficiency can be calculated as

$$Eff = QpK/6,356Hp \qquad (1)$$

where Q is fan capacity in ft³/min, p is total discharge pressure in inches of water column, K is the compressibility factor, dimensionless, and Hp is the fan power consumption in horse power.

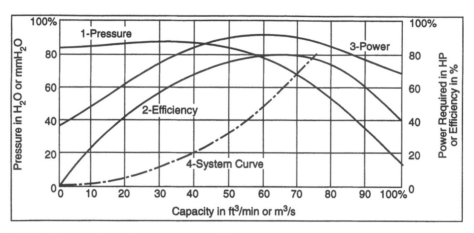

FIGURE 9.1   Characteristics of a constant-speed fan operating on a friction-only system.

Fans are classified into axial flow and radial (or centrifugal) flow types according to the nature of the flow through the blade passages. They move large volumes of air at low pressures with pressure-flow characteristics as shown in Figure 9.2.

If the operating point is to the left of maximum pressure on the fan curve, and particularly if the fan pressure exceeds 10 in. H_2O, pulsations and unstable operation can occur. This maximum point on the curve is referred to as the *surge point* or *pumping limit*. All fans must always operate to the right of the surge point. Under low load conditions the fan can be kept out of surge either by artificially increasing the load through venting or by modifying the fan curve by the use of blade pitch, speed, or vane control.

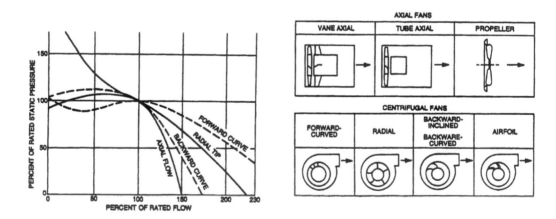

FIGURE 9.2   Different fan types exhibit different pressure-flow characteristics.[1]

C. FAN THROTTLING

Figure 9.3 shows both fan curves and system curves. The system curves in mostly friction transportation systems are parabolic. The operating point for the process is where the fan curve crosses the system curve (point A in Figure 9.3). When the flow and pressure requirements of the process are below the fan curve (points B to E in Figure 9.3), one way to bring them together is to introduce an artificial source of pressure drop. In this figure that source of pressure drop is a suction damper, which modifies the fan curve as it is throttled down (from point A to E). While the introduction

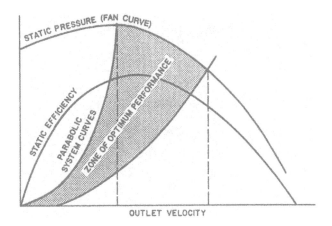

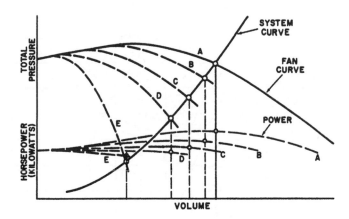

FIGURE 9.3 Upper curve shows optimum operating zone for fans; lower curves describe the effect of suction vane throttling on the fan curve.[1]

of this artificial pressure drop does result in the wasting of transportation energy, this throttling does also shift the surge point to the left and thereby allows for stable operation at lower flows.

Tubeaxial and vaneaxial fans are made with adjustable pitch blades in order to permit the balancing of the fan against the system or to make infrequent adjustments. Vaneaxial fans are also produced with controllable pitch blades (that is, pitch that can be varied while the fan is in operation) for use when frequent or continuous adjustment is needed. Varying pitch angle retains high efficiencies over a wide range of conditions.

Figure 9.4 illustrates both the variable blade pitch design and its performance curves. The efficiency is near maximum at zero pitch angle, and it drops off as the pitch angle is increased. From the standpoint of power consumption, the most desirable method of control is to vary the fan speed to produce the desired performance. If the frequency of change is sufficiently low, belt-driven units may be adjusted by changing the pulley on the drive motor of the fan. Variable-speed motors or variable-speed drives, whether electrical or hydraulic, may be used when frequent or essentially continuous speed modulation is desired.

Figure 9.5 illustrates both the fan curves and the power consumption of fans at partial loads. From the standpoint of noise, variable speed is somewhat better than the variable blade pitch design, and both designs are much quieter and more efficient than the discharge damper or suction vane throttling systems.

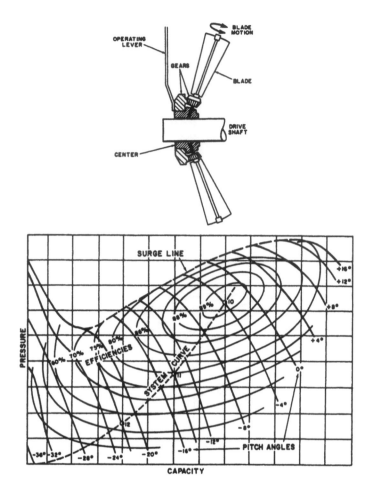

FIGURE 9.4 The design and characteristics of axial-flow fans with variable pitch control are described here. (Lower part is from Reference 2.)

II. BASIC CONTROLS

A. SAFETY INTERLOCKS

All fans should be provided with safety interlocks, such as the ones illustrated in Figure 9.6. Interlock #1 will stop the fan if either excessively high pressures develop on its discharge side (PSH-02) or excessively high vacuums are detected on its suction side (PSL-03). Both of these pressure switches protect the ductwork from bursting or collapsing due to extreme pressure conditions, which can be caused by accidental blockage of the airflow.

Once the fan is shut down, it cannot be restarted until the reset button (R-04) is pressed. This gives the operator an opportunity to eliminate the cause of the abnormal pressure condition before restarting the fan. Whenever the fan is stopped, its discharge damper (XCD-06) is automatically closed to protect it from flow reversal when several fans are connected in parallel. This discharge damper is designed for fast opening and slow closure, to make sure that it is always open when the fan is running.

The time delay (TD-05) guarantees that once the fan is started, it will run for a preset period, unless the safety interlocks turn it off. This protects the motor from overheating as a result of excessive cycling. The fan cycling interlocks (#2) are described in connection with Figure 9.7.

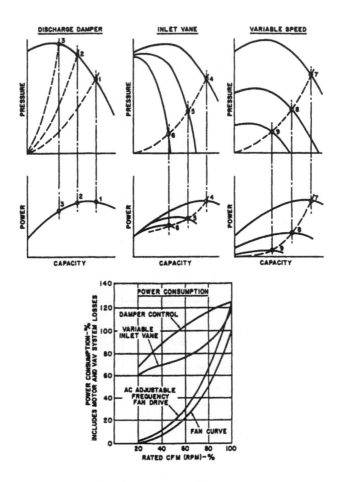

FIGURE 9.5 Turndown efficiency and energy conservation are directly related. (Upper part from Reference 2.)

III. OPTIMIZING CONTROLS

A. CYCLING OF MULTIPLE FANS

The fan station is optimized when it is meeting the process demand for air or other gases at the lowest possible cost. The minimum cost of operation is achieved by first minimizing the number of fans in operation and then minimizing the power consumption of the operating fans. Figure 9.7 illustrates a control system that fulfills both of these goals. The lower part describes the control loops; the fan curves are shown in the upper left and the interlock table in the upper right.

The second fan is started and stopped in response to load variations according to the logic in this interlock table. As the process demand increases, the speed of the operating fan (fan #1) is gradually increased. When its maximum speed is reached (point A on the fan curve) PSH-07 activates interlock #2 to start the second fan. If the speed control signal were unchanged when the second fan was started an upset would occur, because the operating point would instantaneously jump from point A to point C. In order to eliminate this temporary surge of pressure, which otherwise could shut down the station, PY-07 is introduced. This is a signal generator, which upon actuation by interlock #2 drops its output to x. This is the required speed for the two-fan operation at point A. The low signal selector PY-09 immediately selects this signal x for control; thus, the upset is

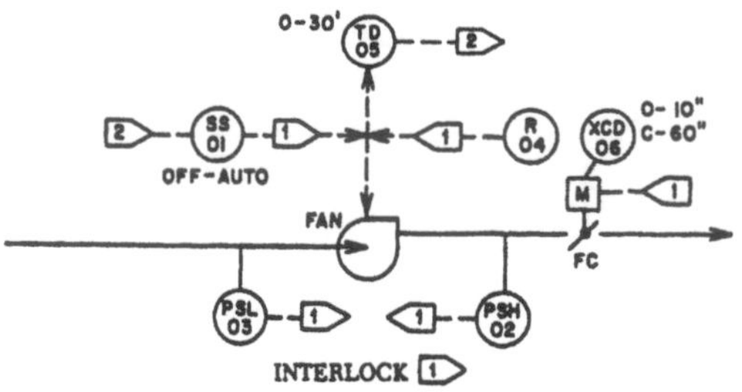

Conditions				Actions		
SS-01	PSH-02	PSL-03	R-04 (Reset)	Fan	XCD-06 (Damper)	TD-06 (Note #1)
Off	—	—	—	Stop	Close	Time
Auto-On	High	—	—	Stop	Close	Time
	Low	—	Stop	Close	Time
	Not high	Not low	Not reset	Stop	Close	Time
			Reset	Run	Open	Reset to start

— = Any state or condition (don't care)
Note #1: Shut-down delay used in interlock [2>

FIGURE 9.6 Safety interlocks should be provided on all fans.

avoided. After actuation, the output signal of PY-07 slowly rises to full scale. As soon as it rises above the output of PIC-10, PY-07's signal is disregarded, and control is returned to PIC-10.

Once both fans are smoothly operating, the next control task is to stop the second fan when the load drops to the point at which it can be met by a single fan. This is controlled by the low flow switch FSL-08, which is set at 90% of the capacity of one fan (point B on the system curve). When the flow drops below the setting of FSL-08, interlock #2 is actuated to stop the second fan. The stopping of the second fan is delayed until time delay TD-05 times out. This shutdown delay protects the fan from overheating due to excessively frequent cycling. Therefore, if TD-05 is set for, say, 20 minutes, the fan cannot be started more than three times an hour.

The fan cycling control described here can be used to cycle any number of parallel fans. For each additional fan, another FSL-08 and PY-07 needs to be added. If n is the number of fans in operation, then FSL-08 is to be set for 90% of the capacity of (n − 1) fans and PY-07 is to be set for x corresponding to the required speed of (n + 1) fans at point A.

B. Optimized Discharge Pressure

The optimum discharge pressure is the minimum pressure that is still sufficient to satisfy all the users. This minimum pressure is found by observing the opening of the most open user damper. If even the most open damper is not fully open, the supply pressure can be safely lowered, whereas if the most open damper is fully open, the supply pressure must be raised. This supply-demand matching strategy not only minimizes the use of fan power but also protects the users from being undersupplied.

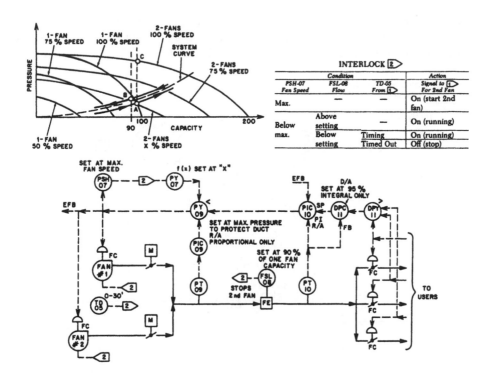

FIGURE 9.7 A fan station is optimized when it is meeting the demand at the lowest possible cost.

The optimization loop functions as follows: DPY-11 in Figure 9.7 selects the opening of the most open damper and sends it as the measurement signal to the damper position controller DPC-11. This controller opens up all the dampers by lowering the supply pressure set point, until one of the dampers reaches 95%. Opening up all the dampers lowers the resistance to flow and saves fan power.

DPC-11 is an integral-only controller with an integral time (minutes/repeat) of 10 times that of PIC-10. This guarantees smooth, stable control, even if the dampers are unstable or cycling. The feedback signal (FB) protects the damper optimizer controller from reset wind-up when PIC-10 is switched to manual set point.

PIC-10 is the load-following controller. It compares the optimized set point with the actual header pressure and adjusts the speed or the blade pitch angle of the fan. When its output signal reaches maximum, PSH-07 acts to start another fan. The role of PIC-09 is to provide overpressure protection at the fan discharge. Under normal conditions the pressure is much below the set point of PIC-09, and its output is saturated at its maximum value. Therefore, under normal conditions PY-09 will select the output signal from PIC-10 for control.

As can be seen from Figure 9.5, if the average load is 60% of full capacity, the above-described optimization strategy can reduce the yearly operating cost to less than 50% of current costs.

C. FAN STATION CONTROL

In most parallel fan installations, all fans except one are constant speed. Therefore, the modulating fan (variable-speed or variable-blade pitch) has control over a narrow load range, and, when the load reaches the limit of that range, one of the constant-speed fans must be stopped or started.

The actual controls are shown in Figure 9.8. The output signal from PIC-10 represents the total demand of the process (A). The system is so calibrated that the output signal is 100% when the total demand matches the total capacity of all fans at full speed.

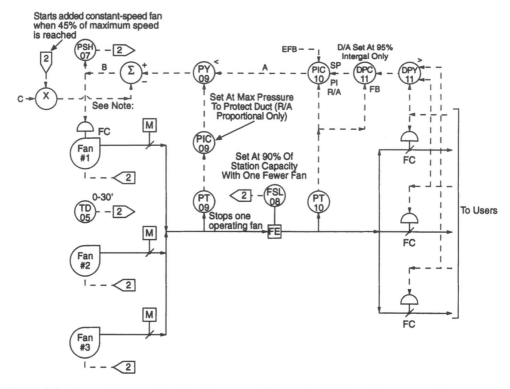

FIGURE 9.8 Fan station control with only one variable speed unit. Note: Summer-amplifier implements equation (2).

If, for example, the maximum station capacity is T, the fan station consists of five fans of equal capacity (C), and the total demand (PIC-10 output) signal is A = 69%, the summer/amplifier will subtract the demand that is met by the three operating constant-speed fans 3(20%) = 60%. This leaves the balance of the total demand to be met by modulating, variable-volume fan as 9% of the station capacity, which is five times that in percentage of the capacity of a single fan. Therefore, if the flow-to-speed relationship is linear, the fan speed would be set at B = 45%. This amplification is also performed by the summer.

In a generalized form, the summer-amplifier performs the following calculation:

$$B = (T/C)[A - (T/N)(\text{Fans on} - 1)] \qquad (2)$$

where A = actual demand (controller output), B = speed setting of modulating fan (if flow is linear with speed), C = capacity of one fan (if all are equal size), N = total number of fans, including variable speed, and T = total capacity of fan station.

As the demand (A) increases, signal B also rises and, when it reaches 100% of the maximum fan speed, interlock #2 starts another constant-speed fan. FSL-08 stops one of the fans when the total flow drops to 90% of the capacity of operation with one fewer fan.

In calibrating and tuning the controls of such a station, the range and dynamic capability of the variable-speed drive (or blade-pitch positioner) should also be considered. This is because, when a constant-speed fan is started or stopped, the signal B will experience a drastic step change.

If the fan drive or blade-pitch positioner is not fast enough to respond, the controlled pressure (PT-10) will be temporarily upset. If the fan surges or becomes unstable below some speed, the value of B must be limited, and the operating range must be narrowed correspondingly, which also changes the loop calibration.

REFERENCES

1. American Society of Heating, Refrigerating, and Air-Conditioning Engineers, *ASHRAE Handbook, 1983 Equipment Volume*, ASHRAE, Atlanta, 1983.
2. Baumeister, T., Ed., *Mark's Standard Handbook for Mechanical Engineers*, 8th ed., McGraw-Hill, New York, 1978, pp 14-53.

BIBLIOGRAPHY

Avery, G., VAV economizer cycle, *Heating, Piping, Air Conditioning*, August 1984.

Brumbaugh, J.E., *Heating, Ventilating and Air Conditioning Library*, rev. ed., Macmillan, New York.

Cooper, F.G., Low pressure sensing and control, *InTech*, September 1992.

Demster, C.S., *Variable Air Volume Systems for Environmental Quality*, McGraw-Hill, New York, 1995.

Haines, R.W. and Wilson, C.L., *HVAC Systems Design Handbook*, McGraw-Hill, New York, 1994.

Hartman, T.B., *Direct Digital Control for HVAC Systems*, McGraw-Hill, New York, 1993.

Lipták, B.G., Applying the Techniques of Process Control to the HVAC Process, ASHRAE Transactions, Paper #2778, Volume 89, Part 2A, 1983.

Lipták, B.G., Optimizing controls for heat pumps, *Chem. Eng.*, October 17, 1983.

Parmley, R.O., *HVAC Design Data Sourcebook*, McGraw-Hill, New York, 1994.

Shinskey, F.G., *Process Control Systems*, 4th ed., McGraw-Hill, New York, 1996.

Stein/Reynolds/McGuinness, *Mechanical and Electrical Equipment for Buildings*, 8th ed., John Wiley & Sons, New York.

Sun, T.-Y., *Air Handling Systems Design*, McGraw-Hill, New York, 1994.

Wang, S.K., *Handbook of Air Conditioning and Refrigeration*, McGraw-Hill, New York, 1994.

10 Heat Exchangers

The transfer of heat is one of the most important operations in the processing industries. Heat can be transferred between the same phases (liquid to liquid, gas to gas), or the transfer of heat can be accompanied by a phase change on either the process side of the exchanger (condenser, evaporator, reboiler), or on its utility side (steam heater). In addition, the heat source can be the heat of combustion of a fuel. This chapter discusses the control and optimization of each of these systems, although, Chapters 2 & 6 provide additional information on condenser controls, as does Chapter 8 for evaporator controls.

I. THE PROCESS

Heat transfer is a complex process. In the steady state, the efficiency of heat transfer is a function of the heat-transfer area (A), the temperature difference between the utility media and the process fluid (ΔT), and the Reynolds number of the flowing streams (Re). While an increase in load (F in Figure 10.1) tends to increase the heat-transfer efficiency by increasing the Reynolds number, the overall effect is a drop because the fixed area (A) of the exchanger makes the transfer of heat less efficient as the load rises.

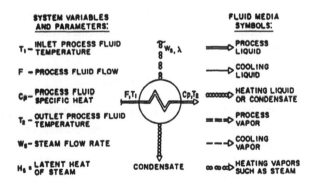

FIGURE 10.1 Temperature and flow are heat exchanger variables; specific and latent heats are parameters.

For most heat-transfer processes, the response to a step change in load (F or T_1 in Figure 10.2) is an S-shaped reaction curve. Such curves are always characterized by the dead time (T_d) and the time constant (T) of the process. The delay, before the outlet temperature (T_2) starts to respond to a step change in load is called the *load side dead time* of the process (T_{dq}) and the delay before T_2 starts to respond to a step change in steam valve opening is called the *manipulated variable side dead time* of the process (T_{dm}).

These dead times are transportation lags caused by the need to displace the process side or utility side volumes of the exchanger before the outlet temperature can start to respond to the changes in flows. Because in most heat-transfer processes the load and the manipulated variable

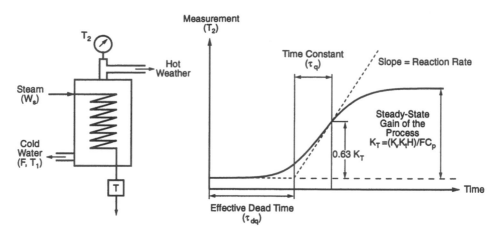

FIGURE 10.2 The process reaction curve caused by a unit size step change in load (cooling water flow –F), showing both the time constant and the dead time of the process. An increase in load reduces both the dead time and the gain (response speed) of the heat-transfer process.

enter at different points, their dead times and other dynamic characteristics are also likely to be different.

A. GAINS AND TIME CONSTANTS

Once the dead time (T_{dq}) has passed, the outlet process temperature (T_2) starts responding at a characteristic speed (reaction rate in Figure 10.2). This process sensitivity can be described by its time constant T_q (the time it takes to reach 63% of the full response to a step change in a single time-constant process) or can be described by the slope of the tangent drawn at the inflection point of the reaction curve, which is related to the *gain* of the process.

The thermal steady-state gain of the heat-transfer process (K_T) determines the full amount of change, which the outlet temperature (T_2) will reach (Figure 10.2) in response to a step change in load. It can be calculated as:

$$K_T = (K_v)(K_t)(H/FC_p) \tag{1}$$

where K_v is the valve gain ($F_{max}/100\%$), K_t is the transmitter gain (100%/span), H is the latent heat of the steam, F is the flow rate of the water, and C_p is the specific heat of the water.

The rate of change between steady states is also a function of the dynamic gain (g_p) of the process. The response of controlled variable (T_2) to manipulated variable (W_s) or to load variables (F or T_1), at a point between steady states is described by the slope of the reaction curve:

$$dc/dm = (G_p)(g_p) \tag{2}$$

B. TUNING THE TEMPERATURE CONTROLLER

As the load on a heat exchanger rises, this tends to lower the dead time (T_{dq}) and increase the time constant (T_q) of the process. Therefore, as the load rises, the reaction curves (Figure 10.2) representing the process will start responding sooner, but will take longer to develop their full response, because the process gain varies inversely with flow. In other words, as the load rises, both the dead time and the gain of the process gain (G_p) drops (Figure 10.3).

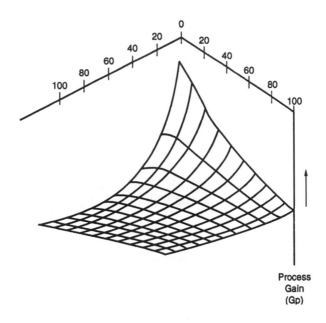

Process
Gain
(Gp)

FIGURE 10.3 The process gain (the change in T_2 resulting from a unit change in either steam or water flow increases as either F or W_s drops.[1]

To properly tune the temperature controller, both the loop gain and the loop dead time should either be made constant or their variations should be compensated for. The variation in dead time can sometimes be minimized by using recirculation to keep the flow through the exchanger constant. This method is widely used on reactor jackets (Chapter 6) but is not considered for heat exchangers except in critical applications.

To keep the temperature loop stable, and to make sure that it will provide 1/4-amplitude damping after an upset, its loop gain should be constant at around 0.5. To guarantee this, the drop in process gain that results from rising loads must be compensated for. If the temperature difference across the exchanger is constant, this can be done by using an equal % control valve, so that G_v will rise as G_p drops (Figure 10.4). If ΔT is a variable, the feedforward adaptation of controller gain is the better choice. One commonly applied method of adaptation is to introduce the load (F in Figure 10.2) into the control algorithm itself, by replacing the controller gain (G_c) with the term (F)(G_c) in the algorithm.

The equation for the flow-adapted three-mode controller is

$$m = K_c F \left(e + \frac{F}{T_I} \int e \, dt + \frac{T_d d}{F \, dt} e \right) + b \tag{3}$$

where F is the fraction of full-scale flow and K_c, T_i, and T_d are the proportional, integral, and derivative settings at full-scale flow.

The open loop gain of an "average" heat exchanger is usually between 0.5 and 5, its time constant is from 1 to 10 minutes, and its dead time is 0.5 to 2 minutes. The total dead time contribution of the instruments in the loop (1 to 2 minutes for thermowells with air-filled annulus, up to 1 minute for I/Ps) must not reach the process dead time. Correctly tuned interacting temperature controllers are usually set for G_c = 0.5 to 5 (PB = 20 to 200%), with integral and derivative both set at about 1/3 of the period of oscillation, or 0.2 to 2 repeats per minute for integral and 0.2 to 2 minutes for derivative.

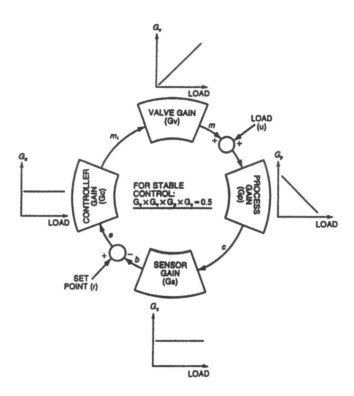

FIGURE 10.4   If the process gain varies with load, such as is the case with heat transfer processes, the gain product of the loop can be held constant by using a valve whose gain variation with load will compensate for the process gain variation.

In the majority of installations, a three-mode controller would be used for heat exchanger service. The derivative or rate action becomes essential in long time-lag systems or when sudden changes in heat exchanger throughput are expected. Because of the relatively slow nature of these control loops, the proportional band setting must be wide to maintain stability (usually between 20% and 200%). This means that the valve will be fully stroked only as a result of a substantial deviation from desired temperature set point. The reset or integral control mode is required to correct for the temperature offsets caused by process load changes. Other variables can also give the appearance of load changes, such as inlet temperature and header pressure changes of the heat transfer medium.

C. Degrees of Freedom

The degrees of freedom of a process represent the maximum number of independently acting automatic controllers that can be placed on that process. Mathematically, the number of degrees of freedom is defined as

$$df = v - e \tag{4}$$

where df = number of degrees of freedom of a system, v = number of variables that describe the system, and e = number of independent relationships that exist among the various variables.

In the case of a liquid-to-liquid heat exchanger, for example Figure 10.5, the total number of variables is six, while the number of defining equations is only one: the first law of thermodynamics, which states the conservation of energy. Therefore, the degrees of freedom of this process are 6 –

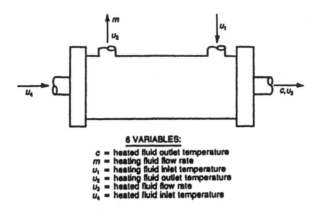

6 VARIABLES:

c = heated fluid outlet temperature
m = heating fluid flow rate
u_1 = heating fluid inlet temperature
u_2 = heating fluid outlet temperature
u_3 = heated fluid flow rate
u_4 = heated fluid inlet temperature

FIGURE 10.5 The degrees of freedom of a liquid-to-liquid heat exchanger.

$1 = 5$. This means that if five variables are held constant, this will result in a constant state for the sixth variable, the outlet temperature (c). Therefore, the maximum number of automatic controllers that can be placed on this process is five.

One would select the *controlled variable (c)* to be the process property which is the most important, because it has the most impact on plant productivity, safety, or product quality. One would select the *manipulated variable (m)* to be that process input variable which has the most direct influence on the controlled variable (c), which in this case is the flow rate of the heating fluid. The other *load variables (u_1 to u_4)* are uncontrolled independent variables. A change in load variables is not responded to until they have upset the controlled variable (c).

When the process involves a phase change, the calculation of the degrees of freedom follows Gibb's phase rule, stated in Equation 5.

$$n = n_c - n_p + 2 \tag{5}$$

where n = number of chemical degrees of freedom, n_c = number of components, and n_p = number of phases.

Figure 10.1 shows a steam heater with its variables and defining parameters. The temperatures and flows are variables, and the specific and latent heats are parameters.

Here, there are four variables and one equation, which is obtained from the first law of thermodynamics, stating the conservation of energy:

$$HsWs = C_pF (T_2 - T_1) \tag{6}$$

Therefore, this system has three degrees of freedom; thus, a maximum of three automatic controllers can be used.

In a steam-heated reboiler or in a condenser cooled by a vaporizing refrigerant (assuming no superheating or supercooling), there are only two flow variables and one defining equation:

$$HsWs = H_1W_1 \tag{7}$$

In this case there is only a single degree of freedom and therefore only one automatic controller can be used.

In the majority of installations, fewer controllers are used than there are degrees of freedom available but every once in a while problems associated with overdefining a system also arise.

D. The Thermal Element

For best performance, both the dead time and the time constant of the temperature sensor should be minimized. The dead time of the sensor (T_d) is minimized by locating it close to the outlet of the heat exchanger. In industrial applications, thermowells are usually required for safety reasons. They should be accessible and so located that the thermal element can be withdrawn without obstruction. When installed in a pipeline, the sensor is immersed to a minimum depth of 3 in. (75 mm). In smaller pipes, this might require pipe expansion, as shown in Figure 10.6.

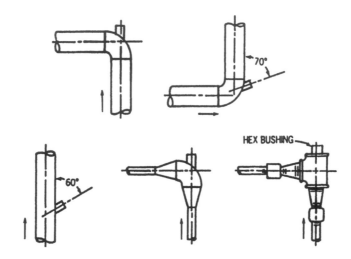

FIGURE 10.6 Alternate ways of installing thermowells in pipes that are 3 in. (75 mm) or smaller in diameter.

The time constant of the sensor increases with its mass. Unprotected miniature sensors are available with a few seconds' time constants. If even a thin protective sheath is added, the time constant rises by at least five seconds. If a thermowell is used, the time constant rises to a minute or more, depending on both the mass of the well and on the thermal contact between the sensor and the well. It is important to eliminate any air gaps in this area by either force-fitting the bulb into the well, or by using a thermally conductive filling material. The time constant of gas temperature measurement is about five times that of liquid temperatures.

One can calculate the time constant or thermal lag of the bulb by calculating the time needed for the heat to pass through the surface area of the bulb.

Calculating the dynamic lag of a typical filled bulb having the area of 0.02 sq ft (0.0018 m²) and heat capacity of 0.005 BTU/°F (9.49 J/°C): If this bulb, which has a bare-bulb diameter of 3/8 in. (9.375 mm), is immersed in a fluid with a heat transfer coefficient (based on flow velocity) of 60 BTU/h/°F/ft² (1.23 MJ/h/°C/m²) and the process temperature is changed at a rate of 25°F/min (13.9°C/min), the dynamic lag can be calculated. First the amount of heat flowing into the element under these conditions is determined:

$$q = \text{(rate of temperature change)(bulb heat capacity)} = (25)(60)(0.005) = 7.5 \text{ BTU/h (7.9 kJ/h)} \quad (8)$$

The dynamic measurement error is calculated by determining the temperature differential across the fluid film surrounding the bulb that is required to produce a heat flow of 7.5 BTU/h (7.9 kJ/h):

$$q = Ah\,\Delta T \quad (9)$$

Therefore,

$$\Delta T = q/Ah = 7.5/(0.02)(60) = 6.25°F \,(3.47°C) \quad (10)$$

If the rate of process temperature change is 25°F/min (14°C/min) and the dynamic error based on that rate is 6.25°F (3.47°C), the dynamic time lag is:

$$t_o = 6.25/25 = 0.25 \text{ minutes} = 15 \text{ seconds} \tag{11}$$

This lag can also be calculated as

$$t_o = \frac{\text{(bulb heat capacity)}}{\text{(bulb area) (heat transfer coefficient)}} = \frac{60 \times 0.005}{0.02 \times 60} = 0.25 \text{ minutes} \tag{12}$$

Bulb time lags vary from a few seconds to minutes, depending on the nature and velocity of the process fluid being detected. Measurement of gases at low velocity involves the longest time lags, and measuring water (or dilute solutions) at high velocity involves the shortest lags.

One method of reducing time lag is by miniaturizing the sensing element. Conventional thermometers are not suitable for all temperature measurement applications. For example, the accurate detection of high-temperature gases (500 to 2000°C) at low velocities (3 to 5 ft/sec) is a problem, even for thermocouples. This is because conductance through the lead wires and radiation both tend to alter the sensor temperature faster than the low velocity gas flow can resupply the lost heat. In such applications, optical fiber thermometry is a good choice. Thermocouples are usually not accurate enough for the precise measurement of temperature differences and are not fast enough to detect high-speed variations in temperature. RTDs can operate with a span for temperature difference of 10°F and can limit the measurement error to ±0.04°F. If high-speed response is desired, thermistors or infrared detectors should be considered.

In the conventional control loop, the measurement lag is only part of the total time lag of the control loop. For example, an air heater might have a total lag of 15 minutes, of which 14 minutes is the *process lag*, 50 seconds is the bulb lag, and 10 seconds is the time lag in the control valve.

II. BASIC CONTROLS

A. LIQUID-TO-LIQUID HEAT EXCHANGERS

The total heat transferred (Q) from the hot to the cold fluid is dependent on the overall heat transfer coefficient (U), the heat transfer area (A), and the log mean temperature difference (ΔTm). Manipulation of any of these three variables can affect the total heat transfer.

$$Q = UA\Delta Tm = UA\frac{(Th_1 - Tc_2)-(Th_2 - Tc_1)}{\ln((Th_1 - Tc_2)/(Th_2 - Tc_1))} \tag{13}$$

$$Q = F_h C_{ph}(Th_1 - Th_2) = F_c C_{pc}(Tc_2 - Tc_1) \tag{14}$$

The dead time of a heat exchanger equals its residence time, which is its volume divided by its flow rate. As process flow increases, the process dead time and the loop gain are decreased. If the controlled variable (Th_2 in Figure 10.7) is differentiated with respect to the manipulated variable Fc, the steady-state process gain is found to be:

$$\frac{dTh_2}{dFc} = \frac{C_{pc}}{C_{ph}F_h} \tag{15}$$

$$Q = UA\Delta Tm = UA\frac{(Th_1 - Tc_2) - (Th_2 - Tc_1)}{\ln \left((Th_1 - Tc_2)/(Th_2 - Tc_1) \right)} \qquad 13$$

$$Q = Fh\,Cph\,(Th_1 - Th_2) = Fc\,Cpc\,(Tc_2 - Tc_1) \qquad 14$$

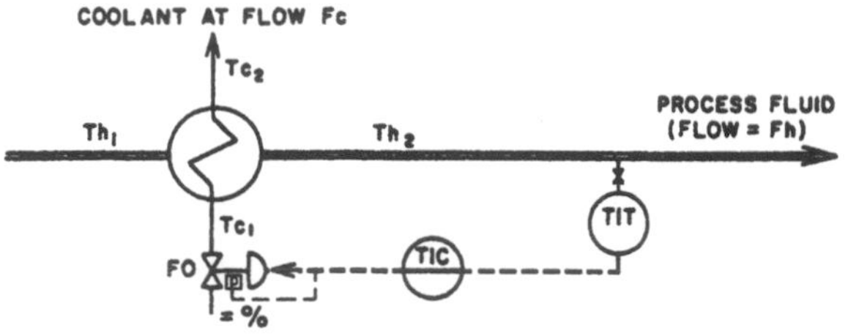

COOLANT AT FLOW Fc

FIGURE 10.7   Feedback control by throttling coolant inlet is a nonlinear, variable gain process.

If $C_{pc} = C_{ph}$, the steady state gain equals $1/F_h$. Therefore, as the process flow (load) drops, the same percent change in the TIC output in Figure 10.7 will have more effect, because the process fluid spends more time in the exchanger. As load is reduced, the exchanger becomes relatively oversized and therefore faster and more effective. As load rises, the exchanger becomes less and less effective, more and more undersized; therefore, it takes more time for the TIC to respond to an upset. An increase in load thus will make the loop sluggish, as the residence and dead times are reduced and the process gain is lowered.

As the process gain of all heat exchangers varies with load, it is not possible to tune the TIC in Figure 10.7 for more than one load. In order to minimize cycling, the TIC is often tuned for the minimum load (maximum process gain). Therefore, it acts in a sluggish manner at higher loads. As the process gain drops with rising load, the total loop gain can be held relatively constant by the introduction of an equal-percentage valve (Figure 10.4). This is a good solution if the temperature difference through the exchanger ($Th_1 - Th_2$) is constant. If it is not, feedforward compensation of the gain is required.

Figures 10.7 and 10.8 illustrate cooler and heater installations with the control valve mounted on the exchanger inlet and outlet, respectively. From a control quality point of view, it makes little difference whether the control valve is upstream or downstream to the heater. The inlet side is usually preferred, because this allows the exchanger to operate at a lower pressure than that of the supply header.

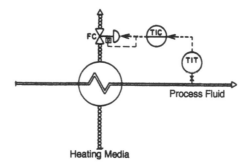

Heating Media

FIGURE 10.8   Feedback control by throttling heating media outlet affects only the operating pressure.

It is generally recommended that positioners for these valves be provided to minimize the valve friction effects. The use of equal-percentage valve trims is also recommended, because it usually contributes to maintaining the control loop gain constant under changing throughput conditions. Equal-percentage trims maintain a constant relationship between valve opening and temperature change (reflecting load variations).

1. Use of Three-Way Valves

Process temperature control quality is a function of the nature of load changes and of the speed of response for the process. In many installations, the process lag time in the heat exchanger is too great to allow for effective control during sudden load changes. In such cases, it is possible to circumvent the dynamic characteristics of the exchanger by partially bypassing it and blending the warm process liquid with the cooled process fluid, as shown in Figure 10.9. The resulting increase in system speed of response, together with some cost savings, are the main motivations for considering the use of three-way valves in such services. The bulb lag has an increased importance in these systems, because it represents a much greater percentage of total loop lag time than in the previously discussed two-way valve installations.

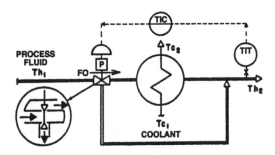

FIGURE 10.9   When a diverter valve is used to control a cooler, it speeds the response and minimizes fouling.

The use of a three-way valve also bypasses the exchanger dynamics and converts the loop dynamics into that of direct mixing of a colder and a warmer stream. This reduces both the dead time and the time constant of the process from minutes to seconds. This increased responsiveness in turn can cut both the gain and the derivative settings of the TIC in half, while doubling the repeats-per-minute setting of its integral action. Therefore, the typical controller settings might range from 0.2 to 2 in gain (PB from 50 to 500%), with 0.5 to 5 repeats per minute for integral and 0.1 to 1 minute for derivative.

The use of a three-way valve such as that shown in Figure 10.9 improves the dynamic response, because the bypass will shorten the time delay between a change in valve position and the response at the temperature sensor. Another benefit from the use of a three-way valve on the process side is that the coolant is not throttled, which keeps the heat transfer coefficient up while eliminating fouling. The disadvantages include increased pumping costs and lower temperature differences between coolant supply and return.

As illustrated in Figures 10.9 and 10.10, either a diverter or a mixing valve can be used as a three-way valve. Stable operation of these valves is achieved by flow tending to open the plugs in both cases. If a mixing valve was to be used for diverting service or if a diverting valve was to be used for mixing, the operation becomes unstable because of the "bathtub effect."

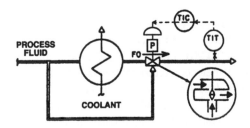

FIGURE 10.10   A mixing valve can be used to control the cooler. It has to flow to open inner valve for stability.

Three-way valves are unbalanced designs and are normally provided with linear ports. The unbalanced nature places a limitation on allowable shutoff pressure difference across the valve, and the linear ports eliminate the potential of compensating for the variable gain of the process.

Misalignment or distortion in a control valve installation can cause binding, leakage at the seats, dead band, and packing friction. Such conditions commonly arise as a result of high-temperature service on three-way valves. The valve, having been installed at ambient conditions and rigidly connected at three flanges, cannot accommodate pipeline expansion because of high process temperature, and therefore distortion can result. Similarly, in mixing applications, when the temperature difference between the two ports is substantial, the resulting differential expansion can also cause distortion. For these reasons, the use of three-way valves at temperatures above 500°F (260°C) or at differential temperatures exceeding 300°F (167°C) is not recommended.

The choice of a three-way valve location relative to the exchanger (Figures 10.9 and 10.10) is normally based on pressure and temperature considerations, with the upstream location (Figure 10.9) usually being favored for reasons of uniformity of valve temperature. When the overriding consideration is the desire to operate the exchanger at a high pressure, the downstream location might be selected.

2. Cooling Water Conservation

Figure 10.11 modifies the system shown in Figure 10.10 to include the additional feature of cooling water conservation. This system tends to maximize the outlet cooling water temperature, thereby minimizing the rate of water usage, while increasing the potential for tube fouling.

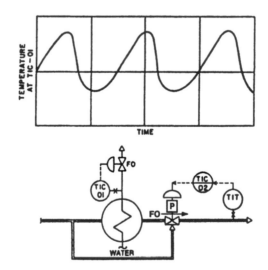

FIGURE 10.11   Conservation of cooling fluid is provided at the price of stability.

Unfortunately, this configuration will not yield stable control, because controlling temperature of a stream by manipulating the flow of the same stream results in limit cycle. The cause of this instability is that the output of TIC-01 affects the dead time of the process which it controls. This comes about because if TIC-01 detects a temperature rise, it opens a valve, causing a sudden drop in temperature as the process dead time is also reduced. As TIC-01 senses this sudden drop in temperature, it closes its valve, but because this also increases the process dead time, the resulting temperature rise will be slow. This limit cycle, consisting of segments of slow rise and fast fall, will also affect the performance of TIC-02 through interaction. The amplitude of the cycle can be reduced severalfold through the doubling of the proportional band of TIC-01, which will also double the period of oscillation.[2] The only way to eliminate this oscillation altogether is to fix the dead time. This can be done by the addition of a recirculating pump, which will return part of the heated water back to the inlet.

As can be seen, there is no easy way to make this configuration stable; therefore, the best solution is to avoid its use.

3. Balancing the Three-Way Valves

It is recommended that a manual balancing valve be installed in the exchanger bypass, as shown in Figure 10.12. This valve is so adjusted that its resistance to flow equals that of the exchanger.

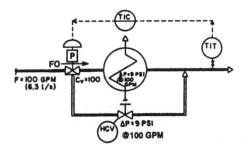

FIGURE 10.12   It is desirable to install a balancing valve in the exchanger bypass.

The resistance to flow in such installations will be maximum when one of the paths is closed and the other is fully open, whereas minimum resistance will be experienced when the valve divides the flow equally between the two paths. If the total flow and exchanger drop is as noted in Figure 10.12, the system drop in one of the extreme positions will be three times that of the pressure drop in the middle position.

4. Using Two Two-Way Valves

If three-way valves cannot be used because the process temperature or the pressure differential is too high or because their linear valve characteristics are unacceptable, improved system response can still be obtained by installing two linear two-way valves having opposite failure positions (Figure 10.13). This, naturally, will increase the cost of the installation. If tight shutoff is required,

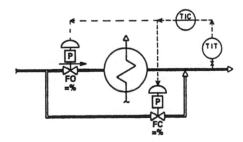

FIGURE 10.13   Exchanger bypass control can be achieved using the two-way valves.

the increase will be more, because only single-seated two-way valves (having smaller C_vs) can be considered. If some valve leakage is acceptable, double-ported two-way valves with larger C_vs can be considered, and the increase in cost will be less.

B. STEAM HEATERS

The heat-transfer rate provided by a steam heater can be calculated as

$$Q = UA\Delta Tm = UA \frac{(T_s - T_2) - (T_s - T_1)}{\ln((T_s - T_2)/(T_s - T_1))} \qquad (16)$$

$$Q = F_s \Delta H_s = FC_p(T_2 - T_1) \qquad (17)$$

where Q = heat-transfer rate, F_s = steam mass flow, ΔH_s = latent heat of vaporization, F = feed rate, C_p = heat capacity of feed, A = area of heat transfer, U = overall heat transfer coefficient, ΔTm = log mean temperature difference, T_1 = inlet temperature, T_2 = outlet temperature, ΔTm = log mean temperature difference, and T_s = condensing steam temperature.

As are all heat-transfer processes, the steam heater is also nonlinear; its process gain and dead time both vary inversely with load. Therefore, if tuned at an intermediate load, its temperature controller will become sluggish as the load rises and unstable as it drops. Those who consider cycling as the less desirable of the two choices will tune the TIC in Figure 10.14 at the minimum load. This will reduce loop responsiveness at higher loads.

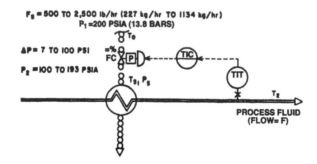

FIGURE 10.14   Feedback control is frequently used on steam-heated exchangers.

As shown in Figure 10.4, the variations in process gain can be compensated by the use of equal percentage valves. This is usually sufficient if the process-side ΔT is relatively constant. If that is not the case, the control algorithm can be adopted by replacing the controller gain (G_c) with the $(F)(G_c)$ product, where F is the actual process flow. In feedforward control systems, this is not necessary, because the gain compensation is always present as a side benefit of the model.

With steam heaters, the desirability of equal-percentage valve trims is even more pronounced than it was with liquid-to-liquid exchangers, because of the high rangeability required in most installations. The need for high rangeability is partially a result of the variations in condensing pressure (steam valve back-pressure) with changes in process load. This can be best visualized by an example.

In Figure 10.14, both the high and low load conditions are shown. When the steam flow demand is the greatest, the backpressure will also be the highest (P_2 = 193 PSIA, or 13.3 bars), leaving the lowest driving force (pressure drop) for the control valve (7 PSI, or 0.5 bar). High flow (2500 lb/hr) and low pressure drop result in a large valve. The backpressure at low loads is only 100 PSIA (6.9 bars), allowing a pressure drop through the valve that is some 16 times greater (100 PSI, or 6.9 bars) than at high loads. The ratio between the required valve coefficients for the high and low load conditions represents the rangeability that the valve has to furnish.

$$\text{rangeability} = S \frac{F_{s,\,max}}{F_{s,\,min}} \sqrt{\frac{\left[\left(P_1 - P_2\right)\left(P_1 + P_2\right)\right]_{min}}{\left[\left(P_1 - P_2\right)\left(P_1 + P_2\right)\right]_{max}}}$$

(18)

$$= 1.5 \times 5 \sqrt{\frac{100 \times 300}{7 \times 393}} = 25.5$$

where S is a safety factor.

The fact that the steam valve ΔP drops with load not only increases the required valve rangeability, but also interferes with the gain-compensating capability of the equal percentage valve, because its gain might not rise with load, as was assumed in Figure 10.4. This is yet another reason for favoring feedforward controls for steam heaters.

1. Minimum Condensing Pressure

The condensing pressure is a function of load when the temperature is controlled by throttling the steam inlet. As long as the heat transfer area is constant, a reduction in load must result in the lowering of the log mean temperature difference across the exchanger. If T_2 is held constant by the TIC, this can occur only if the steam-side temperature (Ts) is lowered. The condensing temperature required can be calculated as follows[3]:

$$Ts = \left(\frac{T_1 + T_2}{2}\right)\left(1 - \frac{L}{100}\right) + \left(\frac{L}{100}\,To\right)$$

(19)

where L is the load in percentage of maximum. To is the steam supply-temperature and Ts is the condensing temperature of the steam. The condensing pressure corresponding to the values of Ts can be plotted as a function of the load. Once this pressure drops below the trap backpressure plus trap differential, the condensing pressure is no longer sufficient to discharge the condensate and it will start accumulating in the exchanger. As condensate accumulation progresses, more and more of the heat transfer area will be covered up, resulting in a corresponding increase in condensing pressure. When this pressure rises sufficiently to discharge the trap, the condensate is suddenly blown out and the effective heat transfer surface of the exchanger increases instantaneously. This can result in cycling as the exchanger surface is covered and uncovered. In addition, noise and hammering can be caused as the steam bubbles collapse on contact with the accumulated cooler condensate. The methods to remedy this situation will be discussed in the following paragraphs.

a. Condensate throttling

Mounting the control valve in the condensate line, as shown in Figure 10.15, is sometimes proposed as a solution to minimum condensate pressure problems. There also is a cost advantage to purchasing a small condensate valve instead of a larger one for steam service.

On the surface this appears to be a convenient solution, because the throttling of the valve causes variations only in the condensate level inside the partially flooded heater and leaves the steam pressure constant, eliminating the problems in condensate removal. Unfortunately, the control characteristics of this loop are not favorable.

Tests have shown that a load decrease takes about three times as long as a load increase.[3] The response time for a 5 to 10% change in load is from a few seconds to a few minutes, whereas for a 50% change it can be from a few minutes to nearly an hour.[3] If the dead time is 5 to 6 seconds, the period of oscillation of the heat exchanger will be close to a half minute.

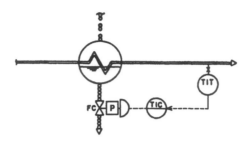

FIGURE 10.15   A control valve mounted in the condensate line can control the rate of condensate removal.

When the load is decreasing, the valve is likely to close completely before the condensate builds up to a high enough level to match the new lower load with a reduced heat transfer area. In this direction, the process is slow, because steam has to condense before the level can be affected. When the load increases, the process is fast, because just a small change in control valve opening is sufficient to drain off enough condensate to expose an increased heat transfer surface. When these "nonsymmetrical" dynamics are present, control is bound to be poor. If the controller is tuned for the fast response speeds of the increasing load direction, sluggish performance and overshoot can result when load is decreasing, whereas if it is tuned for the slow part of the cycle, cycling can occur when the load rises. Therefore, the replacement of the steam trap with an equal-percentage control valve (Figure 10.15) is usually not a good solution.

b. Pumping traps

It is possible to use lifting traps to prevent condensate accumulation in heaters operating at low condensing pressures. This device is illustrated in Figure 10.16 and depends on an external pressure source for its energy.

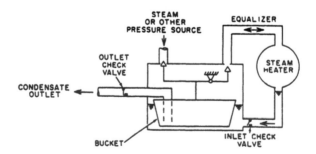

FIGURE 10.16   A lifting or pumping trap, shown here in its filling position, can help prevent condensate accumulation in heaters operating at low condensing pressures.

The unit is shown in its filling position, in which the liquid head in the heater has opened the inlet check valve of the trap. Filling progresses until the condensate overflows into the bucket, which then sinks, closing the equalizer and opening the pressure source valve. As pressure builds in the trap, the inlet check valve is closed; the outlet valve opens when the pressure exceeds that of the condensate header. The discharge cycle follows; in this cycle, the bucket is emptied. When the bucket is near empty, the buoyant force raises the bucket, which then closes off the steam valve and opens the equalizer. Once the pressure in the trap is lowered, the condensate outlet check valve is closed by the header backpressure and the inlet check is opened by the liquid head in the heater, which then is the beginning of another fill cycle.

The pumping trap guarantees condensate removal regardless of the minimum condensing pressure in the heater. If such a trap is placed on the exchanger illustrated in Figure 10.14, it will make temperature control possible even when the heater is under vacuum.

c. Level controllers

The control system in Figure 10.14 provided quick response but was unable to discharge its condensate at low loads. The system in Figure 10.15 eliminated the condensate problem, but at the price of worsening control response.

Because the low condensing pressure is a result of low load and high heat transfer surface area, it is possible to prevent it from developing by reducing the heat transfer area. One method of achieving this is shown in Figure 10.17, in which the steam trap has been replaced by a level control loop.

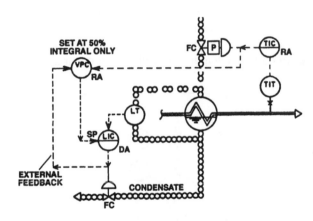

FIGURE 10.17 The level set point can be floated to match the load.

As the required level is a function of the load, it can be adjusted automatically. Load is detected by the valve position controller (VPC) having a set point of, say, 50%. When the load rises, the steam valve opens beyond 50% and the VPC increases the active heat transfer area in the heat exchanger by lowering the set point of the condensate level controller. The VPC is an integral-only controller and therefore will not respond to measurement noise or valve cycling. The integral time is set to give slow floating action that is fast enough to respond to anticipated load changes. The external feedback protects from reset wind-up in the VPC when LIC is switched to local set point.

In less critical applications, in which the cost of an extra level control loop cannot be justified, a continuous drainer trap such as the one shown in Figure 10.18 can serve the same purpose as the level control just described. Its cost is substantially lower, but it is limited to the range within which the level setting can be manually varied and its control point is offset by load variations. For this reason, continuous drainer traps are unlikely to be considered for installations on vertical heaters, reboilers, or where the range of level adjustment can be substantial.

2. Bypass Control

Similarly to liquid-to-liquid exchangers (Figure 10.10), the addition of a three-way valve-operated bypass on the process side of a steam heater can also improve its speed of response. Because the bypass creates one additional degree of freedom, now it is possible to modulate the steam input

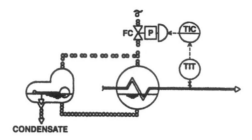

FIGURE 10.18   A continuous drainer trap can serve the same purpose as the level control.

while the condensing pressure is held constant (Figure 10.19). This then solves the condensate removal problem, but introduces another one.

As the demand for heat drops, more and more process fluid will be sent through the bypass of Figure 10.19 and less through the heater. This reduced flow will take longer to displace the heater volume and, the slower it moves, the hotter it will get by the time it leaves. At zero load, the flow is stopped and the temperature of the process fluid approaches the condensing temperature of the steam. If the back pressure on the system is low, boiling can occur at this temperature.

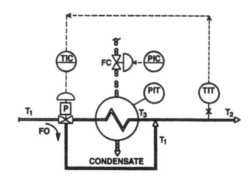

FIGURE 10.19   Bypass control of steam heater provides the added degree of freedom needed for independent control of condensing pressure.

3. Cascade Control

By definition, cascade systems consist of two controllers in series. In heat exchangers, the master detects the process temperature and the slave is installed on a variable that may cause fluctuations in the process temperature. The master adjusts the slave set point, and the slave throttles the valve to maintain that set point. A cascade loop controls a single temperature, and the slave controller is there only to assist in achieving this. In other words, the cascade loop does not have two independent set points. In order for a cascade loop to be successful, the slave must be much faster than the master. A rule of thumb is that the time constant of the primary should be ten times that of the secondary, or that the period of oscillation of the primary should be three times that of the secondary.

Cascade loops are invariably installed to prevent outside disturbances from entering the process. An example of such a disturbance would be the header pressure variations of a steam heater. The conventional single controller system (see Figure 10.14) cannot respond to a change in steam pressure until its effect is felt by the process temperature sensor. In other words, an error in the detected temperature has to develop before corrective action can be taken. The cascade loop, in contrast, responds immediately, correcting for the effect of pressure change before it can influence the process temperature (Figure 10.20).

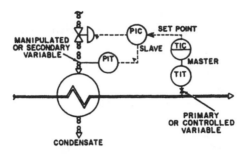

FIGURE 10.20 A temperature-pressure cascade loop on a steam heater increases the speed of response.

The improvement in control quality due to cascading is a function of relative speeds and time lags. A slow primary (master) variable and a secondary (slave) variable that responds quickly to disturbances represents a desirable combination for this type of control. If the slave can quickly respond to fast disturbances, these will not be allowed to enter the process and therefore will not upset the control of the primary (master) variable.

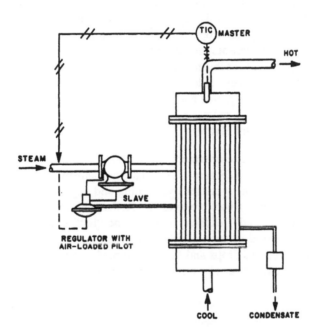

FIGURE 10.21 One of the best cascade slaves is the pressure regulator.

One of the best cascade slaves is the simple and inexpensive pressure regulator. Its air-loaded variety (Figure 10.21) is extremely fast and can correct for steam supply pressure or load variations almost instantaneously.

In Figure 10.20, the controlled variable is temperature and the manipulated variable is the pressure of steam. The primary variable (temperature) is slow, and the secondary (manipulated) variable is capable of responding quickly to disturbances. Therefore, if disturbances occur (a sudden change in plant steam demand, for example), upsetting the manipulated variable (steam pressure), these disturbances will be sensed immediately and corrective action will be taken by the secondary controller so that the primary variable (process temperature) will not be affected.

The addition of a valve-position controller (VPC) can convert the controls in Figure 10.19 into the selective control configuration shown in Figure 10.22. Here, to minimize overheating, the VPC

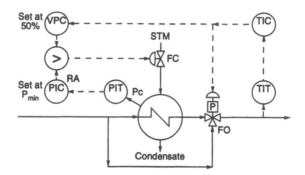

FIGURE 10.22 Balancing flow distribution while maintaining minimum condensing pressure (P_{min}).

is set to evenly distribute the process flow between the heater and its bypass. As the load drops, the VPC gradually reduces the steam valve opening, which in turn lowers the condensing pressure (P_c) within the heater. When P_c has dropped down to the minimum condensing pressure setting (P_{min}), which is required for condensate discharging, the PIC takes over control from the VPC and maintains the condensing pressure at this minimum value. If the temperature corresponding to P_{min} cannot cause boiling on the process side, these controls will fulfill the dual goals of proper condensate discharging on the steam side and no boiling on the process side.

4. Interaction Between Parallel Heaters

If the process fluid is heated to a temperature approaching its boiling point, serious interaction can occur between parallel heaters (Figure 10.23). The mechanism of developing this oscillation is as follows: a sudden drop in flow causes overheating and vaporization in one of the heaters. The vapor formation increases the backpressure and further reduces the flow, eventually forcing all flow through the other ("cold") exchanger, while the "hot" exchanger discharges slugs of liquid and vapor. After a period of such oscillations, when the "hot" exchanger has discharged all of its liquid, the backpressure drops and flow is resumed, drawing feed from the "cold" one. This can cause the "cold" exchanger to overheat, and the cycle is repeated with the roles of the exchangers reversed.

This type of interaction can be eliminated by providing distribution controls. The control system used in making sure that the load is equally distributed between exchangers is the same as the system described for cooling tower balancing (Figure 5.13).

C. CONDENSERS

The control of condensers is also discussed in Chapters 2 and 6. The controlled variable is either the condensate temperature or the condensing pressure (Figure 10.24), while the manipulated variable is the flow rate of either the cooling water or the cooling air. As the cooling load drops, the cooling water valve starts closing, or the blade pitch (or speed) of the air fan is reduced (Figure 10.24).

Slowing the cooling water flow increases its resistance time and therefore its outlet temperature. The outlet temperature should not be allowed to reach 140°F (60°C) because of the probability of calcium carbonate scaling. Therefore, the cooling water is either chemically treated or a small recirculating pump is installed (as P_1 in Figure 2.22), which is started and stopped by FSHL-3, to

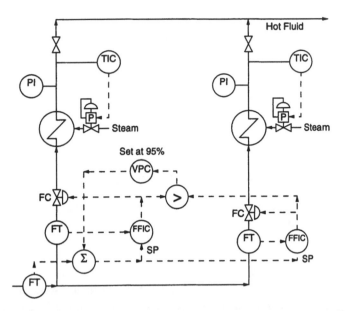

FIGURE 10.23 Parallel heaters operating close to the boiling point of the process fluid can experience serious interaction problems.

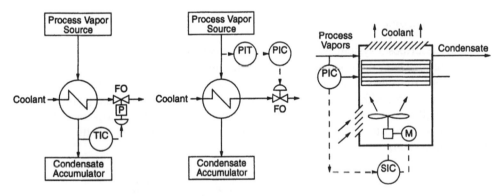

FIGURE 10.24 Condensate temperature or condensing pressure can be controlled by manipulating the flow of either cooling water or cooling air.

make sure that the condenser residence time does not exceed one minute or the water velocity does not drop below 4.5 fps (1.3 m/s).

In the majority of cases, it is preferable to leave the cooling water unthrottled because this eliminates not only the fouling problems but also maximizes heat transfer and therefore minimizes the condensing pressure. In such cases, the condenser is partially flooded and the load is indicated by the condensate level (Figure 10.25) as the exposed tube surface is varied as a function of load. This approach is not applicable to controlling condensate temperature because the degree of subcooling varies with residence time.

When the PIC controls a partially flooded condenser (Figure 10.25), it is tuned as a level controller: PB = 5% to 50%, I = 1 to 10 min/rep, but, because of the heat capacity of the system, a little derivative (1 minute or so) can also be added.

If inerts are present, some purging will also be needed. The main disadvantage of this design is its nonsymmetrical response (the level drops fast, but builds up slowly), as was discussed in connection with Figure 10.15.

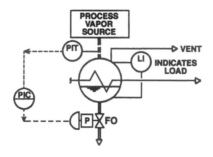

FIGURE 10.25 Condenser control can be achieved by changing the wetted surface area.

The goal of optimization is to maximize heat transfer. This occurs when the coolant flow and the exposed heat transfer surface area are both at their maximums. Therefore, the goal of optimization in Figure 10.24 is to maximize the opening of the coolant valve, while, in Figure 10.25, the goal is to minimize the condensate level. The processes shown in Figure 10.24 can usually be optimized by automatically minimizing the PIC set point. This usually is achieved by detecting the coolant valve opening (by a valve position controller, VPC), which lowers the PIC set point until the valve is nearly fully opened. Similarly, the process shown in Figure 10.25 can be optimized by replacing the level indicator with an LIC and letting it act as a cascade master, reducing the PIC set point, until level is minimized.

1. Refrigerated Condensers

Figure 10.26 describes a system where a compressor is dedicated to serve only one refrigerant-cooled condenser. Here, the condenser heat-transfer area is maximized by the level controller, which keeps all the tubes covered by refrigerant, while the process pressure (or condensate temperature) is controlled by throttling the compressor speed or guide-vane positions. Thereby, compressor loading is automatically matched to process load, without loss of condenser efficiency. Figure 2.12 illustrates how the addition of feedforward can improve the responsiveness of this control system.

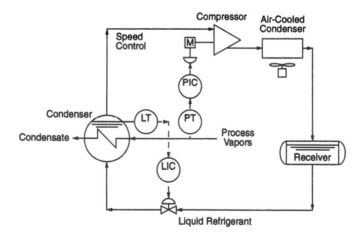

FIGURE 10.26 Refrigerant-cooled condenser controls.

When a refrigerant compressor serves several condensers (Figure 10.27), the individual load controllers can throttle the refrigerant vapor valves on each condenser, while the valve position controller (VPC) lowers the set point of the compressor suction-pressure controller, until one of the vapor valves is nearly fully open.

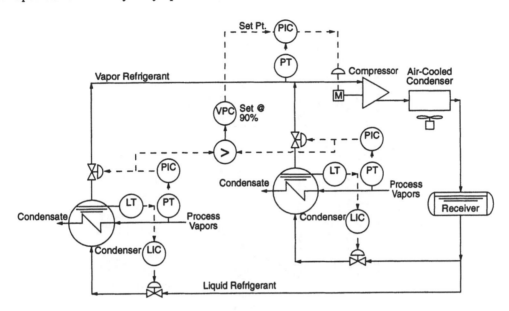

FIGURE 10.27   Multiple refrigerated condensers served by single compressor.

If it is not desirable to add control valves in the refrigerant vapor lines, the configuration shown in Figure 2.11 can be considered. Here, the prices paid for eliminating the valves are reduced responsiveness and lower heat-transfer efficiency because the condenser tubes are not always fully covered.

D. FIRED HEATERS AND VAPORIZERS

When the heat loads and/or the required temperatures are high, direct firing is often used as the heat source. Such furnaces are often used in refineries (Chapter 6) but can also be found in other industrial applications. The firing and draft controls of such furnaces are similar to the ones used on boilers (Chapter 1). Because of this overlap in coverage, the emphasis here will be on the process side controls of industrial furnaces.

1. Coil Balancing Controls

Furnaces or fired heaters, such as an ethylene cracking furnace, usually have a number of parallel heat-transfer tubes. These tubes are protected from melting by the cooling effect of the process fluid as it passes through the heater tubes. Because the heat distribution is not uniform, as localized hot spots develop they cause coking, which in turn accelerates the localized overheating by restricting the flow of the process fluid.

In a large ethylene plant, six to eight identical cracking furnaces may be present, each with from 10 to 20 parallel coils. Maintaining proper flow splits in the coils is a formidable task if done manually.

The objective is to maximize furnace energy utilization per unit feed flow. In general, heat transfer efficiency is highest when the percent vaporization measurements in all the coils are equal. However, without a measurement of percent vaporization, the best that can be done is equalization of all the individual coil outlet temperatures.

Digital control of the total furnace circulation is implemented by feedback manipulation of all furnace coil flow rates (Figure 10.28). The total feed flow (A/D-1) and the individual flows

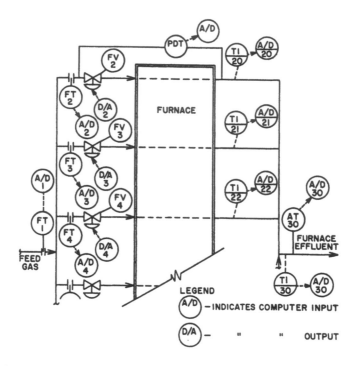

FIGURE 10.28 Digital control of pyrolysis furnace.

(A/D-2, -3, -4, etc.) are monitored by the DCS system, which calculates the proper valve settings based on these flows. The proper valve openings are sent by the DCS system (through signals D/A-2, -3, -4, etc.), and the valves divide the total feed flow properly. A total feed flow rate sets the total charge to the furnace. Each coil flow rate set point is calculated as the total furnace flow (minus the flow of all nonautomated coils, if any), divided by the number of coils being controlled.

Initially the total flow will be divided equally among the passes, but as time progresses, coke builds up at different rates in the individual coils, causing outlet temperatures to vary from coil to coil. The control system keeps the effluent temperatures identical (within some tolerance) by adjusting the flows of the feeds through each coil. The DCS system does this by taking the effluent coil temperature information (A/D-20, -21, -22, etc.), comparing them with the desired temperatures, and modifying the "feed-splitting" computation to correct for deviations. The results of the computation are sent out to reposition the feed valves (FV-2, FV-3, FV-4, etc.).

a. Coil outlet temperature (COT) matrix

The individual COTs can be simultaneously controlled by solving an interaction matrix which relates the change in each coil temperature to the change in valve position of the wall burners. The matrix coefficients determine the gains associated with each COT-firing zone pair. Knowledge of furnace geometry and experimental data are employed to define the matrix. If the matrix accounts for only the steady-state character of the process, the dynamic nature must be handled by feedback trimming of the COT controllers.

The use of a matrix algorithm alone does not necessarily guarantee an even firing pattern, because there is no feedback component in this strategy to correct for errors in the matrix model. Over time, as the effects of coking, air register adjustments, burner fouling, etc., are better understood and quantified, solutions can be implemented by shifting the duty requirements between the firing zones in such a way as to equalize the bias station positions.

The early COT-type furnace-coil balancing strategies were material- and heat balance-based calculations. The more up-to-date control schemes use feedback and feedforward strategies, such as shown in Figure 10.29.

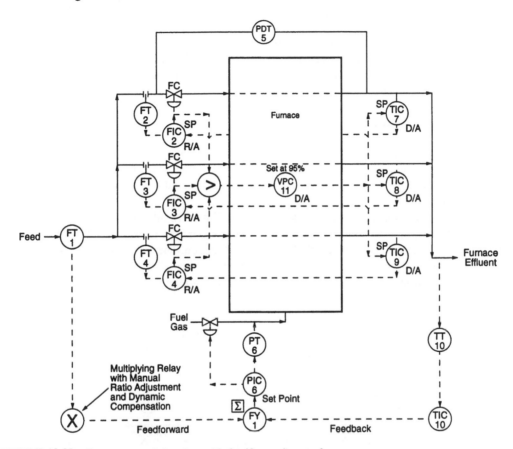

FIGURE 10.29 Furnace pass balancing with feedforward controls.

b. Feedforward bypass balancing

Here the individual pass outlet temperatures are all held at the same set point, which is lowered (by VPC-11 in Figure 10.29) when even the most open valve is not open to 95%. As the set point is lowered, all valves open up, until the most open one reaches 95%. As the set points are reduced, the total stream temperature (TT-10) would also drop if that were not prevented by TIC-10, which increases the firing rate. As the firing rate is increased, the measurements of all the pass TICs rise, the opening of all pass valves increases, and the opening of the most open valve exceeds 95%. Therefore, VPC-11 raises the set points of all the pass TICs.

To minimize the amount of upset in the furnace effluent temperature, a feedforward correction is also provided, which changes the firing rate as soon as the total feed flow has changed. The multiplying block sets the relationship between a change in feed flow and the required corresponding change in fuel header pressure. This is an empirically determined value and can be field adjusted. In critical applications, dynamic compensation is also provided, as shown in the feedforward loop in Figure 10.29.

Changes to the furnace charge rate are normally ramped and limited to avoid rapid upsets in the furnace. Limit constraints, including a differential flow limit between each coil and the average coil flow, are normally included to maintain safe flow rates in all coils, in order to prevent coking or hot spots in the furnace.

Many additional functions can be performed using much of the same input data. Some of these are "off-normal" alarms on feed flow, "off-normal" alarms on effluent temperatures, alarms to signal excessive pressure drop across coils, high coil-metal temperatures, and other scanning functions. Both the control and scanning functions are done on a periodic basis, with the cycle period determined by the nature of the measurement. Flows are usually read more often than temperature or pressure-inputs. Other computations, such as optimization and furnace efficiency calculations, can be performed at much slower rates.

E. Multipurpose Systems

Figure 10.30, for example, depicts a design that uses hot oil as its heat source and water as the means of cooling, arranged in a recirculating system. The points made earlier in connection with three-way valves, cascade systems, and so forth also apply here, but a few additional considerations are worth noting.

In connection with the recirculating design shown in Figure 10.30, it is also important to realize that this is a flooded system; when hot oil enters it, a corresponding volume of oil must be allowed to leave it. The pressure control valve (PCV) serves this function. The same purpose can be achieved by elevational head on the return header, the important consideration being that whatever means are used, the path of least resistance for the oil must be back to the pump suction to keep it always flooded and thereby to prevent cavitation.

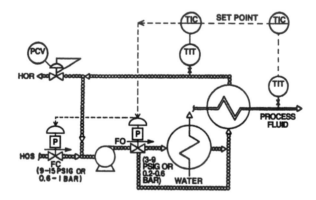

FIGURE 10.30   This recirculating multipurpose heat transfer system uses hot oil as its heat source and water as its coolant.

Probably the most important single feature of this design is that it operates on a *split-range signal*. This means that when the process temperature is above the desired set point, the output signal to the valves will be reduced. When the value of this signal is between 9 and 15 PSIG (0.6 and 1 bar), the three-way valve is fully open to the exchanger bypass and the two-way hot oil supply valve is partially open. If the reduction in the two-way valve opening is not sufficient to bring the process temperature down to set point, the signal will further decrease, thereby fully closing the two-way valve at 9 PSIG (0.6 bar) and beginning to open the flow path through the cooler from the three-way valve. At a 3 PSIG (0.2 bar) signal, the total cooling capacity of the system is applied to the recirculating oil stream, which in that case flows through the cooler without bypass.

The limitations of such a split-range operation include the following.

First, at a signal level near 9 PSIG (0.6 bar), the system can be unstable and cycling, because this is the point at which the three-way valve is just beginning to open to the cooler, and the system might receive alternating slugs of cooling and heating because of the limited rangeability of the valves. (Zero flow is not enough; minimum flow is too much for the particular load condition.)

Second, when the signal is between 9 and 15 PSIG (0.6 and 1 bar), the cooler shell side becomes a reservoir of cold oil. This upsets the controls twice: once when the three-way valve just opens to the cooler and once when the cold oil has been completely displaced and the oil outlet temperature from the cooler suddenly changes from that of the cooling water to some much higher value.

Finally, most of these systems are nonsymmetrical in that the process dynamics (lags and gain reponses) are different for the cooling and heating phases.

To remedy these problems, several steps can be considered. As shown in Figure 10.30, a cascade

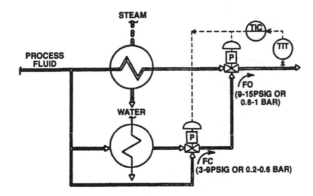

FIGURE 10.31   This multipurpose temperature control system blends process streams at different temperatures.

loop should be used so that upsets and disturbances in the circulating oil loop will be prevented from entering the process and upsetting its temperature. In addition, a slight overlapping of the two valve positioners is desirable. This will offset the beginning of cooling and the termination of heating phases so that they will not both occur at 9 PSIG (0.6 bar). The resulting sacrifice of heat energy can be usually justified by the improved control obtained. Finally, in order to protect against the development of an extremely cold oil reservoir in the cooler, a minimum continuous flow through this unit can be maintained.

Most multipurpose systems represent a compromise of various degrees. Figure 10.31, for example, illustrates a design in which low cost and rapid response to load changes are the main considerations. These characteristics are provided by using the minimum hardware and by circumventing the transient characteristics of the exchangers. The price paid in this compromise involves the full use of both utilities at all times, the development of hot and cold reservoirs, and the necessity for utility disturbances to affect the process temperature before corrective action can be initiated.

Other multipurpose heat transfer systems are used in the control of jacketed chemical reactors. These can be single purpose, double purpose, or configured to operate with three or more different heat transfer fluids as shown in Chapter 6.

III. OPTIMIZING CONTROLS

A. FEEDFORWARD CONTROL

A skillful operator could use a simple feedforward strategy to compensate for changes in inlet water temperature of a direct contact water heater. Detecting a change in inlet water temperature, the operator would increase or decrease the steam rate to counteract the change (Figure 10.32).

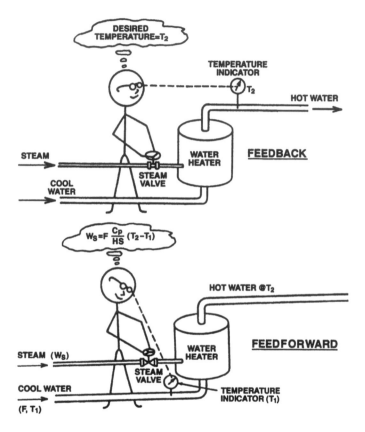

FIGURE 10.32   The feedforward controller has to be "smarter."

The main limitations of feedback control are that it cannot anticipate upsets but can only make corrections after the upsets have occurred, and that it makes its correction in an oscillating, cycling manner. Feedforward control is another basic technique used to compensate for uncontrolled disturbances entering the system. In this technique the control action is based on the state of a disturbance input without reference to the actual system condition. In concept, feedforward control yields much faster correction than feedback control, and in the ideal case compensation is applied in such a manner that the effect of the disturbance is never seen in the process output.

The concept of feedforward control is very powerful, but unfortunately it is difficult to implement in a pure form because it is difficult to generate perfect models or to make perfectly accurate measurements. Because of these limitations, pure feedforward would accumulate the errors in its model and would eventually "self-destruct."

Feedback control measures the controlled variable (c), compares that measurement to a set point (r), and, if there is a difference between the two, changes its output signal to the manipulated variable (m) in order to eliminate the error. This means that feedback control cannot anticipate and prevent errors, it can only initiate its corrective action *after* an error has already developed. Feedback control can be compared to the traditional approach to medicine whereby treatment is started after the symptoms of a sickness have been diagnosed. In this respect feedforward control is inherently different because it is like preventive medicine. Feedforward correction is initiated as soon as a change is detected in a load variable as it enters the process; thus, if the feedforward model is accurate, the load change is prevented from causing an upset in the controlled variable. Because feedforward models and sensors are both imperfect, feedforward loops are usually corrected by feedback trimming.

The output of a feedforward system should be an accurately controlled flow rate. This controlled flow rate cannot be obtained by manipulating the control valve directly, since the valve character-istics are nonlinear and changeable, and their flow is subject to such external influences as upstream and downstream pressure variations. Therefore, most feedforward systems depend on some mea-surement and feedback control of flow to obtain an accurate manipulation of the flow rate. Only when the range of the manipulated variable exceeds that which is available in flowmeters should one consider having the valves positioned directly, and in such cases care must be taken to obtain a reproducible response.

1. Steady-State Model

The first step in designing a feedforward control system is to form a steady-state mathematical model of the process. The model equations are solved for the manipulated variable, which is to be the output of the system. Then the set point is substituted for the controlled variable in the model.

This process will be demonstrated by using the example of temperature control in a heat exchanger (Figure 10.32). A liquid flowing at rate F is to be heated from temperature T_1 by steam, flowing at rate W_s. The energy balance, excluding the losses, is

$$F/C_p(T_2 - T_1) = H_s W_s \qquad (20)$$

Coefficient C_p is the heat capacity of the liquid and H_s is the latent heat given up by the steam in condensing. Solving for W_s yields

$$W_s = F \frac{C_p}{H_s}(T_2 - T_1) \qquad (21)$$

Now T_2 becomes the set point, and W_s and T_1 are load components to which the control system must respond.

Note that this energy balance is imperfect, because it considers only two (major) loads, the flow (F) and the inlet temperature (T_1) of the liquid. Therefore this model disregards a number of (minor) loads, such as the heat losses to the ambient, the heat storage of the exchanger itself, and the variations in the superheat or enthalpy of the steam. Another factor that makes this model imperfect is that even the two major loads that are measured can only be measured imperfectly, as there are no error-free sensors of flow and temperature. The net result is a model which is only an approximation of the energy balance around this process.

Figure 10.33 illustrates a steam heater under feedforward control. This control system consists of two main segments. The feedforward portion of the loop detects the major load variables (the flow and temperature of the entering process fluid) and calculates the required steam flow (W_s) as a function of these variables. When the process flow increases, it is matched with an equal increase in steam flow.

An implementation of Equation 21 is given in Figure 10.33. A control station (TIC) introduces the set point for T_{2^*} into the summing relay and input T_1 is subtracted. The gain of the summing relay is adjusted to provide the constant ratio C_p/H_s. Then, the linear liquid flow signal (F) is

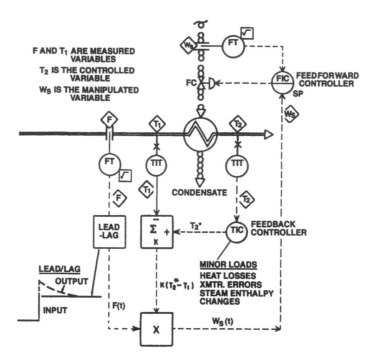

FIGURE 10.33   In feedforward optimization of steam heaters, major load variations (T_1 and F) are corrected by the feedforward portion of the loop, leaving only the minor load variables for feedback correction.

multiplied by the $(C_p/H_s)(T_2 - T_1)$ term to produce the required setting of steam flow (W_s). The flow signals are linear because the temperature difference signal is linear.

The feedback portion of the loop (TIC) has to do much less work in this configuration, as it only has to correct for minor load variables, such as heat losses to the atmosphere, steam enthalpy variations, and sensor errors. The feedback and feedforward portions of the loop complement each other perfectly. Feedforward is responsive, fast, and sophisticated, but inaccurate; feedback is capable of regulating in response to unknown or poorly understood load variations, and although it is slow, it is accurate.

As was already noted in connection with Figure 10.14, the heat transfer process is a variable gain process. Therefore, the process gain varies with flow (F). If the temperature rise ($T_2 - T_1$) is also a variable, even the use of an equal-percentage valve cannot correct for this nonlinearity in the process. In that case, the only way to keep the process gain constant is to use the feedforward system shown in Figure 10.33. There a reduction in process flow causes a reduction in the gain of the multiplier, which exactly cancels the increase in process gain. Thus, the feedforward loop provides gain adaptation as a side benefit, because as the process gain varies inversely with flow, it causes the controller gain to vary directly with flow. The result is a constant total loop gain and therefore stable loop behavior.

During startup, if the steady-state model contains errors, the response to a step change in load (F) and to the resulting instantaneous feedforward correction in steam flow (W_s) will be an under- (or over-) shooting of the T_2 set point, as shown by the dotted line in Figure 10.34. This is corrected by adjusting the gain of the summing relay (K) in Figure 10.33 until the response is that of a "perfect" steady state model (solid line in Figure 10.33). If, at a later time (after this initial adjustment has been made), the feedback controller (TIC in Figure 10.34) notices a deviation from its set point, it adjusts the set point of the feedforward system (T_2^* in Figure 10.33) while the feedforward model adjusts the FIC set point in cascade. The need for such feedback correction can be caused by changes in steam enthalpy, pressure, superheat, by changes in the specific heat of the heated fluid, or by the aging of the loop components.

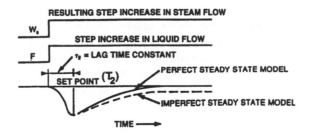

FIGURE 10.34 The response of a dynamically uncompensated feedforward system. T_2 responds faster to changes in F than to changes in W_s.

The feedback TIC should have the same modes but should not be tuned as tightly (gain at least halved, PB doubled), compared with what was recommended in connection with regular feedback tuning (Figure 10.4). This is because, in feedforward control, the feedback controller reacts to an upset by creating another disturbance in the opposite direction one-half cycle later. To not prolong the settling time, the gain of the feedback TIC can be safely reduced, because most of the correction is already done by the feedforward model.

2. Dynamic Model

A step increase in liquid flow results in a simultaneous and proportional step increase in steam flow as dictated by Equation 21.

Under steady-state conditions this step increase would be sufficient to maintain the liquid outlet temperature constant, but under dynamic conditions it is not. This is due to two factors. One is that the increased steam input does not result in an instantaneous increase in heat transfer, because first the steam has to condense, which takes time. The other factor is that in order to increase the heat transfer rate across the fixed heat transfer area of the exchanger, the temperature difference across the tubes must be increased (Equation 16), and this can occur only after the temperature of the exchanger has increased. This takes both time and heat. The heat used to increase the temperature of the heat transfer equipment is unavailable to heat the liquid, and therefore a temporary dip is experienced (Figure 10.34) in that temperature.

This transient error reveals a dynamic imbalance in the system, because the exchanger temporarily needs more heat than the steam flow controller is providing. The reason for this is that the energy level of the process must be increased before an increase in heat transfer level can take place. Some of the added flow of heat from the increased steam flow is diverted in this example to increase the process energy level. Since the balance of the added heat flow is less than that calculated to maintain the exit temperature of the increased liquid flow at steady state, this exit temperature will decrease until the process energy level has been increased adequately. In order to correct the temperature error, an additional amount of energy must be added to the exchanger over what is required for the steady-state balance.

In a more generalized way, we might say that the dynamics on the water side of a steam water heater are faster than on its steam side. In process-control terminology, this means that the sum of load side time constant and dead time ($T_q + T_{dq}$) is less than it is on the manipulated variable side ($T_m + T_{dm}$). Therefore, it is not sufficient in Figure 10.4 to match a change in load (u) with a change in steam flow (m), which is predicted by Equation 16. In effect, what is needed is to bring the sum *of the effects* of load upset (u) and corresponding manipulated variable change (m) to zero. This will be the case if

$$\{(T_m + T_{dm}) - (T_q + T_{dq})\} = 0 \qquad (22)$$

The response in Figure 10.34 indicates that this is not the case there. To set this sum to zero, the mismatch between the time constants on the manipulated and on the load variable sides (T_m and T_q) and the mismatch between the dead times (T_{dm} and T_{dq}) must be eliminated. This is done by dynamic compensation, which can consist of three time-compensator elements: a dead-time compensator (T_d), a lead-time (T_1) and a lag-time (T_2) compensation setting. If the dynamic compensator is correctly tuned, the integrated error in the controlled variable (in this case temperature) will be zero:

$$\{(T_m + T_{dm}) - (T_q + T_{qm}) + (T_d - T_1 + T_2)\} = 0 \tag{23}$$

In the case of a steam heater, the dead-time mismatch (which usually is not substantial and can often be neglected) is eliminated by setting:

$$T_d = T_{dq} - T_{dm} \tag{24}$$

The mismatch in time constants is compensated by setting:

$$T_1 = T_m \text{ and } T_2 = T_q \tag{25}$$

For a more detailed discussion of dynamic compensation, see Reference 4.

As was shown in Figure 10.33, for a steam heater it is usually sufficient to provide a lead-lag compensator, which was mathematically described in Equation 25. Such a unit responds to a step change in load input (F), with an initial inpulse in manipulated variable (oversupply of steam, W_s), as is shown in Figure 10.35. The addition of the dynamic compensator (lead/lag relay) to the control system shown in Figure 10.33 changes the uncompensated response of Figure 10.34 into a compensated one, which is illustrated in Figure 10.36. If the lead/lag compensator is properly tuned, the amount of temperature upset caused by even a drastic load change will still be minimal and about an order of magnitude less than under feedback control (Figure 10.37).

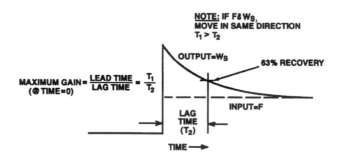

FIGURE 10.35   The step response for a lead-lag unit.

Figure 10.37 describes a major characteristic of a feedforward-controlled process, showing that since most processes do not consist of simple first-order lags, a first-order lead-lag unit cannot produce a perfect dynamic balance. Nevertheless, a major improvement over the uncompensated response is obtained and the error is more or less equally distributed on both sides of the set point.

As shown in Figure 10.37 the temperature upset resulting from the sudden doubling of the load can be reduced by an order of magnitude through the use of feedforward control. Such a tenfold reduction in temperature error can make a great contribution to both safety and quality of production. Therefore, optimized heat exchanger controls are usually justified not on the basis of increased production or reduced energy costs, but on the basis of more stable, accurate, and responsive control.

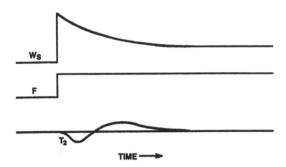

FIGURE 10.36   The dynamic response for a compensated feedforward system.

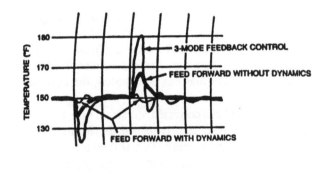

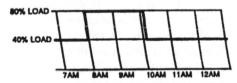

FIGURE 10.37   The addition of a feedforward model, which includes dynamics, can improve the quality of control by an order of magnitude relative to 3-mode feedback control.

3. Adjusting the Lead-Lag Settings

First, a load change should be introduced without dynamic compensation to observe the direction of the error. The compensation can be inhibited by setting the lead and lag time constants at equal values. If the resulting response is in the direction of the load response (this is the case with a steam heater), the lead time should exceed the lag time; if not, the lag time should be greater. Next, measure the time required for the controlled variable to reach its minimum value. This time should be introduced into the lead-lag unit as the smaller time constant (the lag setting). For example, in Figure 10.34 the response of the steam flow (W_s) is in the same direction as the load response (F) and, therefore, the lead time setting should exceed the lag time setting. The time required for the temperature in Figure 10.34 to dip to its minimum should be set equal to the smaller time constant, in this case equal to the lag time setting. Set the lead time constant as twice this value and repeat the load change. If the error curve is still not equally distributed (Figure 10.36) across the set point, increase the lead time constant and repeat the load change.

The outstanding characteristic of the response curve is that its area is equally distributed about the set point if the difference between the lead and lag settings is correct. Once this equalization is obtained, both settings should be increased or decreased with their difference constant until a minimum error amplitude is achieved. When the controller is properly tuned, the steam energy equivalent of the undershoot (area under set point) in Figure 10.34 should match the extra steam introduced (area above input step change) in Figure 10.35.

4. Tuning the Feedforward Loop

At the start of tuning, the lead-lag compensator is off (lead and lag settings are equal) and the feedback controller (TIC in Figure 10.33) is switched to manual, providing a manually generated set point (T_2*) to the feedforward calculation. The value of T_2* can match the desired liquid outlet temperature (T_2), but this correspondence is not essential, as will be seen later.

The steam flow controller (FIC) is switched away from remote (feedforward-cascade) control and the steam flow is manually adjusted until the outlet liquid temperature (T_2) matches exactly its desired set point. This tuning should take place under normal load conditions. "Normal load conditions" means that the liquid flow rate should be what it normally is.

Once the desired outlet temperature is reached, the *actual* steam flow rate which was needed to obtain it is compared to the remote (cascade) steam flow set point calculated by the feedforward model. If there is a difference, the feedforward model is adjusted (by changing the gain K) until its output matches exactly the steam flow that was experimentally determined. When the model has several locations where gain and/or bias can be adjusted, the user can select that which is the most convenient.

Once the feedforward model has been corrected, the steam flow controller (FIC) can be switched to remote (cascade) set point. The next step is to add the lead-lag dynamics. This is done by introducing a load step change into the process. This will result in a response curve as shown in Figure 10.34. If the temperature returns to set point, the steady-state portion of the model is good and only the dynamic tuning remains. This is done as was described in the previous paragraph. Once the initial settings are made, the step-change test is repeated until the temperature response is equally distributed on the two sides of the set point as shown in Figure 10.36. If this is not achieved at the first try, the lead setting is adjusted until equal distribution is reached.

When the dynamics of the feedforward model have been set, the existence of the feedforward portion of the loop can be disregarded, and one can proceed to tune the feedback controller (TIC) as if the "personality" of the inner loop it controls were that of the valve.

The feedforward system is more costly and requires more engineering effort than a feedback system, so, prior to design and installation, the control improvement it brings must be determined to be worthwhile.

B. Furnace Controls

1. Heating Value Calculation

For stable furnace firing, the heating value of the fuel should be kept as constant as possible, or it must be measured and compensated for by the control system. A possible method of accomplishing this is to use a supplemental fuel, having a high and constant heating value, as the control stream. If the heating value of the total stream drops, more of the supplemental fuel is blended in. This instrument can also serve as a safety device by actuating an alarm when the heating value of the fuel drops to such a low point that it will only marginally support combustion. Both the blending of the fuels and emergency shutdown of firing controls can be automated as shown in Figure 10.38. Liquid propane is the supplemental fuel and is under flow control by FRC-1, which is reset by the BTU analyzer controller ARC-1. The level controller (LC-1) on the propane vaporizer adds the amount of steam required to vaporize the supplemental fuel. The knockout drum is required to keep any unvaporized fuel from reaching the burners, which are designed for gas firing only. If the analyzer detects a drop in the heating value, the fuel supply is shut down by ASL-1.

If the heating value of fuel gases #1 and #2 are known, but the percentage of each component in the final mixture is variable, the result is a varying heating value of the total mixture, calculated as

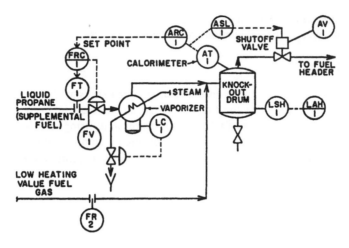

FIGURE 10.38   Calorimeter analyzer application.

$$H_T = \frac{F_1 H_1 + F_2 H_2}{F_1 + F_2} \tag{26}$$

where F_1 = flow rate (lbm/hr) gas no. 1, F_2 = flow rate (lbm/hr) gas no. 2, F_T = flow rate (lbm/hr) mixed gas, H_1 = heating value (BTU/lbm) gas no. 1, H_2 = heating value (BTU/lbm) gas no. 2, and H_T = heating value (BTU/lbm) mixed gas.

This system may be combined into the fuel-firing system shown in Figure 10.39. The fuel gas header pressure for stable burner operation, as controlled by PIC-1, is dependent on the heating values of the fuels consumed, because the two fuels may differ greatly in heating value, and further, the availability of each gas can vary considerably with time. Thus, the heating value of the mixture manipulates the set point of the pressure controller in order to maintain constant heat input to the furnace.

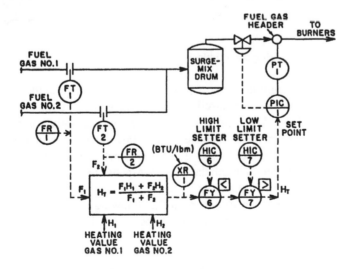

FIGURE 10.39   Header pressure set by heating value of fuel mixture.

The signal-limiting relays FY-6 and FY-7 are installed to maintain the header pressure within safe operating limits even if the calculation would call for extreme pressure settings.

2. Cross-Limited Firing

A cross-limiting firing control technique ensures that air is in excess at all times, to avoid hazardous (fuel-rich) combustible mixtures in the firebox. When an increase in firing rate is needed, this control system first adds air and later fuel. When a decrease in firing rate is desired, the strategy reduces fuel before reducing air. This is performed through a combination of high and low select modules and dynamic exponential lag modules. A control block diagram of the cross-limiting firing circuit is shown in Figure 10.40.

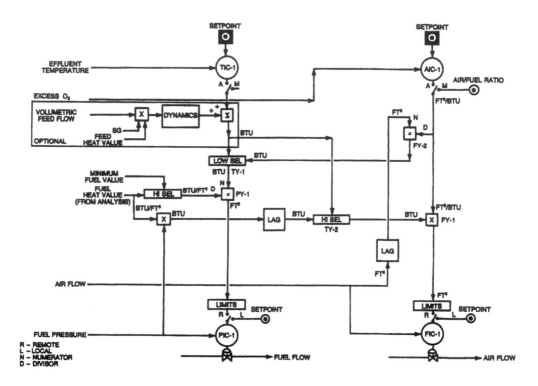

FIGURE 10.40 Cross-limiting firing control block diagram.

The success of the cross-limiting firing is predicated on being able to measure or infer the air and fuel flow rates. If the airflow is controlled only by damper position, this flow rate must be interpolated from that position. Also, fuel is often not metered, but only controlled by a pressure controller, so the header pressure must be converted into flow rates. Such indirect methods of flow approximations are highly inaccurate. Finally, feedforward action based on furnace feed rate can be sluggish and should be adjusted to match the furnace response time. This is because the dynamics of the firing control are normally rather slow.

The cross-limiting firing mechanism works in the following manner. When the effluent temperature controller (TIC-1) calls for additional heat to be supplied by the furnace, the output of the controller increases. After feedforward adjustment, this signal goes to both a high and a low signal selector (TY-1 and TY-2). At the high select, this rising signal will be greater than the signal representing current fuel flow (pressure). The high select (TY-2) will thus call for increased heat by increasing its output, and this increased signal will be multiplied by the current air-fuel ratio in FY-1. The output from this multiplier (FY-1) goes to the set point of the airflow controller (FIC-

1). Therefore the result of an increase in the output of TIC-1 is an immediate increase in the air set point. At the low select (TY-1), the increasing TIC-1 output signal will not be chosen because it is greater than the signal supplied by the air-fuel ratio FY-2. The airflow measurement signal is sent through a lag block and is divided by the air-fuel ratio set point in FY-2, which produces the second input to the low selector. The immediate increase in air, which followed the rise in TIC-1 output, is thus lagged with a first-order lag block prior to the low select (TY-1), resulting in a gradual increase in fuel. Therefore an increase in the firing rate (TIC-1 output) results in a fast increase in air and a slow increase in fuel flow.

When the firing demand drops due to a drop in the feed flow or in the temperature controller output, the reduced signal is sent to both the high and low selector modules (TY-1 and TY-2). The dropping signal is immediately passed through by the low selector (TY-1), resulting in an immediate decrease in the fuel flow set point. However, at the high selector (TY-2), the decreased signal is blocked, because the lag element keeps the apparent flow of fuel higher. The result is a slow (first-order lagged) reduction in the flow of air. A change in the output of the excess air (or O_2) controller (AIC-1) goes through the low selector on the fuel flow (TY-1) in a similar manner. The output of the controller (AIC-1) is the air-fuel ratio setting for the furnace, which is multiplied by the signal from the high selector in TY-1. Therefore the air-fuel ratio directly affects the airflow set point. This same air-fuel ratio is the divisor input to in FY-2. The output of FY-2 is the second input to the low selector TY-1, which selectes the fuel oil flow set point.

REFERENCES

1. McMillan, G.K., *Tuning and Controlling Loop Performance*, 3rd ed., ISA, 1994.
2. Shinskey, F.G., Controlling Unstable Processes, Publication 413-4, The Foxboro Co.
3. Mathur, J., Performance of steam heat-exchangers, *Chem. Eng.*, September 3, 1973.
4. Shinskey, F.G., *Process Control Systems*, 4th ed., McGraw-Hill, New York, 1996.

BIBLIOGRAPHY

Batur, C., et al., Adaptive Temperature Profile Control, ASME Proceedings, Dallas, November 1990.
Dukelow, S.G., Fired Process Heaters, ISA/CHEMPID, April 1980.
Edwards, D., Recent developments in furnace control systems, *Metals Materials*, October 1990.
Forman, E.R., Control dynamics in heat transfer, *Chem. Eng.*, January 3, 1990.
Khalil, E.E., *Modeling of Furnaces and Combustors*, Abacus Press, 1982.
Lipták, B.G., Control of heat exchangers, *Br. Chem. Eng.*, July 1972.
Mathur, J., Performance of steam heat-exchangers, *Chem. Eng.*, September 3, 1973.
McCabe, W.L., et al., *Unit Operations in Chemical Engineering*, 5th ed., McGraw-Hill, New York, 1993.
McMillan, G.K., *Tuning and Controlling Loop Performance*, 3rd ed., ISA, 1994.
O'Callaghan, P.W., *Energy Management*, McGraw-Hill, New York, 1992.
Omata, S., et al., Adaptive Control of Furnace Temperature, Proceedings of the 28th Conference of Society of Instrument and Control Engineers in Tokyo, July 1989.
Shinskey, F.G., Controlling Unstable Processes, Publication 413-4, The Foxboro Co.
Shinskey, F.G., *Process Control Systems*, 4th ed., McGraw-Hill, New York, 1996.
Yokell, S., *A Working Guide to Shell-and-Tube Heat Exchangers*, McGraw-Hill, New York, 1990.

11 HVAC Systems

I. THE PROCESS

The airhandler is the basic unit of operation of space conditioning. It is used to keep occupied spaces comfortable (Figure 11.1) or unoccupied spaces at desired levels of temperature and humidity. In addition to supplying or removing heat and/or humidity from the conditioned space, the airhandler also provides ventilation and fresh air makeup. Depending on the type of space involved, from 75,000 to 300,000 BTU/year (19,000 to 76,000 cal/year) are required to condition one square foot (0.092 m²) of office space. Depending on the energy sources used, this corresponds to a yearly operating cost of a few dollars per square foot of floor space.

Whereas other unit operations have benefited substantially from the advances in process control, airhandlers have not. Airhandlers today are frequently controlled the same way as they were 20 or 30 years ago. For this reason, airhandler optimization can result in much greater percentages of

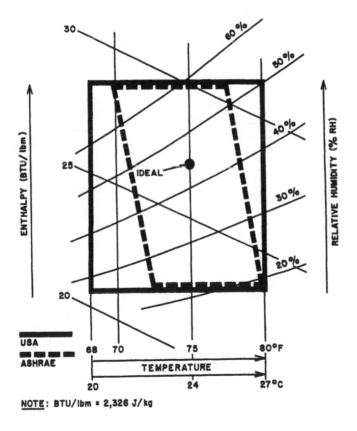

NOTE: BTU/lbm = 2,326 J/kg

FIGURE 11.1 "Comfort zones" are defined in terms of temperature and humidity.[1]

savings than can the optimization of any other unit operation. Optimization can cut the cost of airhandler operation in half—a savings that can seldom be achieved in any other type of unit operation.

Some of the optimization goals and strategies include the following:

Let the building heat itself
Use free cooling and/or free drying
Benefit from gap control or zero energy band (ZEB)
Eliminate chimney effect
Optimize start-up timing
Optimize air makeup (CO_2)
Optimize air supply temperature
Minimize fan energy use
Automate the selection of operating modes
Minimize reheat
Automate balancing of air distribution

A. COMPONENTS OF AN AIRHANDLER

The purpose of heating, ventilation, and air conditioning (HVAC) controls is to provide comfort in laboratories, clean rooms, warehouses, offices, and manufacturing spaces. Supply air is the means of providing comfort in the conditioned zone. The air supplied to each zone must provide heating or cooling, raise or lower humidity, and provide air refreshment. To satisfy these requirements, it is necessary to control the temperature, humidity, and fresh air ratio in the supply air.

Figure 11.2 illustrates the main components of an airhandler. The term *airhandler* refers to the total system, including fans, heat-exchanger coils, dampers, ducts, and instruments. The system operates as follows: Outside air is admitted by the outside air damper (OAD-05) and is

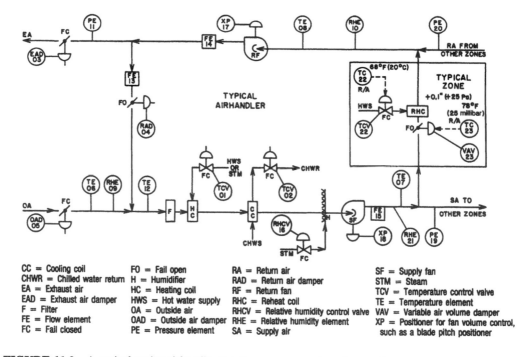

CC = Cooling coil	FO = Fail open	RA = Return air	SF = Supply fan
CHWR = Chilled water return	H = Humidifier	RAD = Return air damper	STM = Steam
EA = Exhaust air	HC = Heating coil	RF = Return fan	TCV = Temperature control valve
EAD = Exhaust air damper	HWS = Hot water supply	RHC = Reheat coil	TE = Temperature element
F = Filter	OA = Outside air	RHCV = Relative humidity control valve	VAV = Variable air volume damper
FE = Flow element	OAD = Outside air damper	RHE = Relative humidity element	XP = Positioner for fan volume control,
FC = Fail closed	PE = Pressure element	SA = Supply air	such as a blade pitch positioner

FIGURE 11.2 A typical major airhandler has these components and controls.

then mixed with the return air from the return air damper (RAD-04). The resulting mixed air is filtered (F), heated (HC) or cooled (CC), and humidified (H) or dehumidified (CC) as required. The resulting supply air is then transported to the conditioned zones (groups of offices) by the variable-volume supply fan station. Variable volume means that the airflow rate generated by the fan(s) is variable.

In each zone, the variable air volume damper (VAV-23) determines the amount of air required, and the reheat coil (RHC) adjusts the air temperature as needed. The return air from the zones is transported by the variable-volume return-air fan station. If the amount of available return air exceeds the demand for it, the excess air is exhausted by the exhaust air damper (EAD-03). The conditioned spaces are typically pressurized to about 0.1 in. H_2O (2.5 millibar), relative to the barometric pressure on the outside. This pressurization results in some air leakage through the walls and windows, which varies with the quality of construction. Therefore, the air balance around the system is

$$OA = EA + \text{pressurization loss} \qquad (1)$$

Under "normal" operation, the airhandler operates with about 10% outside air. In the "purge" or "free cooling" modes, RAD is closed, OAD is fully open, and the airhandler operates with 100% outside air.

As can be seen, the HVAC process is rather simple. Its process material is clean air, its utility is water or steam, and its overall system behavior is slow, stable, and forgiving. For precisely these reasons, it is possible to obtain acceptable HVAC performance using inferior-quality instruments, which are configured into poorly designed loops. Yet there is an advantage in applying state-of-the-art process control to the HVAC process: it can provide a drastic reduction in operating costs, attributable to increased efficiency of operation. Some of the more efficient control concepts are described in the sections below.

B. OPERATING MODE SELECTION

The correct identification and timing of the various operating modes can contribute to the optimization of the building. The *normal* operating modes include start-up, occupied, night, and purge.

Optimizing the time of *start-up* will guarantee that the minimum required cost is invested in getting the building ready for occupancy. This is done by automatically calculating the amount of heat that needs to be transferred and dividing it according to the capacity of the start-up equipment. A computer-optimized control system will serve to initiate the unoccupied (night) mode of operation; it will also recognize weekends and holidays and, in general, provide a flexible means of time-of-day controls.

The *purge* mode is another convenient tool of optimization. Whenever the outside air is preferred to the return air, the building is automatically purged. In this way, "free cooling" can be obtained on dry summer mornings or "free heating" can be provided on warm winter afternoons. Purging is the equivalent of opening the windows in a home. In computer-optimized buildings, an added potential is to use the building structure as a means of heat (or coolant) storage. In this case, the purge mode can be automatically initiated during cold nights prior to hot summer days, thereby bringing the building temperature down and storing some free cooling in the building structure.

1. Summer/Winter Mode Reevaluation

Another important mode selection involves switching from summer to winter mode and vice versa. Conventional systems are switched according to the calendar, whereas optimized ones recognize that there are summerlike days in the winter and winterlike hours during summer days. Seasonal mode switching is therefore totally inadequate. Optimized building operation can be provided only

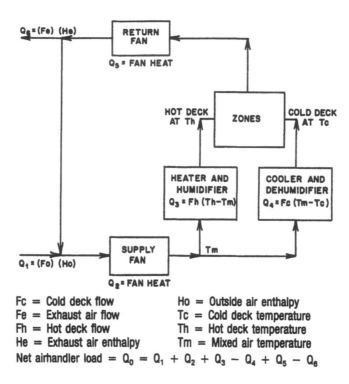

FIGURE 11.3 When the net airhandler load is negative, summer mode is required; when it is positive, winter mode is required.

by making the summer/winter selection on an enthalpy basis: if heat needs to be added, it is "winter"; if heat needs to be removed, it is "summer," regardless of the calendar. In those airhandlers which serve a variety of zones, it is essential to first determine if free cooling (or free heating) by outside air can be used to advantage. Figure 11.3 illustrates the heat balance evaluation that is required to determine the prevailing overall mode of operation. This type of heat balance calculation, which must be reevaluated every 15 to 30 minutes, can be implemented only through the use of computers.

2. Emergency Mode

In addition to the above operating modes, the airhandler can also be placed in *emergency* mode, if fire, smoke, freezing temperature, or pressure conditions require it. Figure 11.4 shows the status of each fan, damper, and valve in each of the operating modes. In a computer-optimized control system, both the mode selection and the setting of the actuated devices is done automatically.

When a smoke or fire condition is detected (S/F-4 and 8), the fans stop, the OADs and RADs close, the EAD opens, and an alarm is actuated. The operator can switch the airhandler into its purge mode, so that the fans are started, OAD and EAD are opened, and RAD is closed. If the smoke/fire emergency requires the presence of fire fighters, the fire command panel is used. From this panel, the fire chief can operate all fans and dampers as needed for safe and orderly evacuation and protection of the building.

In another emergency condition, a freezestat switch on one of the water coils is actuated. These switches are usually set at approximately 35°F (1.5°C) and serve to protect from coil damage resulting from freeze-ups. Multistage freezestat units might operate as follows:

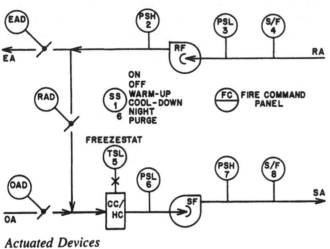

Actuated Devices

Operating Mode or Emergency Condition	Supply Fan	Return Fan	Outside Air Damper	Exhaust Air Damper	Return Air Damper	Coil Control Valves	Alarm
Off	—	—	C	C	O	C	—
On	On	On	←————————Modulating————————→				—
Warm-up	On	On	C	C	O	O(HC)	—
Cool-down	On	On	C	C	O	O(CC)	—
Night	←————————Cycled to maintain required nighttime temperature————————→						
Purge	On	On	O	O	C	Modulating	—
PSH-2	—	Off	—	C	—	—	Yes
PSL-3	—	Off	—	C	—	—	Yes
S/F-4	Off	Off	C	O	C	C	Yes
TSL-5	Off	—	C	—	O	C	Yes
PSL-6	Off	—	C	—	O	—	Yes
PSH-7	Off	—	C	—	O	—	Yes
S/F-8	Off	Off	C	O	C	C	Yes

FIGURE 11.4 Safety and operating mode selection instruments. S/F = Smoke and fire detector(s). FC = Fire command panel, which provides the fire chief with access to all fans and dampers in the building. This panel is used during fire fighting and building evacuation.

- At 38°F (3°C) — close OAD
- At 36°F (2°C) — fully open water valve
- At 35°F (1.5°C) — stop fan

If single-stage freezestats are used, they will stop the fan, close the OAD, and activate an alarm.

Yet another type of emergency is signaled by excessive pressures in the ductwork on the suction or discharge sides of the fans, resulting from operation against closed dampers or from other equipment failures. When this happens, the associated fan is stopped and an alarm is actuated.

II. BASIC CONTROLS

A. FAN CONTROLS

The standard fan controls are shown in Figure 11.5. Each zone shown in Figure 11.2 is supplied with air through a thermostat modulated damper, also called a variable air volume box (VAV-23).

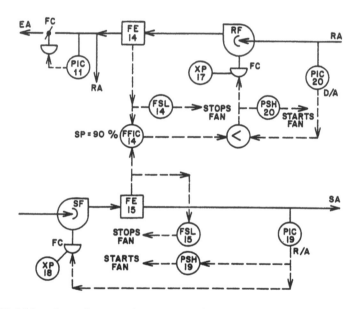

FIGURE 11.5   Variable-volume fan controls operate as shown here.

The VAV box openings in the various zones determine the total demand for supply air. The pressure in the supply modulates the supply-air fan station to match demand (Figure 11.5). When the PIC-19 output has increased the fan capacity to its maximum, PSH-19 actuates and starts an additional fan. Inversely, as the demand for supply air drops, FSL-15 will stop one fan unit whenever the load can be met by fewer fans than the number in operation. The important point to remember is that in cycling fan stations, fan units are started on pressure and stopped on flow control. The operating cost of such a fan station is 20 to 40% lower than if constant-volume fans with conventional controls were used (Figure 11.6).

Because the conditioned zones are pressurized slightly, some of the conditioned air will leak into the atmosphere, creating pressurization loss. Being able to control the pressurization loss is one of the advantages of the control system described in Figure 11.5. The flow controller FFIC-14 is set at 90%, meaning that the return-air fan station is modulated to return 90% of the air supplied to the zones. Therefore, pressurization loss is controlled at 10%, which corresponds to a minimum fresh-air makeup requirement, resulting in a minimum-cost operation.

Because the conditioned zones represent a fairly large capacity, a change in supply airflow will not immediately result in a need for a corresponding change in the return airflow. Thus, PIC-20 (Figure 11.5) is included in the system to prevent the flow-ratio controller from increasing the return air flow rate faster than required. This dynamic balancing eliminates cycling and protects

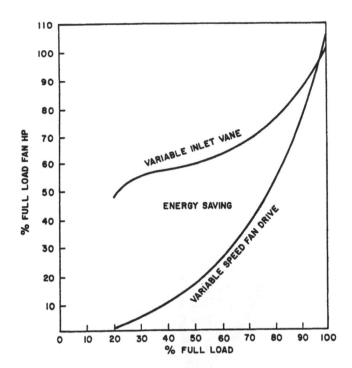

FIGURE 11.6   Using variable-speed fans can save significant amounts of energy. (Courtesy of Dana Corp.)

against collapsing the ductwork under excessive vacuum. Closure of the exhaust air damper by PIC-11 indicates that the control system is properly tuned and balanced and is operating at maximum efficiency. Under such conditions, the outside air admitted into the airhandler exactly matches the pressurization loss, and no return air is exhausted.

To maximize the benefits of such an efficient configuration, the dampers must be of tight shutoff design. With a pressure difference of 4 in. H_2O (100 mm H_2O), a closed conventional damper will leak at a rate of approximately 50 cfm/ft² [15.2 (m³/min)/m²]. In the HVAC industry, a 5cfm/ft² [1.52 (m³/min)/m²] leakage rate is considered to represent a tight shutoff design. Actually, it is cost-effective to install tight shutoff dampers with leakage rates of less than 0.5 cfm/ft² [0.15 (m³/min)/m²], because the resulting savings over the life of the buildings will be much greater than the increase in initial investment for better dampers (Figure 11.7).

In order for dampers to give good control, a fair amount of pressure drop should be assigned to them. They should be sized for a ΔP of about 10% of the total system drop. On the other hand, excessive damper drops should also be avoided, because they will increase the operating costs of the fans. A good sizing basis for outside and return air fans is to size them for 1500 fpm (457 m/min) velocity at maximum flow.

In locations where two air streams are mixed, such as when outside and return airs are ratioed (RAD-04 and OAD-05 in Figure 11.2), it is important that the damper ΔP be relatively constant as the ratio is varied. Figure 11.8 shows that parallel blade dampers give a superior performance in this service.

Figure 11.9 illustrates the typical pressure profiles in airhandlers. It can be seen that the kind of pressure drops that would be required by opposed blade dampers (Figure 11.8) are simply not available. Therefore, if such dampers were installed, the airhandler would be starved for air (the dampers could not pass the design flow) whenever the ratio was near 50–50 (percent).

Figure 11.9 also shows that in traditional airhandlers more fan energy is used than necessary. This is because the return air fan is sized to generate the pressure needed to exhaust the air from

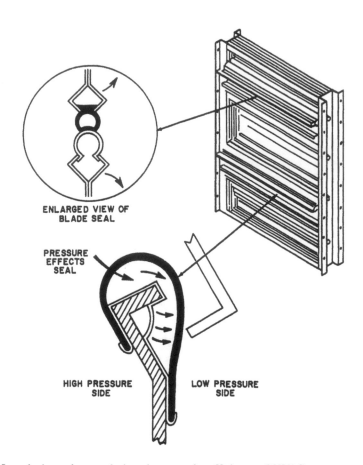

ENLARGED VIEW OF
BLADE SEAL

PRESSURE
EFFECTS
SEAL

HIGH PRESSURE
SIDE

LOW PRESSURE
SIDE

FIGURE 11.7   Low-leakage damper designs increase the efficiency of HVAC systems.

the building. A consequence of this is that the pressure drop of $1^1/_2$ in. H_2O (37.5 millibars) across the return air damper is three times greater than what is necessary ($^1/_2$ in. H_2O, or 12.5 millibars).

The alternate system also shown in Figure 11.9 eliminates this waste of fan energy. Here, only the supply fan (SF) operates continuously, which reduces the pressure drop across the return air damper to $^1/_2$ in. H_2O (12.5 millibars). The return fan (RF) is started only when air needs to be relieved, and its speed is varied to adjust the amount of air to be exhausted. Relocating RF also removes its heat input, which, in the traditional systems, represents an added load on the cooling coil.

B. Temperature Controls

Space temperatures are controlled by thermostats. The traditional thermostat is a proportional-only controller. The pressure of the output signal from a pneumatic "stat" is a near straight-line function of the measurement, described by the following relationship:

$$O = K_c (M - M_o) + O_o \qquad (2)$$

where O = output signal, K_c = proportional sensitivity (K_c can be fixed or adjustable depending on the design), M = measurement (temperature), M_o = "normal" value of measurement, corresponding to the center of the throttling range, and O_o = "normal" value of the output signal, corresponding to the center of the throttling range of the control valve (or damper).

Another term used to describe the sensitivity of thermostats is *throttling range*. As shown in Figure 11.10, this term refers to the amount of temperature change that is required to change the

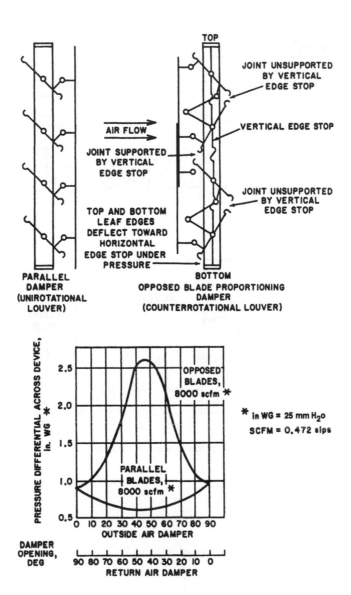

FIGURE 11.8 When outside and return air dampers are throttled to vary their ratio at constant total flow, the required pressure drop varies with damper design (see upper portion of figure). The lower portion of this figure shows the results of American Warming and Ventilating Co. tests (per AMCA Standard 500) of pressure drops across parallel-blade and opposed-blade outside and return air damper sets.[2]

thermostat output from its minimum to its maximum value, such as from 3 to 13 PSIG (0.2 to 0.87 barg). The throttling range is usually adjustable from 2 to 10°F (1 to 5°C).

One important point to remember is that thermostats *do not* have set points, in the sense of having a predetermined temperature to which they would seek to return the controlled space. (Integral action must be added in order for a controller to be able to return the measured variable to a set point after a load change.) M_o does not represent a set point; it only identifies the space temperature that will cause the cooling damper in Figure 11.11 to be 50% open. This can be called a "normal" condition, because relative to this point the thermostat can both increase and decrease the cooling airflow rate as space temperature changes. If the cooling load doubles, the damper will need to be fully open, which cannot take place until the controlled space temperature has risen to 73°F (23°C). As long as the cooling load remains that high, the space temperature must also stay

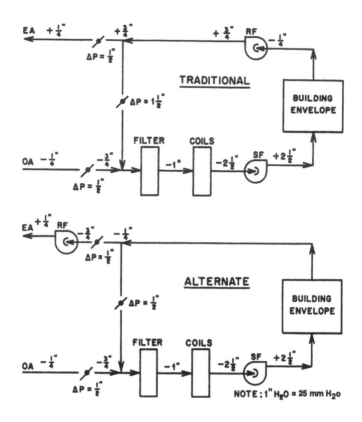

FIGURE 11.9 Pressure profiles and damper drops in various airhandlers.[2]

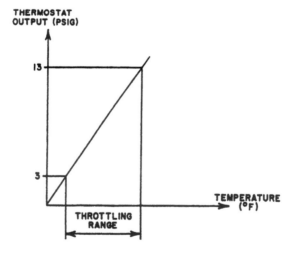

FIGURE 11.10 Throttling range can be defined as the temperature change required to change the thermostat output from its minimum to its maximum value.

up at the 73°F (23°C) value. Similarly, the only way this thermostat can reduce the opening of the cooling damper below 50% is to first allow the space temperature to drop below 72°F (22°C). Thus, thermostats have throttling ranges, not set points. If a throttling range is narrow enough, this gives

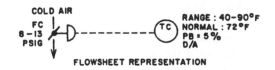

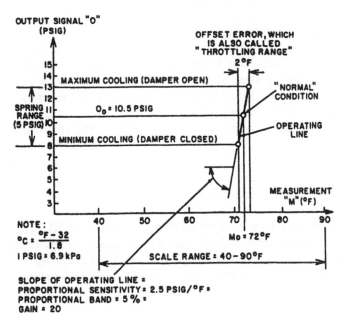

FIGURE 11.11 A fixed proportional band thermostat has a fixed throttling range and no set point.

the appearance that the controller is keeping the variable near set point, when in fact the narrow range only allows the variable to drift within limits.

1. Special-Purpose Thermostats

Night-day, set-back, or *dual room thermostats* will operate at different "normal" temperature values for day and night. They are provided with both a "day" and a "night" setting dial, and the change from day to night operation can be made automatic for a group of thermostats. The pneumatic day-night thermostat uses a two-pressure air supply system, the two pressures often being 13 and 17 PSIG (0.87 and 1.17 barg) or 15 and 20 PSIG (1.0 and 1.35 barg). Changing the pressure at a central point from one value to the other actuates switching devices in the thermostat and indexes them from day to night or vice versa. Supply air mains are often divided into two or more circuits so that switching can be accomplished in various areas of the building at different times. For example, a school building may have separate circuits for classrooms, offices and administrative areas, the auditorium, and the gymnasium and locker rooms. In some of the electric designs, dedicated clocks and switches are built into each thermostat.

The *heating-cooling* or *summer-winter thermostat* can have its action reversed and, if desired, can have its set point changed by means of indexing. This thermostat is used to actuate controlled devices, such as valves or dampers, that regulate a heating source at one time and a cooling source at another. It is often manually indexed in groups by a switch, or automatically by a thermostat that senses the temperature of the water supply, the outdoor temperature, or another suitable variable.

In the heating-cooling design, there are frequently two bimetallic elements, one being direct acting for the heating mode, the other being reverse acting for the cooling mode. The mode is switched automatically in response to a change in the air supply pressure, much as the day-night thermostats operate.

The *limited control range thermostat* usually limits the room temperature in the heating season to a maximum of 75°F (24°C), even if the occupant of the room has set the thermostat beyond these limits. This is done internally, without placing a physical stop on the setting knob.

A *slave* or *submaster thermostat* has its set point raised or lowered over a predetermined range, in accordance with variations in the output from a master controller. The master controller can be a thermostat, manual switch, pressure controller, or similar device. For example, a master thermostat measuring outdoor air temperature can be used to adjust a submaster thermostat controlling the water temperature in a heating system. Master-submaster combinations are sometimes designed as single-cascade action. When such action is accomplished by a single thermostat having more than one measuring element, it is known as *compensated control.*

Multistage thermostats are designed to operate two or more final control elements in sequence.

A *wet-bulb thermostat* is often used for humidity control, as the difference between wet- and dry-bulb temperature is an indication of moisture content. A wick or other means for keeping the bulb wet and rapid air motion to assure a true wet-bulb measurement are essential.

A *dew-point thermostat* is a device designed to control humidity on the basis of dew point temperatures.

A *smart thermostat* is usually a microprocessor-based unit with an RTD type or a transistorized solid-state sensor. It is usually provided with its own dedicated memory and intelligence, and it can also be equipped with a communication link (over a shared data bus) to a central computer. Such units can minimize building operating costs by combining time-of-day controls with intelligent comfort gap selection and maximized self-heating.

2. Zero Energy Band Control

A recent addition to the available thermostat choices is the zero energy band (ZEB) design. The idea behind ZEB control is to conserve energy by not using any when the room is comfortable. As illustrated in Figure 11.12, the conventional thermostat wastes energy by continuing to use it when the area's temperature is already comfortable. The "comfort gap" or "ZEB" on the thermostat is adjustable and can be varied to match the nature of the particular space.

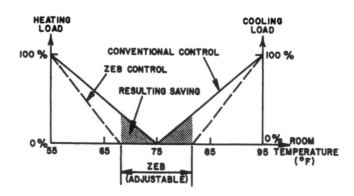

FIGURE 11.12   Zero energy band (ZEB) control is designed to save energy by not using any when the room is comfortable.

ZEB control can be accomplished in one of two ways. The single set point and single output approach is illustrated on the left side of Figure 11.13. Here the cooling valve fails closed and is shown to have an 8 to 11 PSIG (0.54 to 0.74 bar) spring range, while the heating valve is selected to fail open and has a 2 to 5 PSIG (0.13 to 0.34 bar) range. Therefore, between 5 and 8 PSIG (0.34

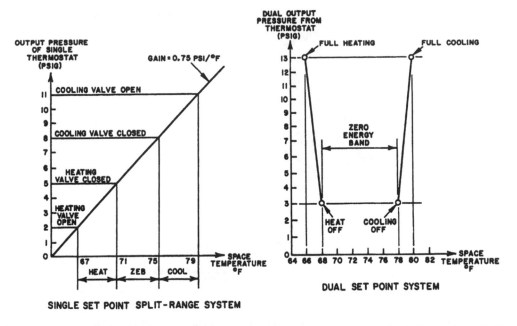

FIGURE 11.13 ZEB control schemes include the single-set point split-range approach (shown on left) and the dual-set point approach (shown on right).

to 0.54 bar), both valves are closed; no pay energy is expended while the thermostat output is within this range. The throttling range is usually adjustable from 5 to 25°F (3 to 13°C). Thus, if the ZEB is 30% of the throttling range, it can be varied from a gap size of 1.5°F (.85°C) to 7.5°F (4.2°C) by changing the throttling range (or gain).

Although the split-range approach is a little less expensive than the dual-set point scheme (shown on the right of Figure 11.13), it is also less flexible and more restrictive. The two basic limitations of the split-range approach are the following:

1. The gap width can be adjusted only by also changing the thermostat gain; maximum gap width is limited by the minimum gain setting of the unit.
2. The heating valve must fail open, which is undesirable in terms of energy conservation.

These limitations are removed when a dual-set point, dual-output thermostat is used. Here both valves can fail closed, and the bandwidth and the thermostat gain are independently adjustable. The gains of the heating and cooling thermostats are also independently adjustable. In Figure 11.13, the heating thermostat is reverse-acting and the cooling thermostat is direct-acting.

The most recent advances in thermostat technology are the microprocessor-based units. These are programmable devices with memory capability. They can be monitored and reset by central computers, using pairs of telephone wires as the communication link. Microprocessor-based units can be provided with continuously recharged backup batteries and accurate room temperature sensors. They can also operate without a host computer (in the "stand-alone" mode). In this case, the user manually programs the thermostat to maintain various room temperatures as a function of the time of day and other considerations.

3. Gap Control and the Self-Heating Building

The winter and summer enthalpy settings of a building are illustrated in Figure 11.14. The concept of gap control is simple: when comfort level in a zone is somewhere between acceptable limits,

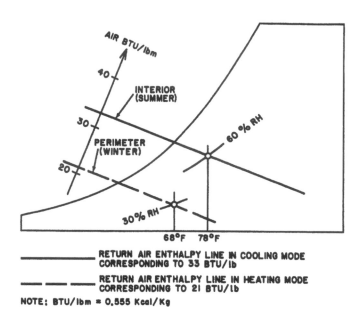

FIGURE 11.14   The air returning from the interior can transport about 10 BTU/lbm to the perimeter, making the building self-heating.[3]

the use of "pay energy" is no longer justified. Allowing the zones to float between limits instead of maintaining them at arbitrarily fixed conditions can substantially reduce the operating cost of the building. The savings come from two sources. First, there is a direct trade-off between the selected acceptable limits of discomfort and the yearly total of required degree days of heating and cooling. Second, there is the added side benefit that the building becomes self-heating during winter conditions. This occurs because during winter conditions, gap control automatically transfers the heat generated in the inside of the building to the perimeter areas, where heating is needed. This can result in long periods of building operation without the use of any pay energy.

Gap control can be looked upon as an override mode of control that is superimposed on the operation of the individual zone thermostats. As such, it can be implemented by all levels of automation, but the flexibility and ease of adjustment of the computerized systems make them superior in those applications in which the gap limits of the various zones are likely to change frequently.

Figure 11.1 illustrates the concept of comfort envelopes. Any combination of temperature and humidity conditions within such envelopes is considered to be comfortable. Therefore, as long as the space conditions fall within this envelope, there is no need to spend money or energy to change those conditions. This comfort gap is also referred to as zero energy band (ZEB), meaning that if the space is within this band, no pay energy of any type will be used. This concept is very cost-effective. When a zone of the airhandler in Figure 11.2 is within the comfort gap, its reheat coil is turned off and its VAV box is closed to the minimum flow required for air refreshment. When all the zones are inside the ZEB, the HW, CHW, and STM supplies to the airhandler are all closed and the fan is operated at minimum flow. When all other airhandlers are also within the ZEB, the pumping stations, chillers, cooling towers, and HW generators are also turned off.

With larger buildings that have interior spaces that are heat-generating even in the winter, ZEB control can make the building self-heating. Optimized control systems in operation today are transferring the interior heat to the perimeter without requiring any pay heat until the outside temperature drops below 10 to 20°F (–12.3 to –6.8°C). In regions in which winter temperature does not drop below 10°F (−12.3°C), ZEB control can eliminate the need for pay heat altogether. In regions farther north, ZEB control can lower the yearly heating fuel bill by 30 to 50%.

4. Supply Air Temperature Control

A substantial source of inefficiency in conventional HVAC control systems is the uncoordinated arrangement of temperature controllers. Two or three separate temperature control loops in series are not uncommon. For example, one of these uncoordinated controllers may be used to control the mixed air temperature, another to maintain supply (SA) temperature, and a third to control the zone-reheat coil. Such practice can result in simultaneous heating and cooling and therefore in unnecessary waste. Using a fully coordinated split-range temperature control system, such as that shown in Figure 11.15, will reduce yearly operating costs by more than 10%.

In this control system, the SA temperature set point (set by the temperature controller, TIC-07) is continuously modulated to follow the load. The methods of finding the correct set point will be discussed under "Optimizing Strategies." The loop automatically controls all heating or cooling modes. When the TIC-07 output signal is low–3 to 6 PSIG (0.2 to 0.4 bar)—heating is done by TCV-01. As the output signal reaches 6 PSIG (0.4 bar), heating is terminated; if free cooling is available, it is initiated at 7 PSIG (0.47 bar). When the output signal reaches 11 PSIG (0.74 bar)—the point at which OAD-05 is fully open–the cooling potential represented by free cooling is exhausted, and at 12 PSIG (0.8 bar), "pay cooling" is started by opening TCV-02. In such split-range systems, the possibility of simultaneous heating and cooling is eliminated. Also eliminated are interactions and cycling.

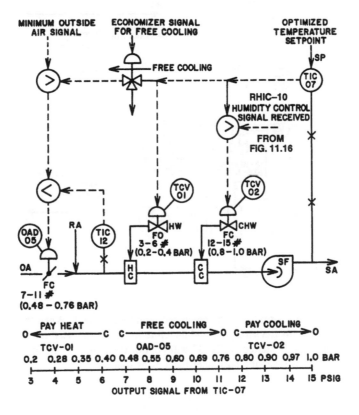

FIGURE 11.15 A fully coordinated split-range temperature control system will reduce yearly operating costs by more than 10%.

Figure 11.15 also shows some important overrides. TIC-12, for example, limits the allowable opening of OAD-05, so that the mixed-air temperature will never be allowed to drop to the freezing point and permit freeze-up of the water coils.

The minimum outdoor air requirement signal guarantees that the outside airflow will not be allowed to drop below this limit.

The economizer signal allows the output signal of TIC-07 to open OAD-05 only when "free cooling" is available. (A potential for free cooling exists when the enthalpy of the outdoor air is below that of the return air.)

Finally, the humidity controls will override the TIC-07 signal to TCV-02 when the need for dehumidification requires that the supply-air temperature be lowered below the set point of TIC-07.

C. HUMIDITY CONTROLS

Humidity in the zones is controlled according to the moisture content of the combined return air (see Figure 11.16). The process controlled by RHIC-10 is slow and contains large dead-time and transport-lag elements. In other words, a change in the SA humidity will not be detected by RHIC-10 until some minutes later. During the winter, it is possible for RHIC-10 to demand more and more humidification. To prevent possible saturation of the supply air, the RHIC-10 output signal is limited by RHIC-21. In this way, the moisture content of the supply air is never allowed to exceed 90% RH.

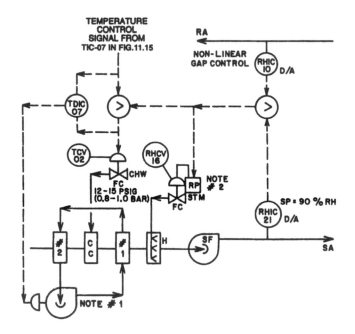

FIGURE 11.16 Humidity is controlled in the combined return air. Note 1: When the need for dehumidification (in the summer) overcools the supply air and therefore increases the need for reheat at the zones, this pump-around economizer loop is started. TDIC-07 will control the pump to "pump around" only as much heat as is needed. Note 2: This reversing positioner fuctions as follows:

Input from RHIC-10	Output to RHCV-16
3	15
9	3

For best operating efficiency, a nonlinear controller with a neutral band is used at RHIC-10. This neutral band can be set to a range of humidity levels–say, between 30% RH and 50% RH. If the RA is within these limits, the output of RHIC-10 is at 9 PSIG (0.6 bar), and neither humidification nor dehumidification is demanded. This arrangement can lower the cost of humidity control during the spring and fall by approximately 20%.

The same controller (RHIC-10) controls both humidification (through the relative humidity control valve, RHCV-16) and dehumidification (through the temperature control valve, TCV-02) on a split-range basis. As the output signal increases, the humidifier valve closes, between 3 and 9 PSIG (0.2 and 0.6 barg). At 9 PSIG (0.6 bar), the reversing positioner closes RHCV-16 and remains so, as the output signal increases to 12 PSIG (0.8 bar). At this condition, TCV-02 starts to open. Dehumidification is accomplished by cooling through TCV-02. This chilled water valve is controlled by humidity (RHIC-10) or temperature (TIC-07). The controller that requires more cooling will be the one allowed to throttle TCV-02.

Subcooling the air to remove moisture can substantially increase operating costs if this energy is not recovered. The dual penalty incurred for overcooling for dehumidification purposes is the high chilled water cost and the possible need for reheat at the zone level. The savings from a pump-around economizer can eliminate 80% of this waste. In this loop, whenever TDIC-07 detects that the chilled water valve (TCV-02) is open more than would be necessary to satisfy TIC-07, the pump-around economizer is started. This loop in coil #1 reheats the dehumidified supply air, using the heat that the pump-around loop removed from the outside air in coil #2 before it entered the main cooling coil.

In this way, the chilled-water demand is reduced in the cooling coil (TCV-02), and the need for reheating at the zones is eliminated. Although Figure 11.16 shows a modulating controller setting the speed of a circulating pump, it is also possible to use a constant-speed pump operated by a gap switch.

D. OUTDOOR AIR CONTROLS

Outdoor air is admitted to satisfy requirements for fresh air or to provide free cooling. Both control loops are shown in Figure 11.17.

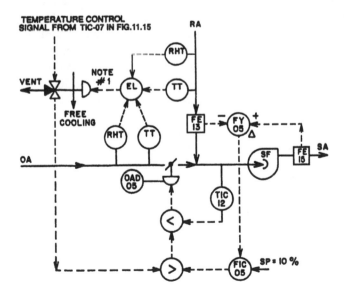

FIGURE 11.17 Outside air control loops provide fresh air or free cooling. The enthalpy logic unit (EL) compares the enthalpies of the outside and return airs and vents its output signal if free cooling is available. Therefore, the economizer cycle is initiated whenever Hoa < Hra.

The minimum requirement for fresh outdoor air while the building is occupied is usually 10% of the airhandler's capacity. In most advanced control systems, this value is not controlled as a fixed percentage but as a function of the number of people in the building or of the air's carbon dioxide content. In the most conventional systems, the minimum outdoor air is provided by keeping 10% of the area of the outdoor air damper always open when the building is occupied. This method is inaccurate, because a constant damper opening does not result in a constant air flow. This flow varies with fan load, because changes in load will change the fan's suction pressure and will therefore effect ΔP across the damper. This conventional design results in waste of air-conditioning energy at high loads and insufficient air refreshment at low loads.

The control system depicted in Figure 11.17 reduces operating costs while maintaining a constant minimum rate of air refreshment, which is unaffected by fan loading. Direct measurement of outdoor airflow is usually not possible because of space limitations. For this reason, Figure 11.17 shows the outdoor airflow as being determined as the difference between FE-15 and FE-13. FIC-05 controls the required minimum outdoor airflow by throttling OAD-05.

1. CO_2-Based Ventilation

In conventional installations the amount of outdoor air admitted is usually based on one of the following criteria:

- 0.1 to 0.25 cfm/ft^2 (30 to 76 lpm/m^2) of floor area
- 10 to 25% of total air supply rates
- About 5 cfm (25 lps) volumetric rate per person

These criteria all originated at a time when energy conservation was no serious consideration; therefore, their aim was to provide simple, easily enforceable rules that will guarantee that the outdoor air intake always exceeds the required minimum. Today the goal of such systems is just the opposite: it is to make sure that air quality is guaranteed at minimum cost. As the floor does not need oxygen–only people do—some of the above rules make little sense.

There is a direct relationship between savings in building operating costs and reduction in outdoor air admitted into the building. According to one study in the United States,[4] infiltration of outdoor air accounted for 55% of the total heating load and 42% of the total cooling load. Another survey[5] showed that 75% of the fuel oil consumed in New York City schools was devoted to heating ventilated air. Because building conditioning accounts for nearly 20% of all the energy consumed in the U.S.,[6] optimized admission of outdoor air can make a major contribution to reducing our national energy budget. This goal can be well served by CO_2-based ventilation controls.

The purpose of ventilation is not to meet some arbitrary criteria, but to maintain a certain air quality in the conditioned space. Smoke, odors, and other air containment parameters can all be correlated to the CO_2 content of the return air.[7] This then becomes a powerful tool of optimization, because the amount of outdoor air required for ventilation purposes can be determined on the basis of CO_2 measurement, and the time of admitting this air can be selected so that the air addition will also be energy efficient. With this technique, health and energy considerations will no longer be in conflict, but will complement each other.

CO_2-based ventilation controls can easily be integrated with the economizer cycle and can be implemented by use of conventional or computerized control systems. Because the rate of CO_2 generation by a sedentary adult is 0.75 cfh (27 lph), control by CO_2 concentration will automatically reflect the level of building occupancy.[8] Energy savings of 40% have been reported[9] by converting conventional ventilation systems to intermittent CO_2-based operation.

2. Economizer Cycles

The full use of free cooling can reduce the yearly air-conditioning load by more than 10%. The enthalpy logic unit (EL) in Figure 11.17 will allow the temperature-controller signal (TIC-07 in Figure 11.15) to operate the outdoor air damper whenever free cooling is available. This economizer cycle is therefore activated whenever the enthalpy of the outdoor air is below that of the return air.

Free cooling can also be used to advantage while the building is unoccupied. Purging the building with cool outdoor air during the early morning results in cooling capacity being stored in the building structure, reducing the daytime cooling load.

The conventional economizers—such as the one shown in Figure 11.17—are rather limited devices for two reasons. First, they determine the enthalpy of the outdoor and return air streams using somewhat inaccurate sensors. Secondly, although they consider the free cooling potential of the outside air, they disregard all the other possibilities of using the outdoor air to advantage.

Advanced, microprocessor-based economizers overcome both of these limitations. They use accurate sensors and memorize the psychrometric chart to evaluate all potential uses of outside air, not only free cooling. Figure 11.18 illustrates the various zones of operation, based on the relative conditions of the outside and return airs.

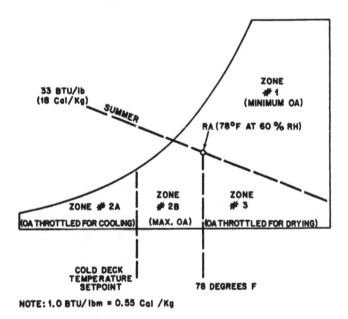

FIGURE 11.18 Free cooling and drying can often be obtained in the summer, depending on the zone where the outside air falls relative to the return air.

If the enthalpy of the outside air falls in zone 1 (that is, if its BTU content exceeds 33), no free cooling is available; therefore, the use of outside air should be minimized in the summer. In the winter or fall, it is possible that the enthalpy of the outside air on sunny afternoons will exceed the return air enthalpy, which in the winter is about 21 BTU/lbm (11.6 cal/kg). Under such conditions, "free heating" can be obtained by admitting the outside air in zone 1.

If the condition of OA corresponds to zone 2 (BTU < 33 and temperature < 78°F), free cooling is available.

If the condition of OA corresponds to zone 3 (BTU < 33 and temperature > 78°F), free dehumidifying (latent cooling) is available.

When there are both cold and hot air ducts in the building (dual duct system), the control system in zone 2 will function differently depending on whether the outside air temperature is above or below the cold deck temperature. If it is above that temperature (zone 2B), maximum (100%) outside air can be used; if it is below that temperature (zone 2A) the use of outside air needs to be modulated or time-proportioned. Therefore, in zone 2A, where the outside air is cooler than the cold deck temperature, free cooling is available, but only some of the total potential can be used. The OA damper therefore must be either positioned or cycled to admit enough cold outside air to lower the mixed air temperature to equal the cold deck temperature set point. This way the need for "pay cooling" is eliminated.

When the outside air is in zone 3 (hot but dry), the control system should admit some of it to reduce the overall need for latent cooling. The amount of outside air admitted under these conditions should be controlled so that the resulting mixed-air dew-point temperature will equal the desired cold deck dew point. This method takes full advantage of the free drying potential of the outside air.

As can be seen from the above, the use of the economizer cycles is similar to opening the windows in a room to achieve maximum comfort with minimum use of "pay energy."

III. OPTIMIZING CONTROLS

The main goal of optimization in airhandlers is to match the demand for conditioned air with a continuous, flexible, and efficient supply. This requires the floating of both air pressures and temperatures to minimize operating costs while following the changing load.

In traditional HVAC control systems, the set point of TIC-07 in Figure 11.19 is set manually and is held as a constant. This practice is undesirable, because for each particular load distribution the optimum SA temperature is different. If the manual setting is less than that temperature, some zones will become uncontrollable. If it is more than optimum, operating energy will be wasted.

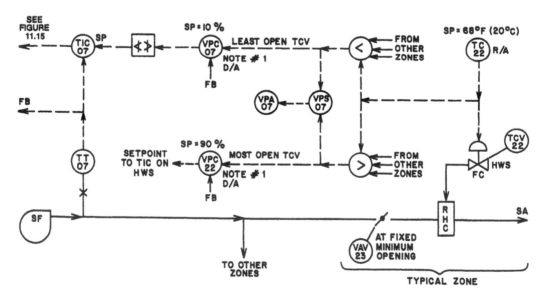

FIGURE 11.19 Optimization of air and water temperatures during heating (winter) mode of airhandler operation is designed to distribute the heat load efficiently between the main heating coil and the zone reheat coils. Valve position controllers (VPCs) are provided with integral action only for stable floating control. The integral time is set to be 10 times the integral time of the associated TIC. External feedback is provided to eliminate reset wind-up when VPC output is overruled by the set point limit on TIC.

A. Temperature Optimization in the Winter

In the winter, the goal of optimization is to distribute the heat load between the main heating coil (HC in Figure 11.15) and the zone reheat coils (RHC in Figure 11.19) most efficiently. The highest efficiency is obtained if the SA temperature is high enough to meet the load of the zone with the *minimum* load. In this way, the zone reheat coils are used only to provide the difference between the loads of the various zones, whereas the base load is continuously followed by TIC-07. In Figure 11.19, the set point of TIC-07 is adjusted to keep the least open TCV-22 only about 10% open. First, the least open TCV-22 is identified and its opening (in the valve position controller, VCP-07) is compared with the desired goal of 10%. If the valve opening is more than 10%, the TIC-07 set point is increased; if less, the set point is lowered. Stable operation is obtained by making VPC-07 an integral-only controller with an external feedback to prevent reset wind-up.

In Figure 11.19, it is assumed that the hot water supply (HWS) temperature is independently adjustable and will be modulated by VPC-22 so that the most open TCV-22 will always be 90% open. The advantages of optimizing the HWS temperature include:

Minimizing heat-pump operating costs by minimizing HWS temperature
Reducing the pumping costs by opening all TCV-22 valves in the system
Eliminating unstable (cycling) valve operation, which occurs in the nearly closed position,
    by opening all TCV-22 valves

In Figure 11.19, a valve position alarm (VPA-07) is also provided to alert the operator if this "heating" control system is incapable of keeping the openings of all TCV-22 valves between the limits of 10% and 90%. Such alarm will occur if the VPCs can no longer change the TIC set point(s), because their maximum (or minimum) limits have been reached. This condition will occur only if the load distribution was not correctly estimated during design or if the mechanical equipment was not correctly sized.

If the HWS temperature cannot be modulated to keep the most-open TCV-22 from opening to more than 90%, then the control loop depicted in Figure 11.20 should be used. In this loop, as long as the most-open valve is less than 90% open, the SA temperature is set to keep the least-open TCV-22 at 10% open (zone A). When the most-open valve reaches 90% opening, control of the least-open valve is abandoned and the loop is dedicated to keeping the most-open TCV-22 from becoming fully open (zone B). This, therefore, is a classic case of herding control, in which a single constraint envelope "herds" all TCV openings to within an acceptable band and thereby accomplishes efficient load following.

B. Temperature Optimization in the Summer

In the cooling mode during the summer, the SA temperature is modulated to keep the most-open variable volume box (VAV-23) from fully opening. Once a control element is fully open, it can no longer control; therefore, the occurrence of such a state must be prevented. On the other hand, it is generally desirable to open throttling devices such as VAV boxes to accomplish the following goals:

1. Reduce the total friction drop in the system.
2. Eliminate cycling and unstable operation (which is more likely to occur when the VAV box is nearly closed).
3. Allow the airhandler to meet the load at the highest possible supply-air temperature.

This statement does not apply if air transportation costs exceed cooling costs (for example, undersized ducts, inefficient fans). In this case, the goal of optimization is to transport the minimum

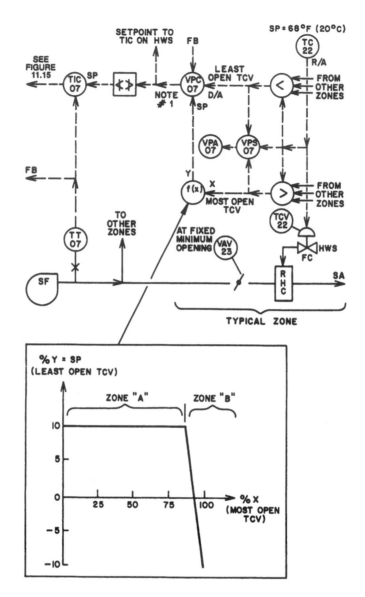

FIGURE 11.20 This alternative method of air supply temperature optimization in winter should be used when the HWS temperature cannot be modulated. Valve position controllers (VPCs) are provided with integral action only for stable floating control. The integral time is set to be 10 times the integral time of the associated TIC. External feedback is provided to eliminate reset wind-up when VPC output is overruled by the set point limit on TIC.

quantity of air. The amount of air required to meet a cooling load will be minimized if the cooling capacity of each unit of air is maximized. Therefore, if fan operating cost is the optimization criterion, the SA temperature is to be kept at its achievable minimum, instead of being controlled as in Figure 11.21.

If the added feature of automatic switchover between winter and summer modes is desired, the control system depicted in Figure 11.22 should be used. When all zones require heating, this control loop will behave exactly as does the one shown in Figure 11.20, when all zones require cooling, the controls of Figure 1.22 will operate as does the system shown in Figure 11.21. In addition, this control system will operate automatically with maximum energy efficiency during the transitional

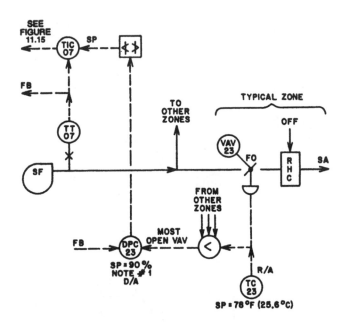

FIGURE 11.21 This method of air supply temperature optimization in summer (cooling mode) should be used if fan operating costs are less than cooling costs. The damper position controller (DPC) has integral action only, with its setting being 10 times the integral time of TIC-07. External feedback is provided to eliminate reset wind-up.

periods of fall and spring. This high efficiency is a result of the exploitation of the self-heating effect. If some zones require heating (perimeter offices) and others require cooling (interior spaces), the airhandler will automatically transfer this free heat from the interior to the perimeter zones by intermixing the return air from the various zones and moving it through the 10-degree zero energy band (ZEB) between the settings of TC-22 and TC-23. When the zone temperatures are within this comfort gap of 68°F (20°C) to 78°F (26°C), no pay energy is used and the airhandler is in its self-heating, or free-heating, mode. This is an effective means of reducing operating costs in buildings. The savings can amount to more than 30% during the transitional seasons.

When the temperature in one of the zones reaches 78°F (26°C), the air supply temperature set point will be lowered by DPC-23 and the air-side controls will be automatically switched to cooling (as depicted in Figure 11.21). If, at the same time, some other zone temperatures drop below 68°F (20°C), requiring heating, their demand will have to be met by the heat input of the zone-reheat coils only. This mode of operation is highly inefficient because of the simultaneous cooling and reheating of the air. Fortunately, this combination of conditions is highly unlikely, because under proper design practices, the zones served by the same airhandler should display similar load characteristics. The advantage of the control loop in Figure 11.22 is that it can automatically handle any load or load combination, including this unlikely, extreme case.

C. Auto-Balancing of Buildings

In computerized building control systems, the optimization potentials are greater than those that have been discussed up to this point. When all zone conditions are detected and controlled by the computer, it can optimize not only the normal operation but also the start-up of the building.

The optimization of airhandler fans is directed at two goals simultaneously. The first goal is to find the optimum value for the set point of the supply air pressure controller (PIC-19 in Figure 11.5). Generally, the supply air pressure is at an optimum value when it is at the lowest possible value, while all loads are satisfied. As the supply pressure is lowered, the fan operating cost is

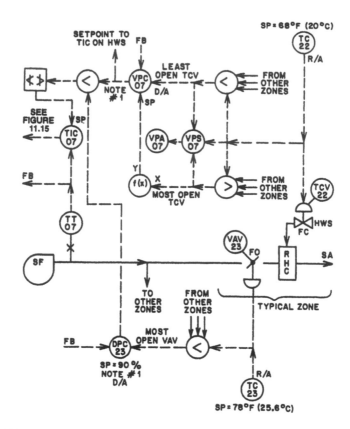

FIGURE 11.22 This control system optimizes the water and air temperatures in both summer and winter. See also Figures 11.20 and 11.21.

reduced, but with lowered supply pressures the VAV boxes serving the individual zones (VAV-23 in Figure 11.2) will have to open up so that the airflow to the zones will not be reduced. Therefore, the optimum setting for PIC-19 is that pressure at which the most-open VAV box is nearly 100% open, while all other VAV boxes are less than 100% open.

The second goal of optimization is to rebalance the air distribution in the building automatically as the load changes. If the VAV boxes (VAV-23 in Figure 11.2) are not pressure independent (are not able to maintain constant airflow when the supply pressure changes), manual rebalancing is required every time the load distribution changes. Naturally this is a very labor-demanding and inefficient operation. The optimization strategy described below serves the multiple purposes of automatic rebalancing and finding the optimum set points for the supply air pressure and temperature.

Figure 11.23 illustrates an airhandler serving several zones. The abbreviations used in that figure and in the algorithm tables that follow are listed below:

AI-1 to AI-N: Analog inputs (zone temperatures)
AI-AT: Analog input (air supply temperatures)
AI-RT: Analog input (return air temperature)
AO-1 to AO-N: Analog output zone (zone VAV opening)
AT: Supply air temperature
EA: Exhaust air
OA: Outside air

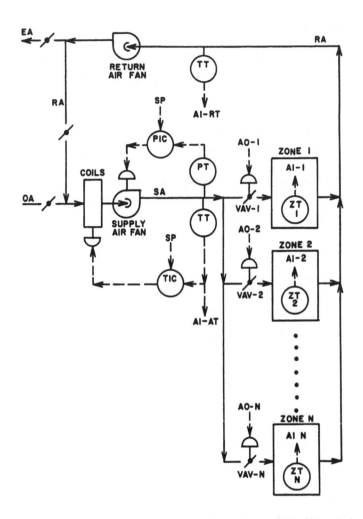

FIGURE 11.23 Airhandler optimization and autobalancing can be handled efficiently by computer.

PIC: Pressure controller (supply air)
PT: Pressure transmitter
RA: Return air
RT: Return air temperature
SA: Supply air
SP: Set point
TH: Upper limit of comfort zone (Figure 11.1) (maximum allowable zone temperature)
TIC: Temperature controller (air supply)
TL: Lower limit of comfort zone (Figure 11.1) (minimum allowable zone temperature)
VAV-1 to VAV-N: Variable air volume boxes
ZT-1 to ZT-N: Zone temperatures
ZT5-1 to ZT5-N: ZT five minutes after start-up
ZT10-1 to ZT10-N: ZT ten minutes after start-up
XMIN: Minimum VAV opening required for ventilation
XSET-1 to XSET-N: Initial VAV opening after "start-up"
XMAX-1 to XMAX-N: Maximum limit on VAV opening during normal operation

D. Start-Up Algorithm

All VAV boxes are set to their minimum openings required for ventilation purposes (XMIN), such as 25%. Therefore, at the time of start-up, AO-1 through AO-N are all set for 25%. PIC is set to midscale; therefore, the start-up value of its SP = 50%. TIC is set for $(TL + 25)°F$ in the heating mode and for $(TH - 25)°F$ in the cooling mode.

After 5 minutes of operation, the zone temperatures are detected (ZT5-1, ZT5-2, etc.), and after 10 minutes of operation they are detected again (ZT10-1, ZT10-2, etc.). At the end of the first 10 minutes of operation, the supply air temperature is also measured as AT10 and the return air temperature as RT10.

Once the above readings are obtained, they are entered into a table such as Table 11.1, which serves as the basis for determining the required start-up openings of each of the VAV boxes (XSET-1, XSET-2, etc.). The purpose of this table is to select the initial opening for each VAV box in a

TABLE 11.1
Algorithm to Determine Start-Up Openings of Individual VAV Boxes (XSET)

Input Conditions			Output
Operating Mode	Approach between zone and supply temperatures	Amount of temperature change during last 5 minutes of start-up	Initial value of XSET to be used for AO-1 to AO-N (%)
		$(ZT10 - ZT5) <0.5°F$	100
		$(ZT10 - ZT5)$ 0.5–1°F	90
		$(ZT10 - ZT5)$ 1–1.5°F	80
Heating $(AT > RT)$	$(TL - ZT10) >5°F$	$(ZT10 - ZT5)$ 1.5–2°F	70
		$(ZT10 - ZT5)$ 2–3°F	60
		$(ZT10 - ZT5)$ 3–4°F	50
		$(ZT10 - ZT5)$ 4–5°F	40
		$(ZT10 - ZT5) >5°F$	30
	$(TL - ZT10) <5°F$	Disregard	25
		$(ZT5 - ZT10) <0.5°F$	100
		$(ZT5 - ZT10)$ 0.5–1°F	90
		$(ZT5 - ZT10)$ 1–1.5°F	80
Cooling $(AT < RT)$	$(ZT10 - TH) >5°F$	$(ZT5 - ZT10)$ 1.5–2°F	70
		$(ZT5 - ZT10)$ 2–3°F	60
		$(ZT5 - ZT10)$ 3–4°F	50
		$(ZT5 - ZT10)$ 4–5°F	40
		$(ZT5 - ZT10) >5°F$	30
	$(ZT10 - TH) <5°F$	Disregard	25

Note: $°C = (°F - 32)/1.8$.

logical manner. Therefore, if the zone temperature after 10 minutes of operation is already within 5°F of reaching the comfort zone, the VAV box can be left at its minimum opening. If comfort zone is not yet within 5°F, a higher opening is needed. The initial VAV opening is increased on the basis of the zone performance during the previous 5 minutes. The larger the temperature change experienced by the zone during the previous 5 minutes, the sooner it will reach the comfort zone and therefore the smaller the opening that is required. By this logic, the VAV boxes on those zones which are furthest from comfort and which are moving most slowly toward comfort will be given the highest openings.

E. Normal Algorithm for VAV Throttling

The initial VAV opening for each zone (XSET), which is determined by the methods above, is then used as the maximum limit on the VAV opening (XMAX) during the first 5 minutes of normal operation. The value of XMAX is reevaluated every 5 minutes, as shown in Table 11.2. The logic here is to increase the maximum limit on VAV opening (XMAX) to any zone in which the VAV has been open to its maximum limit for 5 minutes. Similarly, this logic will lower the XMAX limit if the VAV damper was at its XMIN during the previous 5 minutes. If a VAV damper has been throttled somewhere in between these two limits (XMAX and XMIN), its limit will not be altered. The change increment of 10% shown in Table 11.2 is adjustable for maximum flexibility.

The algorithm described above and illustrated in Table 11.2 guarantees that changes in load distribution will not result in starving some zones; the building will be automatically rebalanced in an orderly manner. The value of XMAX from Table 11.2 and the permanent values of XMIN, determined by ventilation requirements, are used to reevaluate the individual VAV openings every 2 minutes, as described by the algorithm in Table 11.3.

TABLE 11.2
Reevaluation of Value of XMAX

Input Conditions		Output
Has VAV been continuously open to its XMAX during last 5 minutes?	Has VAV been continuously throttled to its XMIN during the last 5 minutes?	Incremental change in value of XMAX at the end of 5-minute period
Yes	Yes	Leave XMAX = XMIN
	No	Increase by 10%
No	Yes	Decrease by 10%
	No	Leave as is

TABLE 11.3
Algorithm to Determine Analog Outputs, Setting the Openings of VAV Boxes

Input Conditions		Output
Operating Mode	Control Criteria	Required VAV opening: AO-1 to AO-N is to be equal
Heating (AT>ZT)	ZT<(TL − 1)	XMAX
	(TL − 1)<ZT<(TL + 1)	No change
	ZT>(TL + 1)	XMIN
Cooling (AT<ZT)	ZT>(TH + 1)	XMAX
	(TH − 1)<ZT<(TH + 1)	No change
	ZT<(TH − 1)	XMIN

The main optimizing and autobalancing feature of this algorithm is that whenever a zone is inside the comfort gap, its VAV opening is reduced to XMIN. This reduces the load on the fans and also provides more air to the zones experiencing the highest loads.

F. Optimization of Air Supply Pressure and Temperature

Optimization means that the load is met at minimum cost. The cost of operating an airhandler is the sum of the cost of air transportation and conditioning. These two cost factors tend to change in opposite directions; minimizing the cost of one will increase the cost of the other. Therefore, it is important to monitor both the transportation and the conditioning costs continuously and to minimize the larger one when optimizing the system. Computerized control systems allow these costs to be calculated readily on the basis of utility costs and quantities.

For example, if the transportation cost exceeds the conditioning cost, the optimization goal is to minimize fan operation. This is achieved by conditioning the space with as little air as possible. The quantity of air transported can be minimized if each pound of air is made to transport more conditioning energy; that is, if each pound of air carries more cooling or heating BTUs. Therefore, when the goal is to minimize fan costs, the air supply pressure is held as low as possible, and the supply temperature is maximized in the winter and minimized in the summer. Fan costs tend to exceed conditioning costs when the loads are low, such as in the spring or fall, or when the economizer cycle is used to provide free cooling.

Table 11.4 describes the algorithm used to achieve this goal. When none of the VAV boxes (Figure 11.23) are fully open, indicating that all loads are well satisfied, the air pressure (PIC set point) is kept at a minimum, and the air temperature (TIC set point) is lowered in the winter and raised in the summer. When more than one VAV box is fully open, the air supply temperature is increased in the winter (lowered in the summer). When its limit is reached, the algorithm will start raising the PIC set point.

Table 11.5 describes the algorithm used when the conditioning costs are higher than the fan operating costs. This is likely to be the case when the loads are high, such as in the summer or the winter. Under such conditions, the supply pressure is maximized before the supply air temperature is increased in the winter or lowered in the summer. When none of the VAV boxes in Figure 11.23 are fully open, the PIC set point is lowered, while the TIC set point is at or near minimum in the winter (maximum in summer). When more than one VAV box is fully open, the PIC set point is increased to its maximum setting. When that is reached, the supply temperature starts to be increased in the winter (decreased in the summer).

The algorithms described above provide the dual advantages of automatic balancing and minimum operating cost. They eliminate the need for manual labor or for the use of pressure-independent VAV boxes, while reducing operating cost by about 30%. They also provide the flexibility of assigning different comfort envelopes (different TL and TH values) to each zone. Thereby as occupancy or use changes, the comfort zone assigned to the particular space can be changed automatically.

G. Elimination of Chimney Effects

In high-rise buildings, the natural draft resulting from the chimney effect tends to pull in ambient air at near ground elevation and to discharge it at the top of the building. Although eliminating the chimney effect can lower the operating cost by approximately 10%, few systems with this capability are yet in operation.

Figure 11.24 shows the required pressure controls. The key element of this control system is the reference riser, which allows all pressure controllers in the building to be referenced to the barometric pressure of the outside atmosphere at a selected elevation. Using this pressure reference allows all zones to be operated at 0.1 in. H_2O (2.5 mm H_2O) pressure (PC-7) and permits this

TABLE 11.4
Optimization of Supply Air Pressure and Temperature, When Fan Costs Exceed Conditioning Costs

(Frequency = 5 minutes)

VAV Status	Airhandler Mode	Is TIC SP at its Limit?	Incremental Ramp Adjustment in the Set Points of	
			TIC	PIC
None at 100% for 15 minutes continuously	Heating (AT>RT)	Yes, at max.	−2°F	N.C.* (at min.)
		No	−1°F	N.C. (at min.)
	Cooling (AT<RT)	Yes, at min.	+2°F	N.C. (at min.)
		No	+1°F	N.C. (at min.)
Not more than one at 100% for 30 minutes continuously	Heating (AT>RT)	Yes, at max.	N.C. (at max.)	N.C.
		No	N.C.	N.C.
	Cooling (AT<RT)	Yes, at min.	N.C. (at max.)	N.C.
		No	N.C.	N.C.
More than one at 100% for more than 30 minutes continuously	Heating (AT>RT)	Yes, at max.	(at max.)	+.25 in. H_2O
		No	+1°F	N.C.
	Cooling (AT<RT)	Yes, at min.	(at min.)	+.25 in. H_2O
		No	−1°F	N.C.
More than one at 100% for more than 60 minutes continuously	Heating (AT>RT)	Yes, at max.	(at max.)	+0.5 in. H_2O
		No	+2°F	N.C.
	Cooling (AT<RT)	Yes, at min.	(at min.)	+0.5 in. H_2O
		No	−2°F	N.C.

Note: N.C. = no change is made at the end of that 5-minute period.

constant pressure to be maintained at both ends of all elevator shafts (PC-8 and 9). If the space pressure is the same on the various floors of a high-rise building, there will be no pressure gradient to motivate the vertical movement of the air, and as a consequence, the chimney effect will have been eliminated. A side benefit of this control strategy is the elimination of all drafts or air movements between zones, which also minimizes the dust content of the air. Another benefit is the capability of adjusting the "pressurization loss" of the building by varying the settings of PC-7, 8, and 9.

Besides reducing operating costs, the use of pressure-controlled elevator shafts increases comfort because drafts and the associated noise are eliminated.

Figure 11.24 also shows the use of cascaded fan controls. The set points of the cascade slaves (PC-2 and PC-5) are programmed so that the air pressure at the fan is adjusted as the square of flow and the pressure at the end of the distribution headers (cascade masters PC-1 and PC-6) remains constant. This control approach results in the most efficient operation of variable-air volume fans.

TABLE 11.5

Optimization of Supply Air Pressure and Temperature, When Conditioning Costs Exceed Fan Costs

(Frequency = 5 minutes)

VAV Status	Airhandler Mode	Is PIC Set Point at its Maximum?	Incremental Ramp Adjustment in the Set Points of	
			TIC	PIC
None at 100% for 15 minutes continuously	Heating (AT>RT)	Yes	$-1°F$	-0.5 in. H_2O
		No	N.C.* (at min.)	$-.25$ in. H_2O
	Cooling (AT<RT)	Yes	$+1°F$	$-.5$ in. H_2O
		No	N.C. (at max.)	$-.25$ in. H_2O
Not more than one at 100% for 30 minutes continuously	Heating (AT>RT)	Yes	N.C.	N.C. (at max.)
		No	N.C.	N.C.
	Cooling (AT<RT)	Yes	N.C.	N.C. (at max.)
		No	N.C.	N.C.
More than one at 100% for more than 30 minutes continuously	Heating (AT>RT)	Yes	$+1°F$	(at max.)
		No	N.C.	$+.25$ in. H_2O
	Cooling (AT<RT)	Yes	$-1°F$	(at max.)
		No	N.C.	$+.25$ in. H_2O
More than one at 100% for more than 60 minutes continuously	Heating (AT>RT)	Yes	$+2°F$	(at max.)
		No	N.C.	$+0.5$ in. H_2O
	Cooling (AT<RT)	Yes	$-2°F$	(at max.)
		No	N.C.	$+0.5$ in. H_2O

Note: N.C. = No change.

If the building is maintained at a constant pressure that corresponds to the pressure at ground elevation plus 0.1″ H_2O, this will result in higher pressure differentials on the higher floors, as the barometric pressure on the outside drops. Therefore air losses due to outleakage and pressures on the windows will both rise. If the pressure reference is taken at some elevation above ground level, these effects will be reduced on the upper floors, but on the lower floors the windows will be under positive pressure from the outside and air infiltration will be experienced.

IV. CONCLUSIONS

The airhandler is just one of the industrial unit operations. The process of air conditioning is similar to all other industrial processes. Fully exploiting state-of-the-art instrumentation and control results

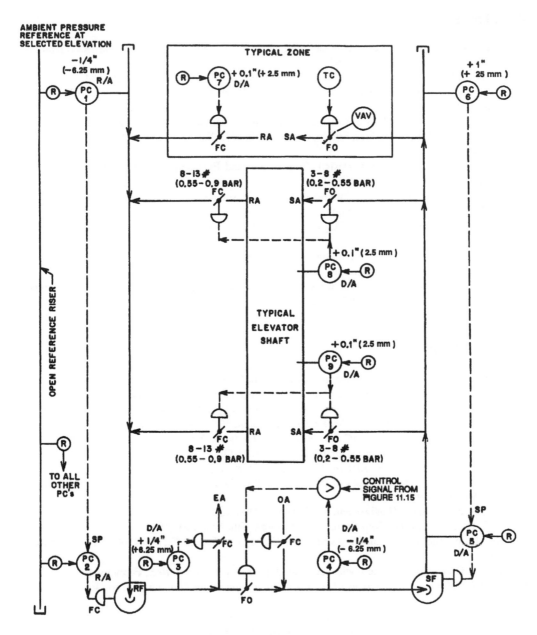

FIGURE 11.24 Chimney effects in high-rise buildings can be eliminated by using the proper pressure controls.

in dramatic improvements. There are few other processes in which the use of optimization and of instrumentation know-how alone can halve the operating cost of a process.

The control and optimization strategies described in this chapter can be implemented by pneumatic or electronic instruments and controlled by analog or digital systems. The type of hardware used in optimization is less important than the understanding of the process and of the control concepts that are to be implemented. The main advantage of digital and computerized systems is their flexibility and convenience in making changes, without the need to modify equipment or wiring.

REFERENCES

1. Lipták, B.G., Applying the techniques of process control to the HVAC process, *ASHRAE Trans.*, Paper No. 2778, Vol. 89, Part 2A, 1983.
2. Avery, G., VAV economizer cycle, *Heating, Piping, Air Conditioning*, August 1984.
3. Lipták, B.G., Reducing the operating costs of buildings by the use of computers, *ASHRAE Trans.*, 83:1, 1977.
4. Kovach, E.G., Technology of Efficient Energy Utilization: The Report of a NATO Science Committee Conference Held at Les Arcs, France, October 8-12, 1973, Scientific Affairs Division, NATO, Brussels, Belgium.
5. Liu, S.T., et al., Research, Design, Construction and Evaluating an Energy Conservation School Building in New York City, NBSIR.
6. Stanford Research Institute, Patterns of Energy Consumption in the United States, prepared for the Office of Science and Technology, Washington, D.C.
7. Nomura, G. and Yamada, Y., CO_2 respiration rate for the ventilation calculation, *Trans. Jpn. Arch. Soc. Meetings*, October 1969.
8. Department of Defense, Environmental Engineering for Shelters, TR-20-Vol. 3, Department of Defense, Office of Civil Defense, May 1969.
9. Kusuda, T., Intermittent Ventilation for Energy Conservation, NBS report of ASHRAE Symposium in Dallas, Texas, February 1976.

BIBLIOGRAPHY

Anderson, E.P., *Air Conditioning: Home and Commercial*, 5th ed., Macmillan, New York.

Brumbaugh, J.E., *Heating, Ventilating and Air Conditioning Library*, rev. ed., Macmillan, New York.

Demster, C.S., *Variable Air Volume Systems for Environmental Quality*, McGraw-Hill, New York, 1995.

Grimm, N.R. and Roasler, R.C., *Handbook of HVAC Design*, McGraw-Hill, New York, 1990.

Haines, R.W. and Wilson, C.L., *HVAC Systems Design Handbook*, McGraw-Hill, New York, 1994.

Hartman, T.B., *Direct Digital Control for HVAC Systems*, McGraw-Hill, New York, 1993.

Huntington, W.C., *Building Construction*, 4th ed., John Wiley & Sons, New York.

Lipták, B.G., Applying the techniques of process control to the HVAC process, *ASHRAE Trans.*, Paper No. 2778, Vol. 89, Part 2A, 1983.

O'Callaghan, P.W., *Energy Management*, McGraw-Hill, New York, 1993.

Parmley, R.O., *HVAC Design Data Sourcebook*, McGraw-Hill, New York, 1994.

Sabatini, J.N. and Smith, R.D., *Building and Safety Codes for Industrial Facilities*, McGraw-Hill, New York, 1993.

Spielvogel, L.G., Exploding some myths about building energy use, *Arch. Rec.*, February 1976.

Stein/Reynolds/McGuinness, *Mechanical and Electrical Equipment for Buildings*, 8th ed., John Wiley & Sons, New York.

Strother, E.F., *Thermal Insulation Building Guide*, R.E. Krieger, 1990.

Sun, T.-Y., *Air Handling Systems Design*, McGraw-Hill, New York, 1994.

Tseng-Yao Sun, *Air Handling System Design*, McGraw-Hill, 1994.

Wang, S.K., *Handbook of Air Conditioning and Refrigeration*, McGraw-Hill, New York, 1994.

Wendes, H.C., *HVAC Retrofits*, Fairmont Press, 1994.

Wendes, H.C., *HVAC Energy Audits*, Fairmont Press, 1996.

Whitman, W.C. and Johnson, W.M., *Refrigeration and Air Conditioning Technology*, Delmar, 1991.

12   Pumps

This chapter begins with a description of the liquid-transportation process. This is followed by a summary of the different pump designs, features, and accessories, such as variable-speed drives. Following this, the basic controls used for single pumps and for multipump stations are covered. The chapter concludes with a review of pumping-system-optimization strategies.

I. THE PROCESS

In the transportation of liquids, impeller velocity or displacement force imparts a head (pressure rise) that moves the liquid through the resistances of the pump cavities and of the process. As the hydraulic power is dissipated through these resistances, the pumping energy is not lost, but is converted into heat.

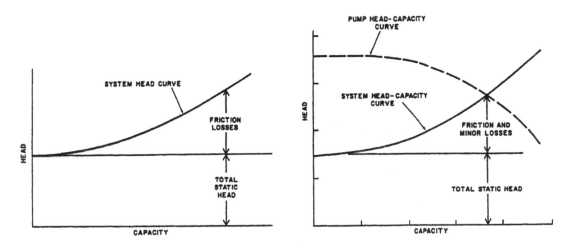

FIGURE 12.1   The system head-capacity curve.

As shown on the left of Figure 12.1, the system curve describes the nature of the process served. The head at any one capacity is the sum of the static and friction heads. The static head does not vary with flow rate, as it is only a function of the elevation or backpressure against which the pump is operating. The friction losses are related to the square of flow and represent the resistance to flow caused by pipe friction. A system curve tends to be flat when the piping is oversized and steep when the pipe headers are undersized. The friction losses also increase with age; therefore, the system curve for old piping tends to be steeper than for new piping.

Figure 12.2 illustrates the system curves of three different types of processes. Curve 1 corresponds to the closed-loop circulation of a fluid in a horizontal plane. Here there is no static head component at all, and the parabola that describes the system starts at zero. Curve 2 is the system

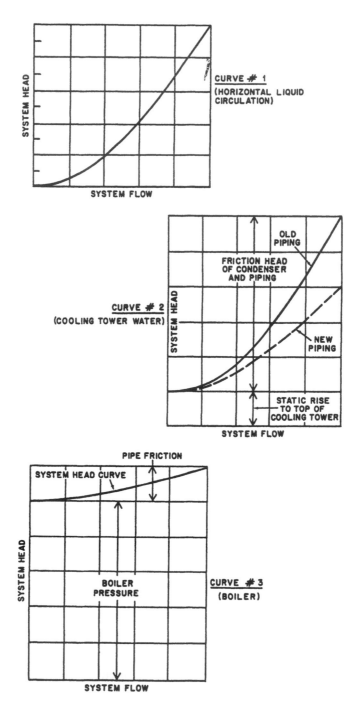

FIGURE 12.2 System curves vary as the static-head components change in various processes.

curve for a condenser water circulation network. Here a limited amount of static head is present, because the pump must return the water to the top of the cooling tower. This curve also illustrates that the friction losses tend to increase when the piping is no longer new. Curve 3 gives an example of a process dominated by static head, such as the feedwater pump of a boiler. This curve is flat and is insensitive to changes in system flow.

As will be discussed later in more detail, when the system curve is flat, there is little advantage to variable- or multiple-speed pumping, and the usual response to system flow variations is the stopping and starting of parallel pumps. Inversely, if the system curve is steep, substantial energy savings can be obtained from the use of booster, multiple-, or variable-speed pumps.

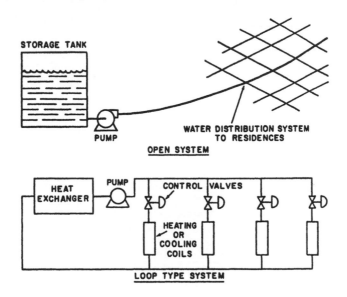

FIGURE 12.3   Pumping systems can be of the open or the loop type.

Hydraulic systems can be open (noncirculating) or loop type, as illustrated in Figure 12.3. Water supply and distribution systems in cities and buildings are typical open systems, whereas hot-and-chilled-water heating and cooling systems are typical loop systems.

Hydraulic systems must also be evaluated as to whether flow is restricted or unrestricted. Restricted-flow systems are those that include valves to regulate the flow through the system. Plumbing systems or hot-and-chilled-water systems are restricted-flow systems, because manual or automatic valves control the flow. Unrestricted-flow systems include sewage and storm water lift stations as well as municipal water flow into elevated storage tanks.

In actual systems, a single system head curve may not exist; often what exists is a system head band, as shown in Figure 12.4. This is because the distribution of active loads will shift the curve, as this figure shows.

A. WATER HAMMER

If a valve opening is suddenly reduced in a moving water column, this causes a pressure wave to travel in the opposite direction to the water flow. When this pressure wave reaches a solid surface (elbow, tee, etc.), it is reflected and travels back to the valve. If, in the meantime, the valve has closed, a series of shocks, sounding like hammer blows, results. An example can illustrate this phenomenon.

Assume that 60°F (20°C) water is flowing at a velocity of 10 ft/s in a 3 in. Schedule 40 pipe, and a valve located 200 ft downstream is suddenly closed. The pressure rise and the minimum acceptable time for valve closure can be calculated. If the valve closes faster than this limit, water hammer will result.

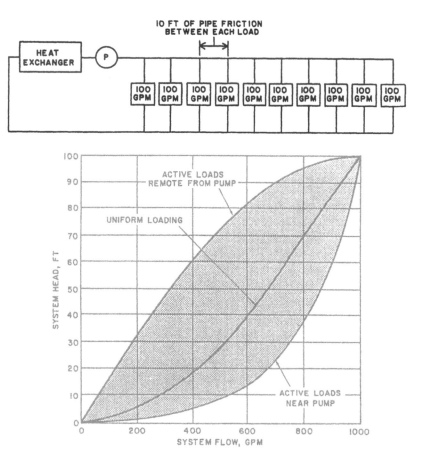

FIGURE 12.4 The system head drops when active loads are near the pump and it rises when the most remote loads are active.

In rigid pipe, the pressure rise (ΔP) is the product of water density (ρ in units of slugs/ft³), the velocity of sound (c in units of ft/s), and the change in water velocity (ΔV in units of ft/s). Therefore, the pressure rise can be calculated as follows:

$$\Delta P = -\rho c\,\Delta V = -(1.937)(4860)(-10)$$
$$= 94{,}138 \text{ lbf/ft}^2 = 653.8 \text{ PSI} \tag{1}$$

In order to prevent water hammer, the valve closure time (t) must exceed the ratio of two pipe lengths (2L) divided by the speed of sound:

$$t = 2L/c = (2)(200)/4860 = 0.0823 \text{ seconds} \tag{2}$$

Therefore, in this example, the valve closure should take more than 0.081 seconds.

The possible methods of preventing water hammer include (1) designing the system with low velocities, (2) using valves with slow closure rates, and (3) providing slow-closing bypasses around fast-closing valves, such as check valves. When water hammer is already present and the cause of it cannot be corrected, its symptoms can be treated (1) by adding air chambers, accumulators, or surge tanks; (2) by using surge suppressors, such as positively controlled relief valves; and (3) when water flows are split or combined, by using vacuum breakers to admit air and thereby cushion the shock resulting from the sudden opening or closing of the second split stream.

B. Process Dynamics and Tuning

In the pressure or flow controllers of pumping systems, the use of the integral mode is essential because these systems are noisy, multi-capacity processes. Usually, proportional and integral (PI) controllers are used, except in self-contained pressure and flow regulators, which are high gain, proportional-only devices. Regulator gains are around 20 (proportional band around 5%).

For PI controllers the typical gain setting is around 0.5 (PB = 200%), although, when individually tuned, the proportional band can range from 50 to 500%. The typical integral time setting is 0.1 minute per repeat (reset setting of 10 repeats per minute), although, if finely tuned, the reset might be between 5 and 60 repeats per minute. To remove noise, the measurement signal is usually filtered by a first-order lag. The filter time constant is set to a period, which is shorter than what the control loop is capable of responding to, for example, 0.02 seconds. When the PI algorithm is digitally implemented, the typical scan time is 1 second, which in the different applications can vary from 0.2 to 2.0 seconds.

When the final control element is a regulator, the inherent characteristic is usually quick opening, and when it is a control valve the recommended valve characteristic is linear.

The pressure-control loop is self-regulating. Its process dead time and time constant are both a fraction of a second (0.05 to 0.5), and therefore, the primary limitations to improved performance are the dead times of the control valve (dead band) and transmitter (smart), or the field bus update times, and the configured cycle time of the controller itself. The open loop gain of a liquid pressure or flow loop is between 0.5 and 2.

C. Pump Types and Characteristics

Table 12.1 gives a summary of the features and capabilities of the various pump designs.

There are two main pump designs: positive displacement pumps and centrifugal pumps. Vertical centrifugal units can pump water from depths up to 2000 feet (600 m), and horizontal units can transport process fluids from clear water to heavy sludge at rates up to 100,000 GPM (6.3 m³/s). The centrifugal designs are of either the radial-flow or the axial-flow type. Liquid enters the radial-flow designs in the center of the impeller and is thrown out by the centrifugal force into a spiral bowl. The axial-flow propeller pumps are designed to push rather than throw the fluid upward. Mixed-flow designs are a combination of the two.

In positive displacement pumps, a piston or plunger inside a cylinder is the driving element as it moves in reciprocating motion. The stroke length and thus the volume delivered per stroke is adjustable within a 10:1 range. Rangeability can be increased to 100:1 by the addition of a variable-speed drive. The plunger designs are capable of generating higher discharge pressures than the diaphragm types because of the strength limitation of the diaphragm. The strain on the diaphragm is reduced if it is not attached directly to the plunger but is driven indirectly through the use of a hydraulic fluid. Because solids will still settle in the pump cavities, these designs are all limited to relatively clean surfaces. For slurry service, the hose-type design is recommended. This design eliminates all the cavities, although the seating of the valves can still be a problem.

Other, less frequently used, pump designs include rotary screw pumps, which are ideally suited for sludge and slurry services, and air pumps, which use compressed air or steam to displace the accumulated liquids from tanks (Figure 1.42). The air system may use plant air (or steam), a pneumatic pressure tank, or an air compressor directly. With large compressors, a capacity of 600 gpm (2.28 m³/min) with lifts of 50 ft (15 m) may be obtained. The advantage of this system is that is has no moving parts in contact with the waste and thus no impellers to clog. Ejectors are normally more maintenance free and longer lived than pumps.

A special variation on pump designs are air lifts. In these, compressed air is blown into the bottom of a submerged updraft tube. As the air bubbles rise, they expand, reducing the density and therefore the hydrostatic head inside the tube. As the head is lower on the inside of the tube, the

TABLE 12.1
Pump Feature Summary

Type Designation	Type of Pump	For Liquid Pumped					For Capacity	For Feet of Head
		Clear Liquids—Low Viscosity	Clear Liquids—High Viscosity	Thin Slurries or Suspensions	Raw or Partially Treated Sewage and Heavy Suspensions	Viscous or Thick Slurries and Sludges	GPM*	PSIG*
A	Radial-flow centrifugals	✓		✓	✓(1)	✓(1)	——	——
B	Axial-flow and mixed-flow centrifugals	✓		✓	✓		——	——
C	Reciprocating pistons and plungers	✓(2)	✓	✓	✓(2)	✓	——	——
D	Diaphragm pumps	✓(2)	✓(2)	✓	✓	✓	——	——
E	Rotary screws		✓(2)		✓(3)	✓	——	——
F	Pneumatic ejectors				✓		——	——
G	Air lift pumps			✓	✓		——	——

GPM scale: 0.1 1.0 10 10^2 10^3 10^4 10^5

Feet of Head scale: PSIG* 0.1 1.0 10 10^2 10^3 / FEET OF HEAD (H_2O)* 0.1 1.0 10 10^2 10^3 10^4

✓ Suitable for normal use. (1) See text for limitations. (2) Not used for this purpose in environmental engineering (with some exceptions).
*GPM = 3.78 l/m, PSIG = 6.9 kPa G, feet = 0.3048 m.
Note: If not checked, either not suitable or not normally used for this purpose.

manometer effect induces the surrounding fluid to enter the updraft tube. Air is blown into the bottom of the submerged updraft tube, and as the air bubbles travel upward, they expand (reducing density and pressure within the tube), inducing the surrounding liquid to enter. Flows as great as 1,500 gpm (5.7 m³/min) may be lifted short distances in this way. This is an efficient and maintenance-free means of lifting large volumes of slurries over short distances of elevation (Figure 12.5).

1. Positive Displacement Pumps

Most frequently a piston or plunger is utilized in a cylinder, which is driven forward and backward by a crankshaft connected to an outside drive. Metering pump flows can readily be adjusted by changing the length and number of strokes of the piston. The diaphragm pump is similar to the piston type, except that instead of a piston, it contains a flexible diaphragm that oscillates as the crankshaft rotates.

Plunger and diaphragm pumps can feed metered amounts of chemicals (acids or caustics for pH adjustment) or can also pump sludge and slurries (Figure 12.6).

Reciprocating pumps, such as the piston and diaphragm types, deliver a fixed volume of fluid per stroke. The control of these pumps is based on changing the stroke length, changing the stroke speed, or varying the interval between strokes. In all cases, the discharge from these pumps is a pulsed flow, and for this reason they are not suited to control by throttling valves. In practice, the volume delivered per stroke is less than the full stroke displacement of the piston or diaphragm.

This hysteresis is a result of high discharge pressures or high viscosity of the fluid pumped. Under these conditions, the check valves do not seat instantaneously. A calibration chart must

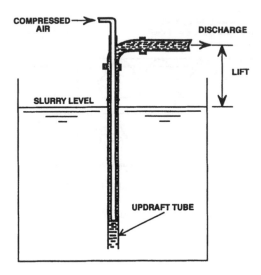

FIGURE 12.5   Airlift pump.

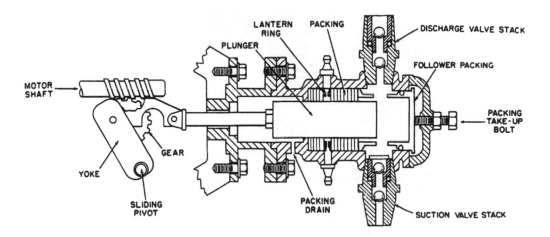

FIGURE 12.6   Plunger- or piston-type metering pump.

therefore be drawn for the pump under actual operating conditions. A weigh tank or level-calibrated tank is usually the reference standard. Since the discharge is a pulsed flow (Figure 12.7), it must be totalized and divided by the time interval to get average flow rate for a particular speed and stroke setting. Metering inaccuracy is approximately ±1% of the actual flow with manual adjustment and ±1.5% with automatic positioning.

In addition to multiple pumping heads, a pulsation dampener can be used on the pump discharge to smooth the flow pulsations. The pulsation dampener is a pneumatically charged diaphragm chamber that stores energy on the pump discharge stroke and delivers energy on the suction stroke, thus helping to smooth the flow pulses. In order to be effective, however, the dampener volume must be equal to at least five times the volume displaced per stroke (Figure 12.8).

When the purpose of a positive displacement pump is to meter the flow rate, certain precautions are needed. These include the removal of all entrained or dissolved gases, which otherwise can destroy metering accuracy. Figure 12.9 shows how entrained gases can be returned to the supply tank.

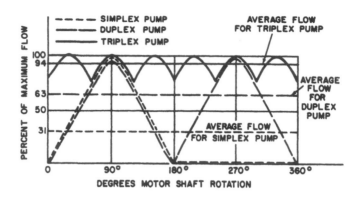

FIGURE 12.7   Flow characteristics of simplex and multiple plunger pumps.

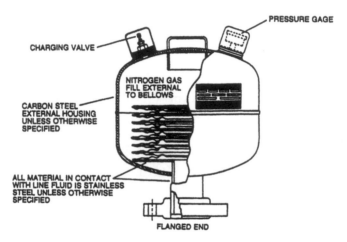

FIGURE 12.8   Pulsation dampener will suppress the pressure surges caused by positive displacement pumps. (Courtesy of The Metraflex Co.)

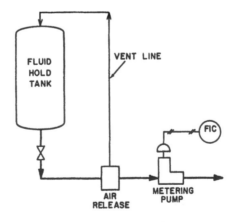

FIGURE 12.9   Elimination of entrained gases in metering pump installations.

Another concern in metering pump installations is to make sure that the discharge pressure is always higher than the suction pressure, so that fluid will not flow unrestricted through the pump. This can be guaranteed either through piping arrangement or by the addition of back pressure regulators to the pump discharge, as shown in Figure 12.10 and 12.11. If it is desirable to dampen

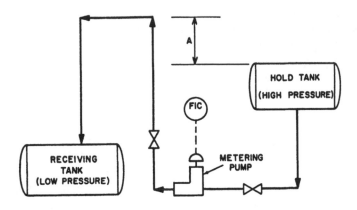

FIGURE 12.10   Piping arrangement to prevent through flow and siphoning.

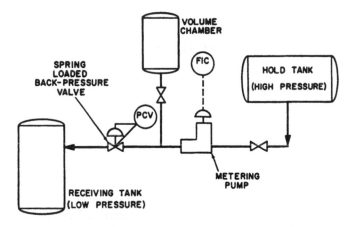

FIGURE 12.11   Metering pump with artificial head created by backpressure valve.

the pulsations, a volume chamber can also be added to the pump discharge. In order to be effective, the dampener volume must be equal to at least five times the volume displaced by each stroke.

It is always desirable to locate the pump below and near the fluid hold tank. Under these conditions the fluid will flow by gravity into the pump suction and loss of prime is unlikely. If the pump cannot be located below the hold tank, other measures must be taken to prevent loss of prime.

2. Rangeability and NPSH

Stroke length, and thus the volume delivered per stroke, is adjustable. The adjustment can be a manual indicator and dial, or for automatic control applications, a pneumatic actuator with positioner can be provided. Stroke adjustment alone offers operating flow ranges of 10:1 from maximum to minimum. Additional rangeability can be obtained by means of a variable-speed drive. A pneumatic stroke positioner used in conjunction with a variable-speed drive provides rangeability of at least 100:1. In the case of automatic stroke adjustment and variable speed, the pumping rate can be controlled by two independent variables, or the controller output can be "split ranged" between stroke and speed adjustment.

For blending two or more streams, several pumps can be ganged to one motor. Stroke length adjustment can be used to control the blend ratio, and drive speed can control total flow. However, in this case rangeability is sacrificed for ratio control.

Pumping efficiency is affected by leakage at the suction and discharge valves. These pumps are therefore not recommended for fluids such as slurries, which will interfere with proper valve seating or settle out in pump cavities.

In order for the pump to operate properly, the net positive suction head must be above the minimum practical suction pressure of approximately 10 PSIA (0.69 bar). The available net positive suction head (NPSHA) is given by Equation 3:

$$NPSHA = P - P_v \pm P_h - \sqrt{\left(\frac{lvGN}{525}\right)^2 + \left(\frac{lvC}{980Gd^2}\right)^2} \tag{3}$$

where P = feed tank pressure (PSIA), P_v = liquid vapor pressure at pump inlet temperature (PSIA), P_h = head of liquid above or below the pump centerline (PSID), l = actual length of suction pipe (ft), v = liquid velocity (ft/sec), G = liquid specific gravity, N = number of pump strokes per minute, C = viscosity (centipoise), and d = inside diameter of pipe (in.).

For liquids below approximately 50 centipoise (0.05 Pa·s), viscosity effects can be neglected, and Equation 3 reduces to

$$NPSHA = P - P_v \pm P_h - \left(\frac{lvGN}{525}\right) \tag{4}$$

The calculated value of NPSHA must be above the minimum suction pressure required by the selected pump.

D. CENTRIFUGAL PUMPS

The centrifugal pump is the most common type of process pump, but its application is limited to liquids with viscosities up to 3,000 centistokes (0.003 m²/s). The head-capacity curve is the operating line for the pump at constant speed and impeller diameter.

In the form of tall, slender, deep well submersibles, centrifugal pumps can pump clear water from depths greater than 2,000 ft (600 m). Horizontal centrifugals with volutes almost the size of a man can pump 9000 gpm (0.57 m³/s) of raw sewage through municipal treatment plants. Few applications are beyond their range, including flow rates of 1 to 100,000 gpm (3.78 lpm to 6.3 m³/s) and process fluids from clear water to all but the densest sludge.

Radial-flow pumps are designed to throw the liquid entering the center of the impeller or diffuser out into a spiral volute or bowl. The impellers may be closed, semi-open, or open, depending on the application (Figure 12.12). Closed impellers have higher efficiencies and are more popular than the other two types. They can readily be designed with nonclogging features. In addition, by using more than one impeller the lift characteristics can be increased. These pumps may be of horizontal or vertical design.

Axial-flow propeller pumps, although classed as centrifugals, do not truly belong in this category since the propeller thrusts rather than throws the liquid upward. Impeller vanes for mixed-flow centrifugals are shaped so as to provide partial throw, partial push of the liquid outward and upward. Axial-flow and mixed-flow designs can handle huge capacities but only at the expense of a reduction in discharge heads. They are constructed vertically.

Most water and wastes can be pumped with centrifugal pumps. It is easier to list the applications for which they are not suited than the ones for which they are. They should not be used for pumping (1) very viscous industrial liquids or sludges (the efficiencies of centrifugal pumps drop to zero, and therefore various positive displacement pumps are used), (2) low flows against very high heads (except for deep well applications, the large number of impellers needed put the centrifugal design

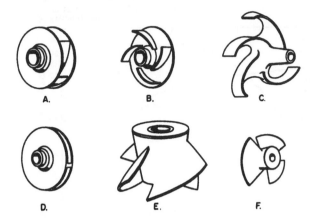

FIGURE 12.12 Types of centrifugal pump impellers: (A) closed impeller, (B) semi-open impeller, (C) open impeller, (D) diffuser, (E) mixed-flow impeller, (F) axial-flow impeller.

at a competitive disadvantage), and (3) low to moderate flows of liquids with high solids contents (except for the recessed impeller type, rags and large particles will clog the smaller centrifugals).

1. Pump Characteristics

Figure 12.13 illustrates the typical pump curves of a single impeller unit and also the more common formulas used in connection with pumping.

Pump efficiency is the ratio of the useful output power of the pump to its input power. It is calculated as follows:

$$E_p = \frac{\text{pump output}}{P_i} = \frac{SPQH_t}{P_i} \text{ (SI units)} \tag{5}$$

$$E_p = \frac{\text{pump output}}{bhp} = \frac{SPQH_t}{bhp \times 550} \text{(U.S. customary units)} \tag{6}$$

where E_p = pump efficiency, dimensionless, P_i = power input, kW (kN · m/s), SP = specific weight of water, lb/ft^3 (kN/m^3), Q = capacity, ft 3/5 (m 3/5) H_t = total dynamic head, ft (m), bhp = brake horsepower, and 550 = conversion factor for horsepower to ft · lb$_f$/s.

The typical range of pump efficiencies is from 60 to 85%.

The affinity laws describe the relationships among changes in speed, impeller diameter, and specific gravity. With a given impeller diameter and specific gravity, pump flow is linearly proportional to pump speed, pump discharge head relates to the square of pump speed, and pump power consumption is proportional to the cube of pump speed. This is why variable-speed pumps can be so highly energy efficient.

2. Design of Systems

In order to choose the proper pump, the conditions that must be known include capacity, head requirements, and liquid characteristics. To compute capacity, one should first determine the average flow rate for the system and then decide if adjustments are necessary. For example, when pumping wastes from a community sewage system, the pump must handle peak flows that are roughly two to five times the average flow, depending on the size of the community. Summer and winter flows

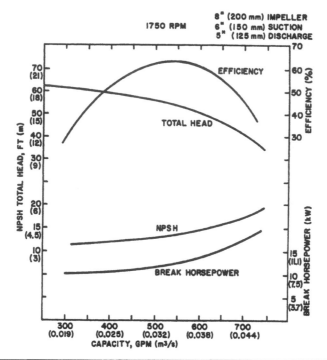

FIGURE 12.13 The operation of centrifugal pumps is described by common formulas.

Formula for	Conventional Units	SI Units
Head	$H = psi \times 2.31/SG^*$ (ft)	$H = kPa = 9.8023/SG^*$ (m)
Output power	$P_o = Q_v \times H \times SG^*/3960$ (hp)	$P_o = Q_v \times H \times SG^*/367$ (kW)
Shaft power	$P_s = \dfrac{Q_v \times H \times SG^*}{39.6 \times E_p}$ (hp)	$P_s = \dfrac{Q_v \times H \times SG^*}{3.67 \times E_p}$ (kW)
Input power	$P_i = P_s \times 74.6/E_m$ (kW)	$P_i = P_s \times 100/E_m$ (kW)
Equipment	(Constant speed pumps)	$E_e = E_p \times E_m \times 10^{-2}$
efficiency, %	(Variable speed pumps)	$E_e = E_p \times E_m \times E_v \times 10^{-4}$
Utilization efficiency, %	Q_D = design flow Q_A = actual flow H_D = design head H_A = actual head	$E_u = \dfrac{Q_D \times H_D}{Q_A \times H_A} \times 100$
System Efficiency Index [see Eq. 12]		$SEI = E_e \times E_u \times 10^{-6}$

*SG = specific gravity.

and future needs may also dictate capacity, and the population trends and past flow rates should be considered in this evaluation.

Head describes pressure in terms of feet of lift. It is calculated by the expression:

$$\text{head in feet} = \frac{\text{pressure (psi)} \times 2.31}{\text{specific gravity}} \qquad (7)$$

The discharge head on a pump is a summation of several contributing factors: static head, friction head, velocity head, and suction head.

Static head (h_d) is the vertical distance through which the liquid must be lifted (Figure 12.14).

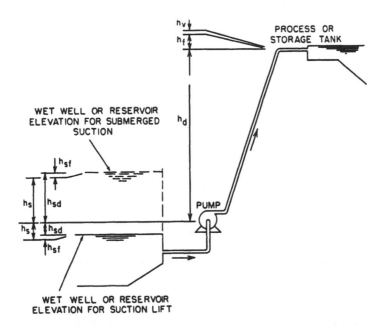

FIGURE 12.14   Determination of pump discharge head requirements. h_v = velocity head; h_f = friction head; h_d = static head; h_s = suction head; h_{sd} = suction side static head; h_{sf} = suction side friction head.

Friction head (h_f) is the resistance to flow caused by the friction in pipes. Entrance and transition losses may also be included. Since the nature of the fluid (density, viscosity, and temperature) and the nature of the pipe (roughness or straightness) affect the friction losses, a careful analysis is needed for most pumping systems, although for smaller systems, tables can be used.

Velocity head (h_v) is the head required to impart energy into a fluid to induce viscosity. Normally this is quite small and may be ignored unless the total head is low.

Suction head (h_s), if there is a positive head on the suction side (a submerged impeller), will reduce the pressure differential that the pump has to develop. If the water level is below the pump, the suction lift plus friction in the suction pipe must be added to the total pressure differential required.

Total head (H) is expressed by

$$H = h_d + h_f + h_v \pm h_s \tag{8}$$

The horsepower required to drive the pump is called brake horsepower. It is found by solving Equation 9:

$$BHP = \frac{\text{capacity (gpm)} \times H \text{ (ft)} \times \text{Sp. Gr.}}{3960 \times \text{pump efficiency}} \tag{9}$$

The shape of the head-capacity curve is an important consideration in pump selection (Figure 12.15). Curve 1 is referred to as a *drooping* curve, curve 2 is called a *flat* curve, and curve 3 would be considered *normal*. For on-off switching control, curves 1 and 2 are satisfactory as long as the flow is above 100 GPM (6.3 l/s). Below this flow rate, curve 1 allows for two flows to correspond to the same head, and curve 2 may drop to zero flow to obtain a small head increase. Both are therefore unstable in this region. Curve 3 is stable for all flows and is best suited for throttling service in cases in which a wide range of flows is desired.

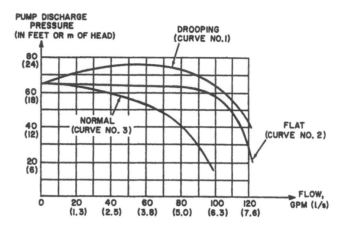

FIGURE 12.15   The shapes of centrifugal pump curves vary with their design.

3. NPSH Calculation

When liquids are being pumped, it is important to keep the pressure in the suction line above the vapor pressure of the fluid. The available head measured at the pump suction is called the *net positive suction head* (NPSH). A pump at sea level that is pumping 60°F water, that has a vapor pressure of $h_{vp} = 0.6$ ft, and that is operating under a barometric pressure of 33.9 ft has an available NPSH of 33.9 − 0.6 = 33.3 ft. As shown in Figure 12.16, the available NPSH increases with barometric pressure and with static head, and it decreases as vapor pressure, friction or entrance losses rise. Figure 12.16 also illustrates the difference between *available* and *required* NPSH. Available NPSH is the characteristic of the process and represents the difference between the existing absolute suction head and the vapor pressure at the process temperature. The required NPSH, on the other hand, is a function of the pump design. It represents the minimum margin between suction head and vapor pressure at a particular capacity that is required for pump operation. If this minimum NPSH is not available, the pump will fail to generate the required suction lift and flow will stop.

In order for the pump to be able to "pull in" the pumped fluid, the net positive discharge head available (NPSHA) for the particular installation must be greater than the NPSH which the pump requires. The NPSH values for the particular pump are obtained from the pump curve (Figure 12.13), while the available NPSH is calculated according to the following equation:

$$\text{NPSHA} = h_b \ (+ \text{ or } -) \ h_s - h_{vp} - h_f \tag{10}$$

where h_b = the absolute barometric pressure (in feet) at the surface of the source of the pumped liquid. If the source is atmospheric, $h_b = 33.96$ feet; h_s = the static head of the installation, which is the vertical distance between the pump inlet and the surface of the liquid on the supply side. It is positive if the liquid level is above the pump inlet and it is negative if it is below; h_{vp} = the vapor pressure of the pumped fluid (in feet) at the operating temperature, and $h_{vp} = h_b$ when the liquid reaches its boiling point; h_f = the suction side friction head in fet. This term increases with the square of flow and reflects the pressure drop through all pipes, valves, and fittings on the suction side of the pump.

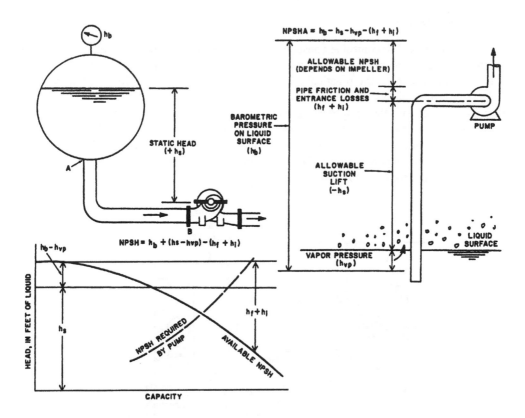

FIGURE 12.16 The available NPSH increases with barometric pressure and static head, and decreases as vapor pressure, friction, or entrance losses rise.

If it is desired to convert NPSHA from feet to PSI, the NPSHA given in feet should be multiplied by the specific gravity of the fluid and should be divided by 2.31. (When converting to metric units, ft = 0.3048 m and PSI = 0.069 bar.)

It is generally sufficient to calculate the available NPSH at maximum flow rate, because at that flow, the suction side friction head (h_f) is maximum and the NPSHA value is likely to be minimum. The required NPSH (NPSHR) of the pump rises with flow (Figure 12.13). Therefore, if NPSHA exceeds NPSHR when the flow is maximum, it will exceed it by an even greater margin as the flow drops.

II. BASIC CONTROLS

A. METERING PUMP CONTROLS

1. On-Off Control

Once the pump has been calibrated, it can be programmed via a timer or counter to deliver a known volume of fluid to a process. The pump may deliver one full stroke and then stop until the next electrical signal is received from the timer or may continuously charge a specified number of pump strokes and then be shut down by a counter. The flow may be smoothed by a pulsation dampener and the system and pump protected from overpressure by a relief valve.

2. Throttling Control

Continuously variable flow control at constant speed may be accomplished by automatic adjustment of the stroke length. The range of flow control by stroke adjustment is 0 to 100%. However, in order to maintain accuracy, the practical range is 10 to 100% of design flow. The flow is related to stroke length through system calibration.

In some applications the reciprocating pump combines the measuring and control functions, receiving no independent feedback to represent flow. In other cases it is used as a final control element, and flow detection is performed by an independent sensor. Figure 12.17 illustrates an installation in which the pump is both the measuring and the control device for ratio control and is provided with automatic calibration capability.

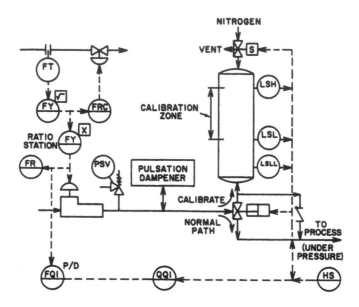

FIGURE 12.17   Ratio and calibration controls for reciprocating pump. When level has reached LSH, the three-way valve returns to the "normal" path and nitrogen enters the tank to initiate discharge. When level drops to LSLL, discharge is terminated by venting off the nitrogen. Counter QQI is running while the rising level is between LSL and LSH. Total count, when compared with known calibration volume, gives total error. Hand switch HS initiates calibration cycle by diverting the three-way valve to the "calibrate" path.

Variable-speed control is usually applied in multiple-head pumps in which all the pumps are coupled mechanically. A control signal may adjust all flows in the same proportion simultaneously. The rangeability and accuracy of flow control depend on the method of speed variation chosen.

B. ROTARY PUMP CONTROLS

The typical pump characteristics for a rotary pump, such as the gear, lobe, screw, or vane type, show a fairly constant capacity at constant speed with large changes in discharge pressure. This is shown in Figure 12.18. These pumps cover the viscosity range from less than 1 centipoise up to 500,000 centipoises (0.001 Pa·s up to 500 Pa·s). The usual application of this type of pump is for the highly viscous liquids and slurries that are beyond the capabilities of centrifugal pumps.

1. On-Off Control

The criterion for the maximum number of starts per hour must be checked carefully for motors on rotary pumps, since the starting torque may be very large (as a result of fluid viscosity), as is the

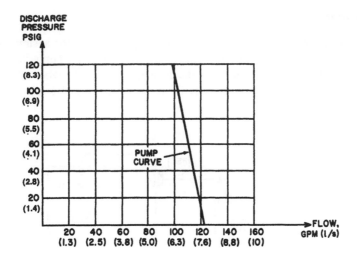

FIGURE 12.18   Rotary pump curve.

inertia load of the column of fluid in the piping, which accelerates under positive displacement each time the pump starts.

In slurry service, on-off control creates problems caused by settling of solids, and is therefore not recommended. Instead, a circulating loop is used with a pressure-controlled bypass back to the feed tank. For example, intermittent flow to feed a centrifuge is obtained by opening an on-off valve via a signal from the cycle timer. Such a loop is shown in Figure 12.19. The pressure-controlled bypass allows the normal pump flow to be maintained in the loop when the centrifuge feed valve is closed.

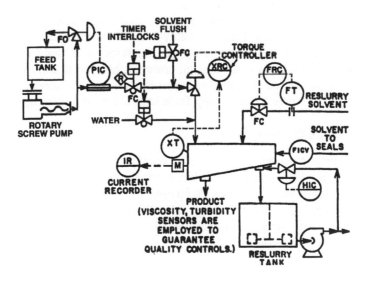

FIGURE 12.19   Pressure-controlled bypass is used on slurry service.

The on-off control in this case is applied to the fluid rather than to the pump motor.

Manual on-off control is often applied to rotary pumps in bulk storage batch transfer services with local level indication.

2. Throttling Control

A safety relief valve is always provided on a rotary pump to protect the system and pump casing from excessive pressure should the discharge line be blocked while the pump is running. The relief valve may discharge to pump suction or to the feed tank. In cases when slurries and viscous materials may not be able to pass through the relief valve, a rupture disk is placed on the discharge line (Figure 12.20).

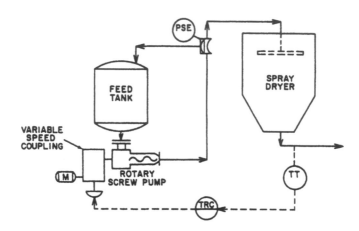

FIGURE 12.20   Speed control of rotary pumps.

The output of a rotary pump may be continuously varied to suit the process demands by use of a pressure-controlled bypass in combination with a flow control valve. The bypass is necessary to accommodate changes in flow to the process, since the total flow through the bypass plus to the process is constant at constant speed.

The capacity of a rotary pump is proportional to its speed, neglecting the small losses due to slippage. The flow can therefore be controlled by speed-modulating devices.

In this case, no bypass is needed. The rangeability is limited by the speed-control device, which for pulleys and magnetic drives is approximately 4:1. The response is slightly slower than with a control valve because of the inertia of the system; however, this type of control would be favored when it is desired to avoid the use of control valves on slurry or gummy services. Such installation is shown in Figure 12.20, where a screw pump feeds latex slurry to a spray dryer.

C. CENTRIFUGAL PUMP CONTROLS

Capacity control of pumps must recognize the incompressibility of liquids, for which reason changes in the volume rate of flow throughout the system occur simultaneously as density is constant at constant temperature, regardless of pressure. Pump capacity may be affected by (1) a control valve in the discharge of a pump, (2) on-off switching, or by (3) variation in the speed of the pump. Flow control by on-off switching provides only zero or full flow, whereas the other control methods provide adjustable flows in the system.

1. Two-Speed

The two-speed motor can be utilized on simple pumping systems in which accurate control of pump pressure is unnecessary. Typical applications of this drive are storm and sewage lift stations that have appreciable pipe friction, as shown in curve 1 in Figure 12.2. They should not be used in systems with high static head and low friction, such as curve 3 in Figure 12.2. This demonstrates that it is necessary to carefully calculate the system head curve before the two-speed motor is used. Where it is acceptable, an efficient, low-cost variable-speed system can be achieved.

Two-speed motors can be utilized on simple pumping systems in which accurate control of pump pressure is unnecessary. For mostly friction systems, such as curve 2 in Figure 12.2, they can offer a reasonably efficient and inexpensive alternative to variable-speed pumping. Standard two-speed motors are available with speeds of 1750/1150 rpm (29 and 19 r/s), 1750/850 rpm (29 and 14 r/s), 1150/850 rpm (19 and 14 r/s), and 3500/1750 rpm (58 and 29 r/s) (Figure 12.21).

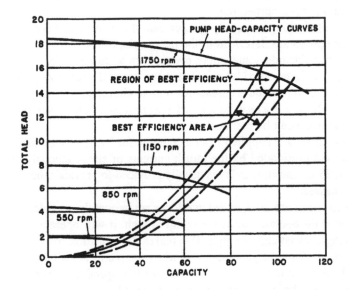

FIGURE 12.21 Several speed combinations are available in two-speed pumps.

a. On-off level control

Figure 12.22 illustrates the use of the level switch for on-off pump control. In order to prevent overheating of the motor, the number of pumps starts should not exceed 15 per hour. This can be checked by calculation of the filling and emptying times. At a feed rate of 75 GPM (4.7 l/s), it takes 13 minutes to fill the 1,000-gallon (3,785 l) volume between LSL and LSH, when the discharge pump is off. The pump capacity being 100 GPM (6.3 l/s), it takes 40 minutes to discharge this same volume while the feed is on. Therefore the pump will start approximately once an hour and will run 75% of the time.

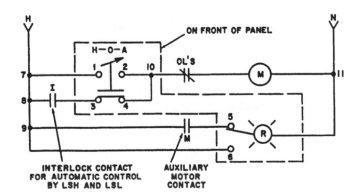

FIGURE 12.22 On-off pump-down interlocks often utilize two-probe conductivity switches.

Therefore the 1,000-gallon (3785 l) tank is satisfactory. The switch prevents tank overflow and dry pumping, which would damage the seals. Since the pump starts once an hour, it will not overload the motor by excessive starts and stops.

The bottom portion of Figure 12.22 describes a simple pumps starter. It is controlled by the three-position hand-off-automatic switch. When placed in automatic (contacts 3 and 4 connected), the status of the interlock contact I determines whether the circuit is energized and whether the pump is on. The purpose of the auxiliary motor contact M is to energize the running light R whenever the pump motor is on. The parallel hot lead to contact 6 allows the operator to check quickly to see if the light has burned out.

The amount of interlocking is usually greater than that shown in Figure 12.22. Some of the additional interlock options are described in the paragraphs below.

Most pump controls include some safety overrides in addition to the overload (OL) contacts shown in Figure 12.22. These overrides might stop the pump if excessive pressure or vibration is detected. They usually also provide a contact for remote alarming.

Most pump controls include a reset button that must be pressed after a safety shutdown condition is cleared before the pump can be restarted.

When a group of starters is supplied from a common feeder, it may be necessary to make sure that the starters are not overloaded by the inrush currents of simultaneously started units. If this feature is desired, a 25-second time delay is usually provided to prevent other pumps from starting until that time has passed.

When two-speed pumps are used, added interlocks are frequently provided. One interlock might guarantee that even if the operator starts the pump in high, it will operate for 0 to 30 seconds in low before advancing automatically to high. This makes the transition from off to high speed more gradual. Another interlock might guarantee that when the pump is switched from high to low, it will be off the high speed for 0 to 30 seconds before the low-speed drive is engaged. This will give time for the pump to slow before the low gears are engaged.

If the same pumps can be controlled from several locations, interlocks are needed to resolve the potential problems with conflicting requests. One method is to provide an interlock to determine the location that is in control. This can be a simple local/remote switch or a more complicated system. When many locations are involved, conflicts are resolved by interlocks that either establish priorities between control locations or select the lowest, highest, safest, or other specified status from the control requests. In such installations with multiple control centers, it is essential that feedback be provided so that the operators are always aware not only of the actual status of all pumps but also of any conflicting requests coming from the other operators or computers.

Large pumps should also be protected from overheating caused by frequent starting and stopping. This protection can be provided by a 0- to 30-minute time delay guaranteeing that the pump will not be cycled at a frequency faster than the delay time setting.

When dual pumps are used an alternator is interposed between the level switches and the motors. The alternator places the pumps in service in an alternating sequence. Thus, if two pumps are used, each will start half as many times per hour, permitting the tank size to be reduced. This application is shown in Figure 12.23, where a cooling water return sump is illustrated. The minimum sump size is to be found allowing 10 starts per pump per hour. In this system, each pump is designed to handle the total feed flow into the sump by itself. The pumps do not operate together except when abnormally high flow rates are charged into the sump.

The design is for 10 starts per hour for each pump. Because there are two pumps, the sump capacity can allow 20 starts per hour, or 3 minutes per empty and refill cycle. Let x = sump capacity. The minimum sump capacity is calculated by allowing the sum of feed (4000 GPM) and discharge (5000 − 4000 = 1000 GPM) times to equal 3 minutes.

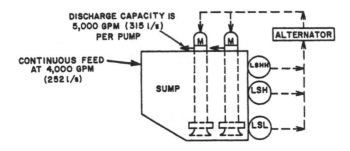

FIGURE 12.23   On-off level control of dual pump station.

$$\frac{x}{4000} + \frac{x}{1000} = 3 \text{ minutes}$$

$$x = 2400 \text{ gallons} \left(9.09 \text{ m}^3\right)$$

(11)

In addition to the reduction of sump volume, a further advantage is that a spare pump is available for emergency service. The purpose of a LSHH level switch is to put both pumps on line at once should the continuous feed flow exceed 5000 GPM (0.32 m³/s), resulting in the rise of the level to this point. The combined pump flow of 10,000 GPM (0.63 m³/s) will then prevent overflow of the sump. When the level drops to LSL, both pumps will be off, and the normal alternating cycle is resumed.

b.   On-off flow control

Illustrated in Figure 12.24 is a tandem pump arrangement that responds to varying flow demands on pump outlet. Pump I is normally operating at point (1) (at 80 GPM and 36 ft, or 5 l/s and 10.8 m). When flow demand increases to 120 GPM (7.6 l/s), the head drops to 22 ft at point (2), and FSH starts pump II. The combined characteristic gives 120 GPM at 40 ft (7.6 l/s at 12 m) at point (3). In this control scheme, a wide range of flows is possible at high pump efficiency without serious loss in pressure head.

When two or more pumps operate in parallel, the combined head-capacity curve is obtained by adding up the capacities at the same discharge head, as illustrated in Figure 12.25. The total capacity of the pump station is found at the intersection of the combined head-capacity curve with the system head curve. This point also gives the head at which each of the pumps is operating. If the selection is to be very accurate, the head-capacity curves should be modified by substituting the station losses (the friction losses at the suction and discharge of the individual pumps) so that the resulting "modified head" curve will represent the pump plus its valving and fittings.

When constant-speed pumps are used in parallel, the second pump can be started and stopped automatically on the basis of flow. In this case, an adjustable dead band is provided in the flow switch (FSH in Figure 12.24); this dead band starts the second pump when the flow rises to 120 GPM (7.6 l/s) but will not stop it until it drops to 100 GPM (6.3 l/s). This prevents excessive cycling. The addition of an alternator can equalize the running times of the two pumps. In Figure 12.24, the normal operation of the system is represented by point 1. If the load rises, the single pump meets it until point 2 is reached. At this load, the second pump is started and the system operates at point 3. When the load drops off, if will pass through point 3 without any effect and the second pump is not stopped until the load drops to point 4. At this load, the second pump is turned off and the system operation is returned to point 5 using only one pump.

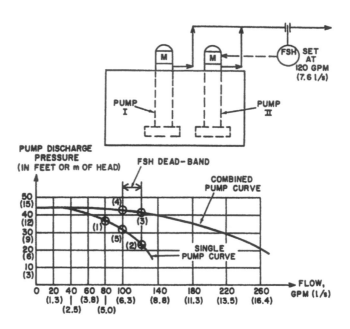

FIGURE 12.24 On-off flow control can be used with parallel pumps.

c. On-off pressure control: booster pumps

A pressure switch may be used to start a spare pump in order to maintain pressure in a critical service when the operating pump fails. In this case, a low pressure switch would be used to actuate the spare pump, piped in parallel with the first pump. A second function, as illustrated in Figure 12.26, is to boost pressure. Pump I is normally operating at point (1). When the discharge pressure rises to 50 ft (15 m) at point (2), the flow is reduced from 53 to 20 GPM (3.3 to 1.3 l/s). At this point the pressure switch will start pump II and close the bypass valve. The system will now operate at point (3) on the combined characteristic, delivery 60 GPM at 50 ft (3.8 l/s at 15 m) of head.

When a booster pump is added to a main fed by several *parallel* pumps, the total head-capacity curve is obtained by adding the booster curve to the modified head of the parallel pumps at each capacity point. Series pumping is most effective when the system head curve is steep, such as in Figure 12.27. With such mostly friction loads, series pumping can substantially reduce the over-pressure at low loads. Therefore, booster pumps or two-speed pumps (Figure 12.21) can both be considered for the same kind of system curve. Multiple pumps in series are preferred from an operating cost point of view, but a single two-speed pump represents a lower capital investment.

When constant-speed pumps are used, the booster pump can be started and stopped automatically on the basis of pressure. In this case, an adjustable dead band is provided in the pressure switch. The normal system operation can be represented by point 1 in Figure 12.27. As the load increases, the pump discharge head drops to the set point of PSL at point 2 and the booster pump is automatically started. Now the system operates to the right of point 3 until the load drops off again. The booster stays on as the load drops below point 3 until the PSL turns it off at point 4. At this point, the system is returned automatically to the single pump operation at point 5. The dead band in the PSL prevents the on-off cycling of the booster pump at any particular load. The width of the dead band is a compromise: as the band is narrowed, the probability of cycling increases, and widening the band results in extending the periods when the booster is operated unnecessarily. If the pumps are identical, their running times can be equalized by alternating between them so that the pump with the higher running time will be the one that is stopped.

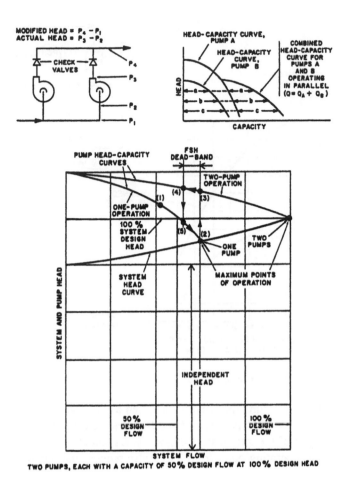

FIGURE 12.25 Multiple pumps in parallel can eliminate overpressure at low flows.

2. Throttling Control

For good controllability, the control valve is usually sized to pass the design flow with a pressure drop equal to the system dynamic friction losses excluding the control valve but not less than 10 PSID (0.69 bar) minimum.

Pump flow is controlled by varying the pressure drop across the valve. This relationship is shown at point (1) in Figure 12.28. When the flow is throttled to 15 GPM (0.95 l/s) at point (2), the control valve must burn up a differential of 50 PSID (3.5 bars). However, the pump must not be run at zero flow or overheating will occur and the fluid will vaporize, causing the pump to cavitate. To avoid this, a bypass line can be provided with a backpressure regulator. It is set at a pressure that will guarantee minimum flow as the pump is throttled toward zero flow.

One can calculate the minimum flow needed through the pump to prevent vaporization from a heat balance on the pump by assuming that the motor horsepower is converted to heat. If this flow were calculated to be 20 GPM (1.3 l/s), then the PCV in the bypass line would be sized to pass 20 GPM with a corresponding set pressure of 63 PSIG (4.35 bars).

If there is no throttling valve in the system, a constant-speed pump will always operate at the intersection of the pump and the system curve (point 2 in Figure 12.29). If the flow is controlled by throttling a control valve on the pump discharge, the pump will operate at point 1, and the differential head between point 1 and 3 will be burned up in the form of pressure drop. If the system flow is reduced, the energy wasted in the form of valve pressure drop will increase.

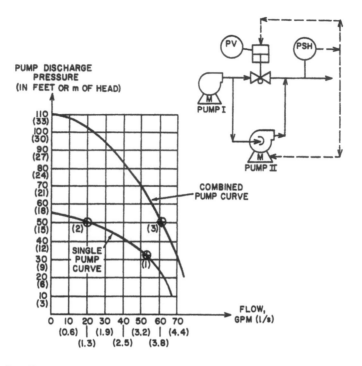

FIGURE 12.26 On-off pressure control of pumps.

In Figure 12.29, the useful pumping head is identified as H_s, and the actual pumping head of the throttled system is given as H_p. The ($H_p - H_s$) difference identifies the energy wasted through throttling. This is only part of the total loss of efficiency. As can also be seen in the lower part of Figure 12.29, moving the operating point from 2 to 1 also reduces the pump efficiency from 81% to 71%.

a. Energy cost of throttling

If the unit cost of electricity is 6¢/kWh, to calculate the energy cost of throttling control, the following rule of thumb can be used: It costs $0.5/year for a flow of 1.0 GPM to overcome a differential of 1.0 PSID. The example illustrated in Figure 12.30 calculates the yearly waste of pumping energy at different loads. If the FRC is set for 625 GPM (set point 2), the pump operates at point 2 on its curve and the kickback PCV is closed, because the pump discharge head at point 2 is less than the PCV set point of 90 PSIG. Therefore, the cost of throttling between points 2 and 3 can be calculated as follows:

$$\text{(FRC set point)}(\Delta P_{2-3})(0.5 \text{ \$/yr}) = (625)(40)(0.5) = \$12,500/\text{yr}$$

If, in this same example, the FRC set point is reduced to 250 GPM, the pump is throttled back on its curve to point 4. Because the pump head at point 4 is higher than the set point of the PCV, it will open and bypass enough flow to reduce the pump head to 90 PSIG. Therefore, in this case, the pump will operate at point 1. As the flow at point 1 is below that at point 2, it is fair to assume that the energy waste will be greater under these conditions. To calculate the amount of waste, the energies burned up in both the PCV and FCV must be added as shown below:

$$\text{(FRC set point) } (\Delta P_{1-5})(0.5 \text{ \$/yr}) + \text{(PCV flow) } (\Delta P_{0-1}) (0.5 \text{ \$/yr})$$
$$= (250) (73) (0.5) + (300) (90) (0.5) = 9,125 + 13,500 = \$22,625/\text{yr}$$

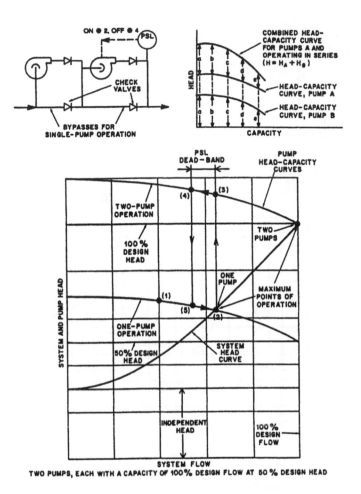

FIGURE 12.27 Multiple pumps in series are effective when the system head curve is steep. The two pumps illustrated by the lower graph are each capable of generating 100% design flow at 50% design head.

For good controllability, the control valve is usually sized to pass the design flow with a pressure drop equal to the system friction losses excluding the control valve but not less than 10 PSID (0.7 bar) minimum. Pump flow is controlled by varying the pressure drop across the valve. However, the pump must not be run at zero flow or overheating will occur and the fluid will vaporize, causing the pump to cavitate. To avoid this, a bypass line can be provided with a backpressure regulator. This regulator is set at a pressure that will guarantee minimum flow as the pump is throttled toward zero flow.

The minimum flow required to prevent overheating is calculated by assuming that the motor horsepower is converted to heat. One rule of thumb calls for a flow of 1.0 GPM for each pump horsepower to prevent overheating. If a minimum flow bypass cannot be provided, then a safety interlock should be furnished. In this interlock, a minimum flow switch is used to stop the pump on low flow.

b. Pump speed throttling

Flow control via speed control is less common than throttling with valves, because most AC electric motors are constant-speed devices. If a turbine drive is used, speed control is more convenient. An instrument air signal to the governor can control speed within ±.5% of the set point. In order to vary pump speeds with electric motors, it is generally necessary to use a variable-speed device in the power transmission train. Variation of the pump speed generates a family of head-capacity

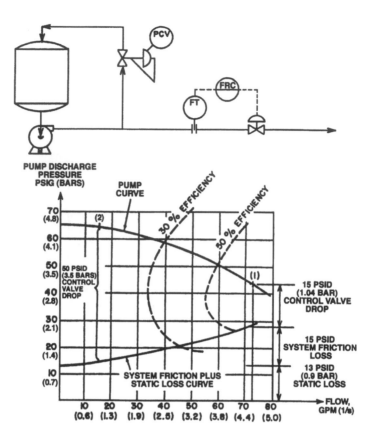

FIGURE 12.28 Throttling control of centrifugal pump.

curves, as shown in Figure 12.31, in which the volume flow is proportional to speed if the impeller diameter is constant. The intersection of the system curve with the head curve determines the flow rate at points 1, 2, or 3.

Because the area of peak pump efficiency falls on a parabolic path (Figure 12.31), speed throttling will usually not reduce the pump efficiency as much as valve throttling (Figure 12.29). This increases the total energy saving obtained from pump speed control. As shown in Figure 12.32, when the flow is reduced from F_1 to F_2, instead of wasting the excess pump head of $(P_1 - P_2)$ in pressure drop through a valve, that pump head is not introduced in the first place. Thereby, speed throttling saves energy that valve throttling would have wasted.

c. When to consider variable-speed pumps

The shape of the system curve determines the saving potentials of variable-speed pumps. All system head curves are parabolas ($H \sim Q^2$), but they differ in the steepness of these curves and in the ratio of static head to friction drop. As shown in Figure 12.33, the value of variable-speed pumping increases as the system head curve becomes steeper.

Studies indicate that in *mostly friction* systems (such as zone 4 in Figure 12.33), the savings represented by variable-speed pumping will increase with reduced pump loading. If, on the yearly average, the pumping system operates at not more than 80% of design capacity, the installation of variable-speed pumps will result in a payback period of approximately three years.

The zones in Figure 12.33 are defined by the H_t/H_s ratio. The higher this number, the higher the zone number and the more justifiable the use of variable-speed pumps. Figure 12.34 illustrates how the H_t/H_s ratio is calculated. The shaded areas identify the energy-saving potentials of variable-speed pumps. The values of H_t and H_s are identified by determining the average yearly flow rate

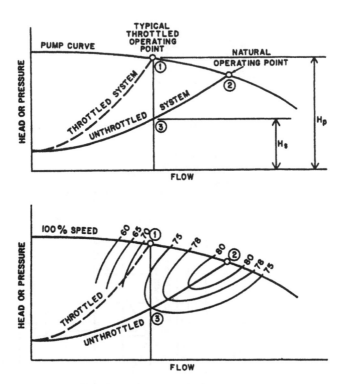

FIGURE 12.29 The throttled system not only wastes pumping energy through valve pressure drop but also operates at a less efficient point on the pump curve.

(Fa) and determining its intersections with the pump and system curves. The larger the shaded area in Figure 12.34, the higher the H_f/H_s ratio will be and, therefore, the shorter the payback period for the use of variable speed pumps is likely to be.

d. Calculating the savings

As was shown in Figure 12.30, the savings resulting from variable-speed pumping can be calculated at any operating point on the pump curve. Based on this information, the relationship between the demand for flow and the power input required can be plotted. Figure 12.35 shows these curves for both control valve throttling and variable-speed pumping systems. The difference between the two is the savings potential.

Once the savings curve shown in Figure 12.35 has been established, the next step is to determine the operating cycle. The operating cycle identifies the percentages of time when the load is 10%, 20%, etc., up to 100% (Figure 12.36). When both the savings and the operating curves are established, all that needs to be done is to incrementally calculate the savings as follows:

Knowing the total horsepower of the pumps and the cost of electricity makes it possible to convert the resulting percentages into yearly costs. For example, if the cost of electricity is 6¢/kWh, then the yearly savings of each 100-horsepower increment has the value of $39,000.

e. Variable-speed drives

All variable-speed drives should be reliable and serviceable and should provide repeatable control and wide turndown at all loads. Variable-speed pumps must operate within the speed range of 50 to 100%, but the drive must also be able to function at speeds down to zero without damage or overheating.

Variable-speed pump drives are either electrical or electromechanical. The electrical ones can be of a number of types: two-speed, direct current, variable voltage, variable frequency (current

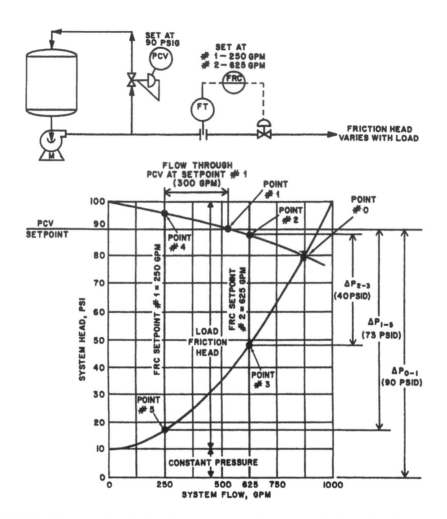

FIGURE 12.30   This example calculates the energy waste due to pump discharge throttling.

source or pulse width modulated), and wound rotor regenerative. The common feature of all of these designs is that they depend on altering the characteristics of the electricity to the motor in order to change its speed. The electromechanical drives, on the other hand, utilize a speed-changing device that is interposed between the motor and the pump. These designs include eddy-current couplings, hydraulic or hydroviscous drives, and V-belt drives. The electromechanical drives are generally less efficient than their electrical counterparts. The variable voltage electric drive using high slip motors is an exception to this rule, as it is less efficient than the mechanical drives.

The *variable frequency drives* are among the most reliable and efficient units, having turndown ratios of up to 3:1. They convert alternating current at 60 Hz to direct current and back to alternating at 0 to 120 Hz. As the frequency sent to the standard squirrel-cage type induction motor changes—say, from 60 Hz to 40 Hz—the pump speed is correspondingly reduced from 100% to 60%. These drives are relatively expensive and sophisticated devices (Tables 12.2 and 12.3). They are normally available in sizes up to 250 HP (185 kW), with specially designed units being available in sizes up to 2000 HP (1491 kW).

The *wound rotor regenerative drives* give one of the best efficiencies, although their rangeability is limited to 2:1. In the conventional wound rotor design, the motor speed is altered by varying the resistance in the rotor. This is rather inefficient. In the regenerative design, the resistors are eliminated, and through the use of rectifiers and inverters the excess power is returned to the supply.

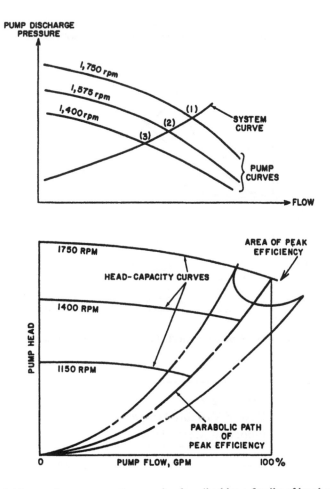

FIGURE 12.31   Variable-speed pump operation can be described by a family of head-capacity curves.

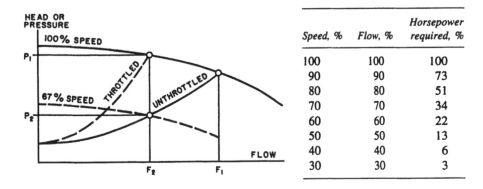

Speed, %	Flow, %	Horsepower required, %
100	100	100
90	90	73
80	80	51
70	70	34
60	60	22
50	50	13
40	40	6
30	30	3

FIGURE 12.32   Instead of wasting the unnecessarily introduced pump energy, speed is reduced so that such energy is not introduced in the first place.

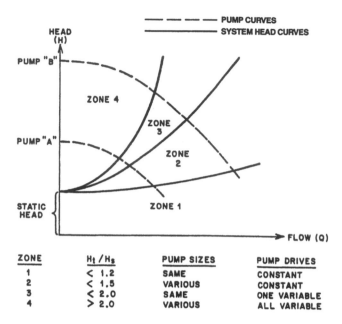

FIGURE 12.33 Pumps and drives should be selected as a function of the steepness of the system curve.

TABLE 12.2
Cost of Variable-Frequency Induction Motor Drives

Power Rating	PWM	Current-fed ASCI
10 Hp	$3500	
20 Hp	$4500	
50 Hp	$7000	
100 Hp	$11,000	$12,000
200 Hp	$15,000	$15,000
500 Hp		$32,000
1000 Hp		$75,000

TABLE 12.3
Cost of Tyristor DC Drives

Power Rating	Single Converter	Dual Converter
100 Hp	$7300	$8600
200 Hp	$10,000	$13,000
500 Hp	$22,000	$28,000
1000 Hp	$24,000	$32,000

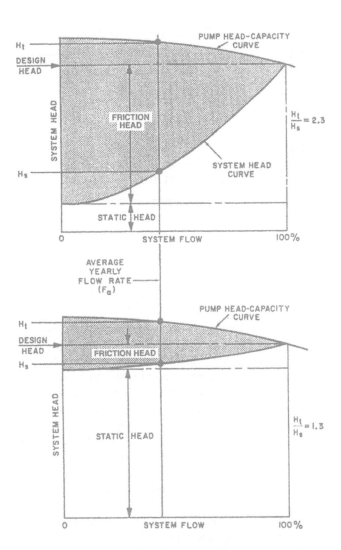

FIGURE 12.34 The higher the H_f/H_s ratio, the more justifiable the use of variable-speed pumps.

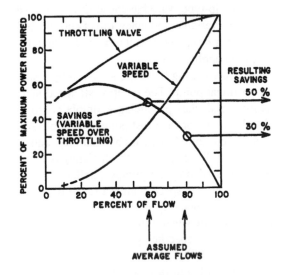

FIGURE 12.35 The savings due to variable-speed pumping increase as the flow rate drops off.

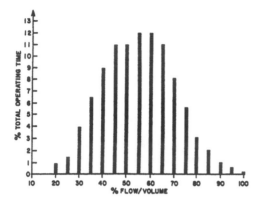

% Flow	% of Time	% Savings	Total Savings (%)
Below 25	2.5	57	1.4
25–35	10	60	6.0
35–45	20	56	11.2
45–55	23	52	11.9
55–65	23	50	11.5
65–75	14	40	5.6
75–85	5.5	30	1.6
85–95	1.5	15	0.2
95–100	0.5	0	0
	100%		49.4%

FIGURE 12.36   The pump operating cycle identifies the percentages of time when load is 10 percent, 20 percent, etc.

Sizes vary from 25 to 10,000 HP (20 to 7,455 kW). The wound rotor type motor is more expensive than the induction type, and the brushes used in the regenerative design require replacement every three to four years.

Direct current drives were among the first variable-speed drives using wound rotors. In combination with silicon-controlled rectifiers (SCRs), they provide stepless speed control at high efficiency and modest first cost. As the armature rotates at right angles to the magnetic field established by the field coils, rotation can be reversed easily by reversing polarity. An advantage of these drives is the ability to maintain constant speed while the load varies. Disadvantages include the limited availability of DC pump motors, the difficulty in preventing explosions resulting from sparking, and the approximately yearly replacement requirement of the carbon brushes through which the power is applied. Units are available in sizes of 1 to 500 HP (0.75 to 373 kW).

In *variable voltage drives*, a specially designed squirrel-cage, high-slip induction motor is used to reduce speed as the primary voltage is reduced. This is one of the least desirable drive designs because of poor efficiency, limited turndown, and high rates of heat generation. This design is not recommended for centrifugal pumps.

In an *eddy-current coupling*, the motor rotates a drum or rotating ring. Inside, but free from the drum, is an electromagnetic pole-type rotor assembly that is connected to the pump. The speed-control system regulates the amount of flux that exists between the drum and rotor assemblies. The amount of slip or speed difference between motor and pump increases and decreases with the flux density in the coupling. Being a slip-type, variable-speed device, the eddy-current coupling does not have as high an efficiency as the variable frequency, direct current, 2nd wound rotor regenesative types of variable-speed drives. Otherwise, however, it is a good design. It is rugged and used widely, particularly in large turbine pump applications.

Fluid couplings consist of two coupling halves, one driven by a standard electrical motor and the other connected to the centrifugal pump. The pump shaft speed increases as the amount of oil supplied to the coupling is increased, thus producing a simple control system for varying pump speed. The fluid coupling has been one of the most popular variable-speed devices for centrifugal pumps because of its relatively low initial cost, high reliability, and ease of maintenance. This design is particularly useful for installations in which efficiency is of no serious concern.

Mechanical drives have been used only on pumps smaller than 100 HP (75 kW). They use rubber belts or metal chains to adjust the speed within a range of 6:1. Because of their relatively low reliability and efficiency, they are not very popular.

The above brief descriptions provide general information on the various types of variable-speed drives. Actual selection of a drive for a specific application requires an evaluation of (1) wire-to-shaft efficiency, (2) first cost, (3) reliability, (4) serviceability, (5) maintenance costs, (6) need for special motor, (7) speed range, and (8) control repeatability.

No variable-speed drive should be selected for a centrifugal pump without careful evaluation of the process load itself to determine the feasibility of variable speed.

Variable-Speed Drive Efficiencies — The efficiency of variable-speed drives changes substantially with their design (Figure 12.37). Electronic designs eliminate dead bands, require less maintenance, and give better turndowns and simpler interface with DCS controls.

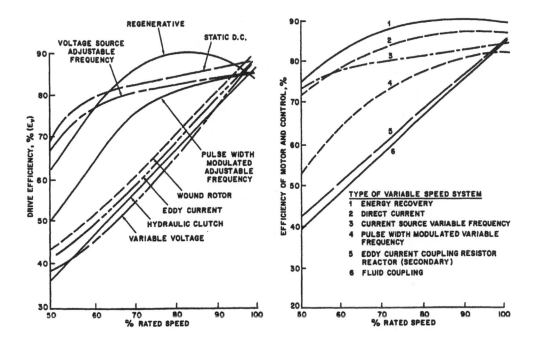

FIGURE 12.37   Variable-speed drives in the 100-HP and larger sizes offer a wide range of efficiencies.

When compared to valve throttling, the energy savings are less with electromechanical or slip control drives than with solid-state electrical drives (Figure 12.38). At 50% of rated speed, the variable-speed drive efficiency can be as low as 40% or as high as 70%, depending on the design selected (Figure 12.37). Naturally, the less efficient variable-speed drives are also the less expensive ones. For a 100 HP motor, the cost of a variable-speed drive can range from less than $10,000 to more than $25,000 as a function of efficiency (Tables 12.2 and 12.3).

3. Overall System Efficiency Index (SEI)

The overall system efficiency index (SEI) of a variable-speed pump installation is determined as follows:

$$SEI = (E_p \times E_m \times E_v \times E_u)10^{-6} \tag{12}$$

where E_p = the pump efficiency (%), E_m = the motor efficiency (%), E_v = the variable-speed drive efficiency (%), and E_u = the efficiency of utilization (%).

Figure 12.39 gives some typical efficiency values for E_p, E_m, and E_v.

The efficiency of utilization (E_u) is an indicator of the quality of the specific water distribution system design. It reflects both on safety factors and on bad design practices. For example, in the piping distribution system illustrated in Figure 12.40, Q_r might represent the required water flow and H_r the pressure head at the pump required to transport Q_r. Because three-way valves are used

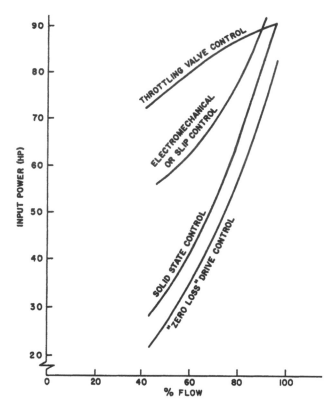

FIGURE 12.38   The energy savings vary with the type of variable-speed drive used.

in this figure, the actual flow (Q_a) is much higher than the required flow, and the head required to transport this actual flow (H_a) is also greater than the required head. Therefore, the efficiency of utilization is defined as follows:

$$\overline{} \atop \text{*SG = specific}$$ (13)

The numerical value of E_u can be accurately obtained only by testing. Once all four efficiencies are determined, they can be represented by a single system efficiency index (SEI) curve (Figure 12.41). By combining different pumps, motors, drives, and system designs, it is possible to arrive at a number of SEI curves. The relative advantages of different devices and designs can be evaluated quantitatively by comparing these curves.

III. OPTIMIZING CONTROLS

A pumping system is optimized when it meets the process demand for liquid transportation at minimum pumping cost and in a safe and stable manner. Once the equipment is installed, the potential for optimization is limited by the capabilities of the selected equipment and piping configuration. As an example, Figure 12.42 illustrates the optimization of a pump station consisting of a variable-speed and a constant-speed pump, which is the correct equipment selection if the system curve falls in zone 3 in Figure 12.33.

Figure 12.42 illustrates the instrumentation requirements of an optimized, variable-speed pumping system. PDIC-01 maintains a minimum of 10 PSID (0.7 bar) pressure difference between supply

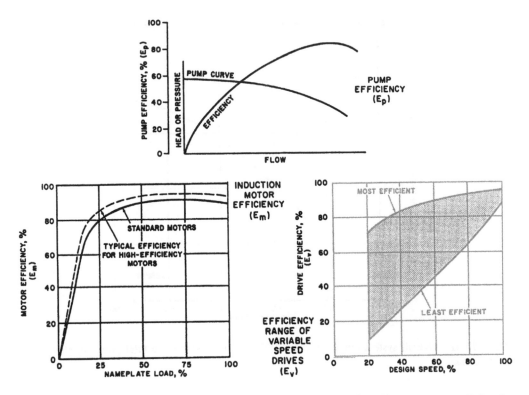

FIGURE 12.39   Overall pump efficiency is the result of the motor and variable-speed drive efficiencies.

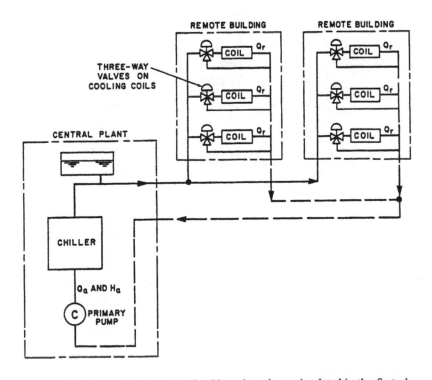

FIGURE 12.40   The flow bypassing the coils should not have been circulated in the first place.

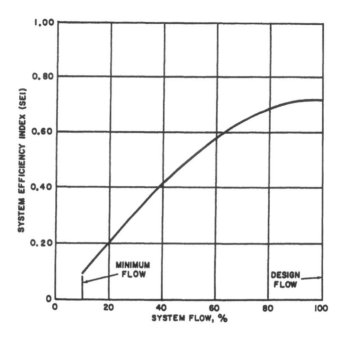

FIGURE 12.41   Overall system efficiency index (SEI) curves describe the total pumping efficiency as a function of load.

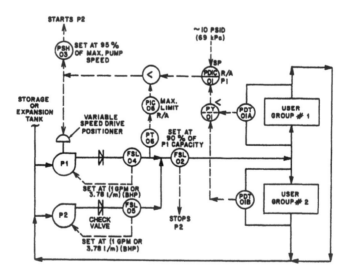

FIGURE 12.42   Pump station controls can be optimized for efficient, safe, and stable operation.

and return from each user. Therefore, no user will be denied more fluid if its control valve opens further. On the other hand, if demand drops, PDIC-01 will slow down the variable-speed pump to keep the differential at the users from rising beyond 10 PSID (0.7 bar).

When the variable-speed pump approaches its maximum speed, PSH-03 will automatically start pump 2. When the load drops down to the set point of FSL-02, the second pump is stopped. Extra increments of pumps are started by pressure but stopped by flow controls.

FSL-04 and FSL-05 are safety devices that will protect the pumps from overheating or cavitation, which might occur if the pump capacity is very low. As long as the pump flow rate is greater than 1 GPM (3.78 l/m) per pump break horsepower, the operation is safe. If the flow rate drops below this limit, the pumps are stopped by the low flow switches.

Variable-speed pumping will never completely replace the use of control valves. However, its use will increase because variable-speed pumps do represent a more energy-efficient method of fluid transportation and distribution than did the constant-speed pumps used in combination with control valves. Therefore, if the fluid is being distributed in a mostly friction system and the yearly average loading is under 80% of pump capacity, the installation of variable-speed pumps will reduce operating costs.

As is shown in Figure 12.42, the pump speed is set to keep the lowest user pressure drop above some minimum limit; thus, it is important that the pressure drop across the users (PDT) be accurately detected. Figure 12.43 illustrates both the correct and the incorrect location of these transmitters. The differential pressure transmitter should measure the head loss not only of the load but also of the control valve, hand valves, and piping.

Similarly, if the header pressure is to be controlled (or limited, as in Figure 12.42), the pressure transmitters should be located on main raisers or headers. They should not be near major on-off loads, as shown in Figure 12.43.

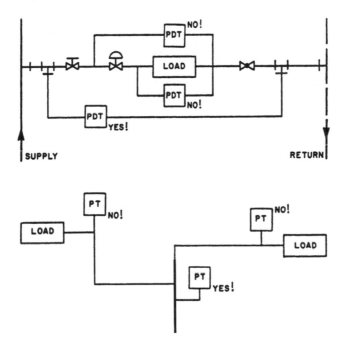

FIGURE 12.43 The pressure and differentiated pressure transmitters must be correctly located in the pipe distribution network.

A. VALVE POSITION OPTIMIZATION

One of the best methods of finding the optimum pump discharge pressure is illustrated in Figure 12.44. The optimum pressure is that which will keep the most open user valve at a 90% opening. As the pressure rises, all user valves close; as it drops, they open. Therefore, opening the most open valve to 90% causes all others to be opened also. This keeps the pump discharge pressure and the use of pumping energy at a minimum.

As the valve position controller (VPC-02) opens the valves, it not only minimizes the valve pressure drops but also reduces valve cycling and maintenance. This is because cycling is more

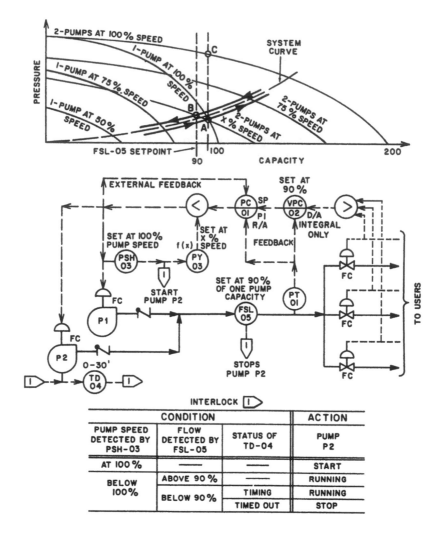

FIGURE 12.44 A pump station optimization system can be based on valve position control of two equal-size variable-speed pumps

likely to occur when the valve is nearly closed, and maintenance is high when the pressure drop is high. The other important advantage of this control system is that no user can ever run out of coolant, as no valve is ever allowed to be 100% open. This increases plant safety.

In order to make sure that the pressure controller (PC-01) set point is changed slowly and in a stable manner, the valve position controller (VPC-02) is provided with integral action only, and its integral time is set to be about 10 times that of PC-01. In order to prevent reset wind-up when the PC-01 is switched from cascade to automatic, the valve position controller is also provided with an external feedback.

The pump station in Figure 12.44 consists of two variable-speed pumps. Their speed is set by PC-01. When only one pump is in operation and the PC-01 output reaches 100%, PSH-03 is actuated and the second pump is started, as shown in the interlock table in Figure 12.44. When both pumps are in operation and the flow drops to 90% of the capacity of one pump, the second pump is stopped if TD-04 has timed out. The purpose of the time delay (TD-04) is to make sure that the pump is not started and stopped too often.

The cycling of the second pump is also illustrated on the pump curves on the top of Figure 12.44. As the demand for water rises, the speed of the one operating pump increases until 100% is reached at point A. Here PSH-3 starts the second pump. However, if the speed-control signal was unchanged when the second pump was started, an upset would occur, as the operating point would instantaneously jump from point A to C. In order to eliminate this temporary surge in pressure, which otherwise could shut down the station, PY-03 is introduced. This is a signal generator that, upon actuation by interlock 1, drops its output to x. This is the required speed for the two-pump operation at point A. The low signal selector immediately selects this signal x for control, thereby avoiding the upset. After actuation, the output signal of PY-03 slowly rises to full scale. As soon as it rises above the output of PC-01, it is disregarded and control is returned to PC-01.

Once both pumps are operating smoothly, the next control task is to stop the second pump when the load drops to the point where it can be met by a single pump. This is controlled by the low flow switch FSL-05, which is set at 90% of the capacity of one pump (point B on the system curve).

The pump-cycling controls described here can be used for any number of pumps. For each additional pump, another PY and FSL must be added, but otherwise, the system is the same.

B. Alternative Optimization Strategies

In pump distribution systems in which the number of users served is large, Shinskey suggests a simpler system than the one described in Figure 12.44.

The logic behind the control loop shown in Figure 12.45 is that the terminal pressure will be

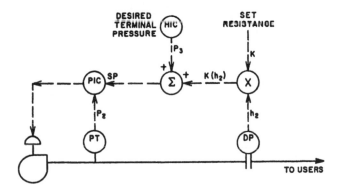

FIGURE 12.45 This system saves energy by reducing discharge pressure with reduced flow, while maintaining a constant terminal pressure.

kept constant if pump discharge pressure is varied in proportion to the orifice drop. The pressure drop through an orifice varies in the same way as does the pressure drop in a mostly friction system. Therefore, if the orifice drop (h_2) is measured, this measurement can be related to the pipeline pressure losses by multiplying the drop by a set resistance ratio (K). Therefore, a set terminal pressure at the user can be maintained under variable load conditions by adjusting the pump speed, as shown in Figure 12.45.

It is also advisable to monitor the electric power consumption of the pumps continuously, so that pump efficiency data will be empirical and up to date. This allows any load to be met with the most efficient pump or pump combination. In addition, the continuous monitoring of efficiency can also be used to signal the need for pump maintenance.

C. Water Distribution System Efficiencies

The overall efficiency of the water distribution system shown in Figure 12.40 is shown in Figure 12.46 by curve 1. A substantial increase in efficiency (reduction in operating cost) can be obtained

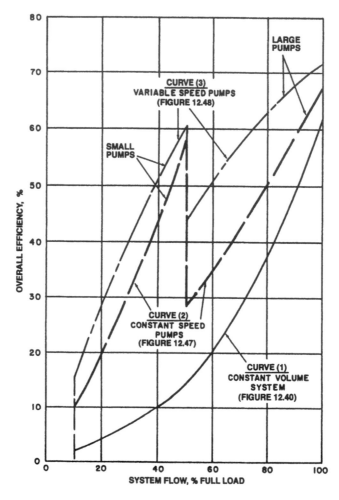

FIGURE 12.46 Overall efficiency is maximum if two variable-speed pumps are used and is minimum with a constant volume installation.

by replacing the three-way valves with two-way ones and by replacing the single large pump with several smaller ones. Figure 12.47 shows such a system. Here a small and a large primary pump are provided in the central plant and a small and a large booster pump are furnished in each of the user buildings. When the load is low, the small pumps are operating; when it is high, the large pumps take their place. The minimum flow requirements of the chiller are guaranteed by a bypass valve, and the chilled water makeup into the recirculating loop of each building is under temperature control (TC). The resulting improvement in overall efficiency is shown by curve 2 in Figure 12.46.

The highest overall efficiency can be obtained through the use of variable-volume load-following. Figure 12.48 illustrates such a system, utilizing variable-speed pumps in two sizes. In this system, all waste is eliminated except the small minimum flow bypass around the chiller, which is guaranteed by a small constant-speed pump. The resulting increase in overall efficiency is illustrated by curve 3 in Figure 12.46.

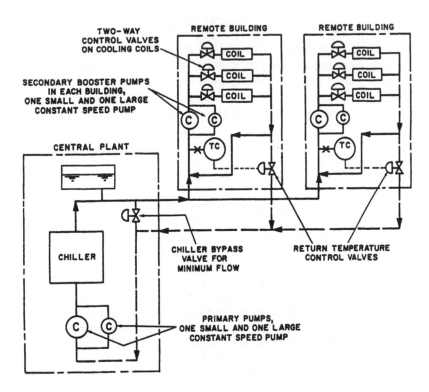

FIGURE 12.47 Supply-demand matching can be achieved using constant-speed pumps of different sizes.

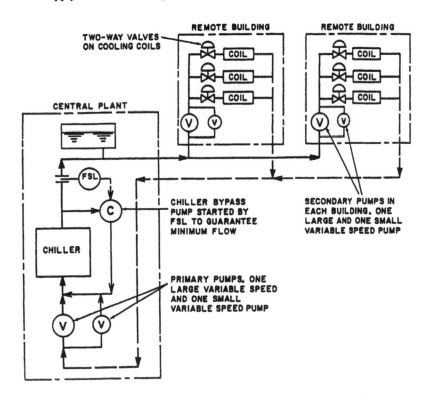

FIGURE 12.48 Variable-volume water distribution systems provide maximum efficiency.

IV. CONCLUSIONS

As was shown in Figures 12.35 and 12.46, the use of optimized variable-speed pumps can cut the yearly operating costs in half when the average load is low. In addition, optimized supply-demand matching (Figure 12.44) can protect from loss of control while reducing maintenance and cycling. The full automation of pumping stations—including automatic start-up and shut-down—not only will reduce operating costs but will also increase operating safety as human errors are eliminated.

BIBLIOGRAPHY

Baumann, H.D., Control Valves vs. Speed Controlled Pumps, 1981 Texas A&M Symposium.

Bose, B.K., *Power Electronics and AC Drives*, Prentice Hall, 1986.

Conzett, J.C., Adjustable Speed Drives, Reliance Electric Bulletin, latest edition.

Gibson, I.H., Variable Speed Pumps for Flow Control, ISA/93 Technical Conference, Chicago, September 19-24, 1993.

Hall, J., Motor drives keep pace with changing technology, *Instrum. Control Sys.*, September 1982.

Karassik, Krutzsch, Fraser and Messing, *Pump Handbook*, McGraw-Hill, New York.

Langfeldt, M.K., Economic Considerations of Variable Speed Drives, ASME Paper 80-PET-81.

Lipták, B.G., Save energy by optimizing your boilers, chillers, and pumps, *InTech*, March 1981.

Murphy, J.M.D. and Turnbull, F.G., *Power Electronic Control of AC Motors*, Pergamon Press, 1988.

O'Callaghan, P.W., *Energy Management*, McGraw-Hill, New York, 1992.

Rishel, J.B., Wire to Water Efficiency of Pumping Systems, 1975 Central Chilled Water Conference, Purdue University.

Schroeder, E.G., Choose variable speed drives for pump efficiency, *InTech*, September 1981.

Index

logic controls, 200
split-range, 188
Service core, 97
Set point(s)
adaptation, 145, 146
optimized, 166
Shaft power, 372
Shrink/swell effect, 10
Shutdown
delay, 288
sequence, 217
Signal generator, 287
Silicon-controlled rectifiers (SCRs), 392
Single-effect evaporation, 254
Siphoning, 369
Smart thermostat, 340
Sodium brines, 49
Solids moisture content, 248
Sparking, explosions resulting from, 392
Speed
control, 116
governor classification, 112
modulation, 111
throttling, 386
Split-range
operation, limitations of, 317
system, 341
Spray dryers, 245, 246, 378
Stack temperature, 36
Stand-alone air compressor, 120, 121
Standard air dry, 157
Standard ton, 49
Start-up
algorithm, 354
openings, 354
Static head, 373
Steady-state
gain, 178, 300
model, 273, 319
Steam
atomization, 16
availability, 276
drum, 1
enthalpy, 266
extraction, 266
-to-feed ratio, 271
flow, 13
governor, 89, 90
heaters, 304, 320
jet type ejector, 214
management flexibility, 267

pressure optimization, 37
supply, 267
temperature, 23
water heater, 321
Storage optimization, 86
Strainer cycle, 84
Stripping controls, 211
Subfreezing weather, 168
Subzone(s)
controls of, 96
pressure, 99
Sucrose solutions, 263
Suction
head, 373
vane throttling, 285
Sugar juice concentration, 258, 259
Summer
-amplifier, 290
-winter thermostat, 339
Sump, 157
Superheat
control valve, 52
of evaporated vapors, 52
tubes, 1
Superheater
feeler bulb, 53
secondary, 14
Supply
air pressure, optimization of, 358
-demand matching, 152, 288, 401
temperature, 69, 71, 166
Surface moisture, evaporation of, 227
Surge
beginning of, 141
controls, 88, 127
loops, 140
valve, 134 137
curge, 145
curve variations, 130
flow controller, 137
line, 61, 114, 129, 286
phenomenon of, 128
point, 284
protection backup, 139
set points, 143
spike detection, 128
valves, 135, 136
Swell or shrink period, 21
Switching, optimal, 196
System
curves, 362